ASP.NET Core 应用开发入门教程

周志刚　编著

北京航空航天大学出版社

内 容 简 介

本书假设读者已经熟悉 C# 和 .NET 的开发,并且对关系型数据库有所了解,但不要求对 C# 有太深入的了解。初学 C# 的读者可能会发现有些代码难以阅读,因为本书不会详细讲解 C# 的内容。

ASP.NET Core 是一个开源跨平台框架,用于构建 Web 应用、IoT 应用和移动后端应用。ASP.NET Core 应用程序可以运行于 .NET Core 和完整的 .NET Framework 之上。全书共分 9 章,深入浅出地介绍了 ASP.NET Core 的基础知识和前端 UI 选择方面的知识,主要包括.NET Core 的基础知识及其安装、dotnet 命令、Visual Studio 开发工具的安装和使用、ASP.NET Core 的原理及其组件介绍、Entity Framework Core 框架、ASP.NET Core MVC 框架和依赖注入等。

本书面向从未使用过 ASP.NET Core 和 EF Core 的初、中级用户,以及想了解 ASP.NET Core 和 EF Core 的读者,也可作为各初、高等院校师生的教学和自学丛书,以及社会相关领域培训班的教材。

图书在版编目(CIP)数据

ASP.NET Core 应用开发入门教程 / 周志刚编著. ——北京:北京航空航天大学出版社,2020.1
 ISBN 978 - 7 - 5124 - 2987 - 1

Ⅰ. ①A… Ⅱ. ①周… Ⅲ. ①网页制作工具—程序设计—教材 Ⅳ. ①TP393.092.2

中国版本图书馆 CIP 数据核字(2019)第 260644 号

版权所有,侵权必究。

ASP.NET Core 应用开发入门教程
周志刚　编著
责任编辑　宋淑娟

*

北京航空航天大学出版社出版发行

北京市海淀区学院路 37 号(邮编 100191)　http://www.buaapress.com.cn
发行部电话:(010)82317024　传真:(010)82328026
读者信箱: emsbook@buaacm.com.cn　邮购电话:(010)82316936
涿州市新华印刷有限公司印装　各地书店经销

*

开本:710×1 000　1/16　印张:27.5　字数:586 千字
2020 年 1 月第 1 版　2020 年 1 月第 1 次印刷　印数:3 000 册
ISBN 978 - 7 - 5124 - 2987 - 1　定价:89.00 元

若本书有倒页、脱页、缺页等印装质量问题,请与本社发行部联系调换。联系电话:(010)82317024

前　言

自2000年6月微软提出.NET战略,经过16年的推广之后,在2016年推出了升级版.NET Core。这是一个真正跨平台的框架,如今,.NET Core已经得到了比较广泛的认可。使用.NET开发B/S应用程序当然首推ASP.NET Core,因为在ASP.NET Core 2.1中内置了更多的功能。相对于ASP.NET Core 1.0,ASP.NET Core 2.x降低了学习门槛,引进了新的Razor Page用户界面设计方式,可以很容易地与.NET Framework 4.6以上版本兼容,便于用户轻松地把旧项目迁移到.NET Core环境上来。

ASP.NET Core由模块化的组件构成,是一个比ASP.NET更为精简且模块化的框架。ASP.NET Core不再基于System.Web.dll,而是基于一系列颗粒化的、且有良好架构的NuGet包。ASP.NET Core并不只是前端技术,也是后端技术。ASP.NET Core是ASP.NET开发人员需要了解的一种技术,是在多种平台上进行Web开发时可供使用的另一种全栈解决方案,以便在Windows、Mac和Linux上跨平台地开发和运行自己的ASP.NET Core应用程序。

本书对ASP.NET Core 2.x进行了全面讲解,包括.NET Core、ASP.NET Core的原理及其组件、ASP.NET Core MVC框架和Entity Framework Core框架等,通过本书的学习可以帮助开发者走进ASP.NET Core跨平台开发的世界。本书侧重于各种小功能的实现,并在实现过程中揭示ASP.NET Core 2.x的特性。

全书共分9章,内容如下:

第1章介绍.NET Core的基础知识,并通过一个简单的控制台应用介绍.NET Core。

第2章介绍.NET Core CLI的dotnet命令,并通过一些简单示例来学习如何使用这些命令。

第3章介绍Visual Studio 2017和NuGet包,并通过一些简单示例来学习如何使用Visual Studio 2017进行应用开发及引用NuGet包。

第4章介绍ASP.NET Core框架。

ASP. NET Core 应用开发入门教程

第 5 章介绍 Entity Framework（EF）Core 框架和配置特性，并通过构建使用 EF Core 的 ASP. NET Core 应用程序示例来具体应用 EF Core 功能。

第 6 章介绍 ASP. NET Core MVC 使用"模型-视图-控制器"设计模式构建 Web 应用和 API 的丰富框架。

第 7 章介绍依赖注入，并通过示例来学习如何使用依赖注入。

第 8 章介绍 Razor 标记语言。

第 9 章介绍设计一个优秀 Web 应用程序应注意的前端 UI 选择问题，以及现在比较流行的几种 UI，这些 UI 为统一整个 Web 应用程序的风格和减少冗余代码提供了很好的解决方案。

读者在阅读完本书后能够了解 ASP. NET Core 的基础知识和原理，学会使用 ASP. NET Core 并结合 EF Core 开发简单的跨平台应用程序。由于笔者能力有限，本书只能抛砖引玉，有未尽如人意之处，希望读者海涵并提出宝贵意见，以期共同进步。

作 者

2019 年 2 月

目 录

第 1 章 .NET Core ··· 1

1.1 .NET Core 介绍 ·· 1
1.1.1 什么是.NET ·· 1
1.1.2 什么是.NET Framework ·· 1
1.1.3 什么是.NET Core ·· 2
1.2 .NET Core 跨平台 ·· 2
1.2.1 .NET Standard ··· 3
1.2.2 .NET Core 的特点 ·· 4
1.3 .NET Core SDK 下载安装 ·· 7

第 2 章 dotnet 命令 ·· 12

2.1 dotnet 命令结构 ··· 12
2.1.1 dotnet 命令结构介绍 ··· 13
2.1.2 dotnet 命令示例 ··· 14
2.2 dotnet new ··· 16
2.2.1 介绍 ·· 16
2.2.2 示例 ·· 20
2.3 dotnet restore ··· 22
2.4 dotnet sln ·· 24
2.4.1 介绍 ·· 24
2.4.2 示例 ·· 25
2.5 dotnet build ··· 27
2.5.1 介绍 ·· 27
2.5.2 示例 ·· 28
2.6 dotnet pack ··· 28
2.6.1 介绍 ·· 28
2.6.2 示例 ·· 29

2.7 dotnet run ··· 30
　2.7.1 介　绍 ··· 30
　2.7.2 示　例 ··· 32
2.8 dotnet publish ··· 32
　2.8.1 介　绍 ··· 32
　2.8.2 示　例 ··· 34
2.9 dotnet add package ·· 35
　2.9.1 介　绍 ··· 35
　2.9.2 示　例 ··· 35
2.10 dotnet add reference ··· 37
　2.10.1 介　绍 ·· 37
　2.10.2 示　例 ·· 37
2.11 dotnet 命令综合示例 ··· 38

第3章 Visual Studio 2017 与 NuGet ·· 44

3.1 安装 Visual Studio 2017 ··· 44
　3.1.1 检查计算机安装环境 ··· 44
　3.1.2 下载 Visual Studio 2017 ·· 44
　3.1.3 运行 Visual Studio 2017 安装程序 ·································· 45
　3.1.4 选择工作负载 ·· 46
　3.1.5 逐个选择组件(可选) ·· 47
　3.1.6 安装语言包(可选) ··· 47
　3.1.7 更改安装位置(可选) ·· 48
　3.1.8 起始页介绍 ··· 48
3.2 使用 Visual Studio 创建程序 ··· 49
3.3 Visual Studio 功能简介 ·· 53
　3.3.1 菜　单 ··· 53
　3.3.2 解决方案资源管理器 ··· 53
　3.3.3 快速启动 ··· 56
　3.3.4 编辑器 ··· 58
　3.3.5 运行和调试应用程序 ··· 61
　3.3.6 调试代码 ··· 62
　3.3.7 使用重构和 IntelliSense ·· 66
3.4 NuGet 简介 ·· 68
　3.4.1 包、创建者、主机和使用者之间的关系 ····························· 68
　3.4.2 包的兼容性 ··· 69
　3.4.3 NuGet 工具 ··· 70
　3.4.4 管理依赖项 ··· 70
　3.4.5 跟踪引用和还原包 ··· 71

3.5 在 Visual Studio 中安装和使用包 …… 72
   3.5.1 程序包管理器 UI …… 72
   3.5.2 程序包管理器控制台 …… 74
   3.5.3 在应用中使用 Newtonsoft.Json API …… 75

第4章 ASP.NET Core 简介 …… 77

4.1 为何使用 ASP.NET Core …… 77
4.2 ASP.NET Core 启动的秘密 …… 78
   4.2.1 ASP.NET Core 启动流程 …… 78
   4.2.2 宿主构造器：WebHostBuilder …… 79
   4.2.3 UseStartup⟨Startup⟩() …… 80
   4.2.4 WebHostBuilder.Build() …… 82
   4.2.5 WebHost.Initialize() …… 85
   4.2.6 WebHost.Run() …… 87
   4.2.7 构建请求处理管道 …… 87
   4.2.8 启动 WebHost …… 90
   4.2.9 启动 Server …… 92
   4.2.10 启动 IHostedService …… 93
4.3 ASP.NET Core 中间件 …… 94
   4.3.1 什么是中间件 …… 94
   4.3.2 中间件的运行方式 …… 94
   4.3.3 中间件排序 …… 96
   4.3.4 Use、Run 和 Map 方法 …… 98
   4.3.5 内置中间件 …… 103
4.4 ASP.NET Core 中的静态文件 …… 104
   4.4.1 如何将静态文件注入到项目中 …… 105
   4.4.2 自定义静态文件夹 …… 107
   4.4.3 添加默认文件支持 …… 109
   4.4.4 设置 HTTP 响应标头 …… 110
   4.4.5 启用目录浏览 …… 112
4.5 ASP.NET Core 中的配置 …… 113
   4.5.1 配置相关的包 …… 114
   4.5.2 文件配置 …… 115
   4.5.3 XML 配置 …… 120
   4.5.4 按环境配置 …… 121
   4.5.5 在 Razor 页面中访问配置 …… 122
   4.5.6 其他配置方式 …… 122
4.6 ASP.NET Core 中的日志记录 …… 123
   4.6.1 日志模型三要素 …… 124

    4.6.2 日志记录级别 ·· 125
    4.6.3 将日志写入不同的目的地 ·································· 126
    4.6.4 添加筛选功能 ·· 130
    4.6.5 根据等级过滤日志消息 ···································· 131
    4.6.6 设置文件配置 ·· 133
    4.6.7 作用域 ·· 135
    4.6.8 日志记录建议 ·· 137
  4.7 在 ASP.NET Core 中使用多个环境 ······························ 137
    4.7.1 环　境 ·· 137
    4.7.2 在运行时确定环境 ·· 137
    4.7.3 开发环境 ·· 139
    4.7.4 生产环境 ·· 141
    4.7.5 基于环境的 Startup 类和方法 ······························ 142
  4.8 Session 详解 ··· 142
    4.8.1 什么是 Session ··· 142
    4.8.2 理解 Session 机制 ·· 143
    4.8.3 ASP.NET Core 中的 Session ······························· 144
    4.8.4 ASP.NET Core 中如何使用 Session ························· 145
  4.9 ASP.NET Core 中的缓存 ·· 149
    4.9.1 缓存的基础知识 ·· 149
    4.9.2 将数据缓存在内存中 ······································ 149
    4.9.3 基于 SQL Server 的分布式缓存 ····························· 158

第 5 章 Entity Framework Core ······································ 167

  5.1 先决条件 ·· 167
  5.2 Visual Studio 开发 ·· 167
    5.2.1 使用 NuGet 的包管理器用户界面 ··························· 167
    5.2.2 使用 NuGet 的包管理器控制台 ····························· 168
  5.3 创建数据库 ·· 168
  5.4 EF Core 的两种编程方式 ······································· 170
  5.5 EF Core 2.0 Database First 的基本使用 ··························· 170
  5.6 Entity Framework Core 的实体特性 ······························ 177
    5.6.1 数据注释特性——Key ····································· 178
    5.6.2 数据注释特性——Timestamp ······························ 182
    5.6.3 数据注释特性——ConcurrencyCheck ······················· 183
    5.6.4 数据注释特性——Required ································ 184
    5.6.5 数据注释特性——MaxLength ······························ 186
    5.6.6 数据注释特性——MinLength ······························ 187
    5.6.7 数据注释特性——Table ··································· 188

5.6.8 数据注释特性——Column	191
5.6.9 数据注释特性——ForeignKey	193
5.6.10 数据注释特性——NotMapped	196
5.7 EF Core 2.0 Code First	199
5.8 EF Core 2.0 Code First 创建数据库	199
5.8.1 创建实体	199
5.8.2 创建数据库	203
5.8.3 数据库修改	204
5.8.4 还原迁移	206
5.8.5 删除迁移	207
5.8.6 生成 SQL 脚本	208
5.8.7 创建存储过程	209
5.8.8 给数据库添加初始数据	209
5.9 用 EF Core 2.0 Code First 查询数据	214
5.9.1 查询的工作原理	215
5.9.2 执行查询	215
5.9.3 基本查询	216
5.9.4 异步查询	217
5.9.5 加载所有数据	218
5.9.6 加载单个实体	218
5.9.7 条件查询	219
5.9.8 使用 SQL 语句查询	220
5.9.9 基本 SQL 查询	221
5.9.10 传递参数	221
5.9.11 使用 SQL 查询,用 LINQ 编写条件排序	222
5.9.12 跟踪与非跟踪查询	223
5.10 EF Core 2.0 Code First 保存数据	224
5.10.1 添加数据	225
5.10.2 修改数据	228
5.10.3 删除数据	230
5.10.4 单个 SaveChanges 中的多个操作	232
5.10.5 异步保存	235
5.10.6 使用事务	235
5.10.7 默认事务	235
5.10.8 显式事务	235
5.11 EF Core 2.0 Code First 处理并发冲突	239
5.11.1 并发冲突	239
5.11.2 乐观并发	239
5.11.3 检测并发冲突	241

5.11.4 解决并发冲突 ·· 243
5.11.5 使用时间戳和行级版本号 ······································ 246

第 6 章 ASP.NET Core MVC ·· 253

6.1 ASP.NET Core MVC 概述 ·· 253
6.1.1 什么是 MVC 模式 ·· 253
6.1.2 什么是 ASP.NET Core MVC ·· 254
6.2 ASP.NET Core 中的路由 ·· 259
6.2.1 路　由 ·· 259
6.2.2 路由基础知识 ·· 259
6.2.3 路由模板 ·· 263
6.2.4 路由约束 ·· 263
6.2.5 正则表达式 ··· 265
6.3 ASP.NET Core 中的模型绑定 ·· 266
6.3.1 模型绑定简介 ·· 266
6.3.2 模型绑定的工作原理 ··· 266
6.3.3 数组绑定 ·· 269
6.3.4 返回带格式的数据 ·· 270
6.4 ASP.NET Core MVC 中的模型验证 ······································ 271
6.4.1 模型验证简介 ·· 271
6.4.2 验证特性 ·· 272
6.4.3 自定义验证 ··· 273
6.4.4 客户端验证 ··· 275
6.4.5 远程验证 ·· 277
6.5 ASP.NET Core MVC 中的视图 ··· 279
6.5.1 Razor 视图引擎 ·· 279
6.5.2 使用视图的好处 ·· 280
6.5.3 创建视图 ·· 281
6.5.4 控制器如何指定视图 ··· 281
6.5.5 向视图传递数据 ·· 283
6.6 ASP.NET Core 中的布局 ·· 288
6.6.1 什么是布局 ··· 288
6.6.2 指定布局 ·· 290
6.6.3 导入共享指令 ·· 291
6.6.4 在呈现每个视图之前运行代码 ······································ 292
6.7 ASP.NET Core 中的标记助手 ·· 293
6.7.1 什么是标记助手 ·· 293
6.7.2 标记助手的功能 ·· 294
6.7.3 管理标记助手的作用域 ··· 296

6.7.4　标记助手的智能提示支持 ……………………………………………… 297
6.8　ASP.NET Core 中的分部视图 ………………………………………………… 299
　　6.8.1　什么是分部视图 …………………………………………………………… 299
　　6.8.2　何时使用分部视图 ………………………………………………………… 300
　　6.8.3　声明分部视图 ……………………………………………………………… 300
　　6.8.4　分部视图访问示例 ………………………………………………………… 300
6.9　ASP.NET Core 中的视图组件 ………………………………………………… 304
　　6.9.1　什么是视图组件 …………………………………………………………… 304
　　6.9.2　如何创建视图组件类 ……………………………………………………… 304
　　6.9.3　创建一个简单的视图组件 ………………………………………………… 305
　　6.9.4　调用视图组件作为标记助手 ……………………………………………… 308
　　6.9.5　在控制器方法中直接调用视图组件 ……………………………………… 309
　　6.9.6　指定视图名称 ……………………………………………………………… 310
6.10　在 ASP.NET Core MVC 中使用控制器处理请求 …………………………… 312
　　6.10.1　什么是控制器 …………………………………………………………… 312
　　6.10.2　定义操作 ………………………………………………………………… 313
　　6.10.3　控制器响应返回的方法 ………………………………………………… 313
6.11　ASP.NET Core 中的过滤器 …………………………………………………… 314
　　6.11.1　过滤器 …………………………………………………………………… 314
　　6.11.2　过滤器的工作原理 ……………………………………………………… 314
　　6.11.3　授权过滤器 ……………………………………………………………… 315
　　6.11.4　资源过滤器 ……………………………………………………………… 315
　　6.11.5　操作过滤器 ……………………………………………………………… 317
　　6.11.6　异常过滤器 ……………………………………………………………… 319
　　6.11.7　结果过滤器 ……………………………………………………………… 321
　　6.11.8　内置过滤器特性 ………………………………………………………… 322
　　6.11.9　取消和设置短路 ………………………………………………………… 323
　　6.11.10　依赖关系注入 …………………………………………………………… 324
　　6.11.11　过滤器示例 ……………………………………………………………… 325
6.12　ASP.NET Core 中的区域 ……………………………………………………… 331

第 7 章　依赖注入 ……………………………………………………………………… 337

7.1　什么是依赖注入 ………………………………………………………………… 337
　　7.1.1　什么是依赖 ………………………………………………………………… 337
　　7.1.2　什么是注入 ………………………………………………………………… 338
　　7.1.3　为什么要反转 ……………………………………………………………… 338
　　7.1.4　何为容器 …………………………………………………………………… 340
7.2　.NET Core DI ……………………………………………………………………… 341
　　7.2.1　构造函数注入行为 ………………………………………………………… 341

  7.2.2 实例的注册 ……………………………………………………………… 341
  7.2.3 实例的生命周期 …………………………………………………………… 342
 7.3 DI 在 ASP.NET Core 中的应用 ……………………………………………………… 349
  7.3.1 在 Startup 类中初始化 …………………………………………………… 351
  7.3.2 在控制类中使用 …………………………………………………………… 352
  7.3.3 通过 HttpContext 来获取实例 …………………………………………… 353
 7.4 在 ASP.NET Core 中将依赖项注入到视图中 ……………………………………… 354
  7.4.1 简单示例 …………………………………………………………………… 354
  7.4.2 填充查找数据 ……………………………………………………………… 356
 7.5 如何替换其他的 IoC 容器 …………………………………………………………… 358
  7.5.1 Autofac 的基本使用 ……………………………………………………… 359
  7.5.2 用 Autofac 代替原来的 IoC ……………………………………………… 364
  7.5.3 一个接口对应多个实现的情况 …………………………………………… 369

第 8 章 Razor 视图 …………………………………………………………………… 373

 8.1 什么是 Razor ………………………………………………………………………… 373
 8.2 Razor 保留关键字 …………………………………………………………………… 374
  8.2.1 Razor 关键字 ……………………………………………………………… 374
  8.2.2 C# Razor 关键字 ………………………………………………………… 374
 8.3 使用 Razor 语法编写表达式 ………………………………………………………… 375
  8.3.1 隐式 Razor 表达式 ………………………………………………………… 375
  8.3.2 显式 Razor 表达式 ………………………………………………………… 377
  8.3.3 表达式的编码 ……………………………………………………………… 378
 8.4 Razor 代码块 ………………………………………………………………………… 379
 8.5 Razor 逻辑条件控制 ………………………………………………………………… 382
  8.5.1 if 和 switch 条件语句 ……………………………………………………… 382
  8.5.2 循环语句 …………………………………………………………………… 383
  8.5.3 复合语句@using ………………………………………………………… 385
  8.5.4 异常处理语句@try、catch、finally ……………………………………… 386
  8.5.5 加锁语句@lock …………………………………………………………… 386
  8.5.6 注　释 ……………………………………………………………………… 387
 8.6 指　令 ………………………………………………………………………………… 387
 8.7 ASP.NET Core 中的 Razor 页面介绍 ……………………………………………… 389
  8.7.1 启用 Razor 页面 …………………………………………………………… 389
  8.7.2 Razor 页面介绍 …………………………………………………………… 390
  8.7.3 编写基本窗体 ……………………………………………………………… 391
  8.7.4 页面的 URL 生成 ………………………………………………………… 405
  8.7.5 针对一个页面的多个处理程序 …………………………………………… 406

第 9 章　Web UI 框架的选择 ……………………………………………………… 412
9.1　以 JQuery 为核心的前端框架 …………………………………………… 413
9.1.1　EasyUI ………………………………………………………………… 413
9.1.2　DWZ JUI ……………………………………………………………… 414
9.1.3　LigerUI ………………………………………………………………… 415
9.2　以 Bootstrap 为核心的前端框架 ………………………………………… 416
9.2.1　HUI …………………………………………………………………… 416
9.2.2　H+ UI ………………………………………………………………… 416
9.2.3　Ace Admin …………………………………………………………… 417
9.2.4　Metronic ……………………………………………………………… 418
9.2.5　AdminLTE …………………………………………………………… 419
9.2.6　INSPINIA ……………………………………………………………… 420
9.3　以 ExtJS 为核心的前端框架 ……………………………………………… 421

参考文献 ………………………………………………………………………… 423

第1章 .NET Core

1.1 .NET Core 介绍

在介绍.NET Core 之前先来了解一下什么是.NET。

1.1.1 什么是.NET

.NET 是一个微软搭建的开发者平台,是微软的新一代技术平台,可为敏捷商务构建互联互通的应用系统,这些系统是标准的、联通的、适应变化的、稳定的和高性能的。从技术角度讲,一个.NET 应用是一个运行于.NET Framework 之上的应用程序。更精确地讲,一个.NET 应用是一个使用.NET Framework 类库来编写,并运行于公共语言运行时(Common Language Runtime)之上的应用程序。其特点主要包括:

1) 跨语言,即只要是面向.NET 平台的编程语言(C♯、Visual Basic、Visual C++/CLI、F♯、IronPython、IronRuby、Delphi 和 Visual COBOL),用其中一种语言编写的类型就可以无缝地用在使用另一种语言编写的应用程序中,即具有互操作性。

2) 可用于该平台下开发人员的技术框架体系(.NET Framework、.NET Core、Mono、UWP 等),包括:

① 定义了通用类型系统和庞大的 CTS 体系;

② 用于支撑.NET 下的语言运行时的环境 CLR;

③ .NET 体系技术的框架库 FCL。

3) 可用于支持开发人员开发的软件工具(即 SDK,如 Visual Studio 2017 和 Visual Studio Code 等)。

1.1.2 什么是.NET Framework

1. .NET Framework

.NET Framework 是微软 2002 年发布的 Windows 操作系统下的.NET 技术框架,与 Windows 操作系统高度耦合,当前被.NET 开发人员经常使用。.NET Framework 4.5 版及其更高版本实现.NET Standard,因此,面向.NET Standard 的代码都可在这些版本的.NET Framework 上运行。.NET Framework 还包含一些特定于 Windows 的 API,如通过 Windows 窗体和 WPF 进行 Windows 桌面开发的 API。

.NET Framework 非常适合于生成 Windows 桌面应用程序,如图 1.1 所示。

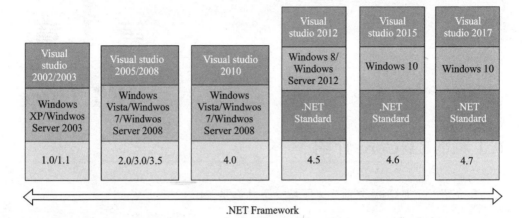

图 1.1　.NET Framework 演绎图

2. Mono

Mono 是在.NET Core 发布之前运行于 Linux 上的开源的.NET 框架,是由 Novell 公司(现在是 Xamarin)发起主持的开源项目。它包含一个 C♯语言的编译器,一个 CLR 的运行时和一组类库,并实现了 ADO.NET 和 ASP.NET。开发人员在 Linux 环境中使用 C♯开发程序。Mono 不仅可以运行于 Windows 操作系统上,还可以运行于 Linux、Android、Unix、Mac OS X、IOS 和 Solaris 上。Mono 还支持使用 Unity 引擎生成的游戏。它支持所有当前已发布的.NET Standard 版本。

1.1.3　什么是.NET Core

简单地说,.NET Core 是其他操作系统的.NET Framework 实现。

操作系统不止有 Windows,还有 Mac 和类 Linux 等系统,之前的.NET Framework 只是 Windows 操作系统下的.NET 的实现。.NET Core 是微软在其他操作系统上的.NET 实现,是.NET Framework 的新一代版本,是微软开发的具有跨平台(Windows、Mac OS X、Linux)、跨语言(C♯、Visual Basic、F♯)能力的应用程序开发框架,可用于设备、云和嵌入式/IoT 方案,以及处理大规模的服务器和云工作负荷。

.NET Core 是.NET Framework 基于.NET Framework 基类库(BCL)的跨平台版本,是根据.NET Standard 规范进行编写的。.NET Core 提供了一个可用于.NET Framework 或 Mono/Xamarin 的 API 子集。

1.2　.NET Core 跨平台

跨平台是软件开发中的一个重要概念,既不依赖于操作系统,也不依赖于硬件环境,在一个操作系统下开发的应用,放到另一个操作系统下依然可以运行。

一个.NET应用程序能运行的关键在于.NET CLR,为了能让.NET程序在其他平台上运行,一些非官方社区和组织为此开发了在其他平台上的.NET实现(最具代表性的是Mono),但因为不是官方,所以在一些方面多少有些缺陷(如FCL),后来微软官方推出了.NET Core,并开源在GitHub中,由微软和GitHub上的.NET社区共同维护。

要想实现真正的.NET跨平台,所有平台上的运行时(Runtime)都要根据一个统一的标准来实现,为此,微软推出了.NET Standard规范。

1.2.1 .NET Standard

.NET Standard是一种API规范,用于描述开发者在每个.NET实现代码中可以使用的一组标准化的.NET API。更正式地说,它是构成协定统一集(这些协定是编写代码的依据)的特定的.NET API组,这些协定要在每个.NET实现中来实现。.NET实现包括.NET Framework、.NET Core和Mono。

.NET Standard也是一个目标框架。如果开发者开发的.NET应用是面向.NET Standard版本的,则它可在支持该.NET Standard版本的任何.NET实现上运行,这是在每个.NET实现中都可用的.NET API的正式规范。

在没有.NET Standard之前,.NET的世界是:各个不同的.NET各自实现自己的基类库,如图1.2所示。

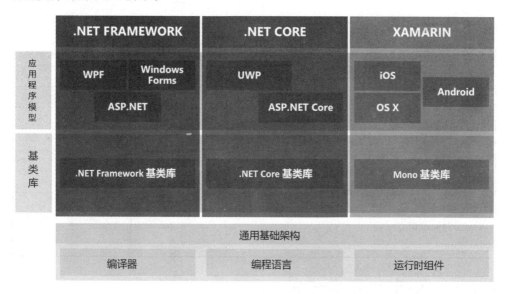

图1.2 .NET Framework框架图

在有了.NET Standard之后,.NET的世界是:各个不同的.NET有了统一的基类库,如图1.3所示。

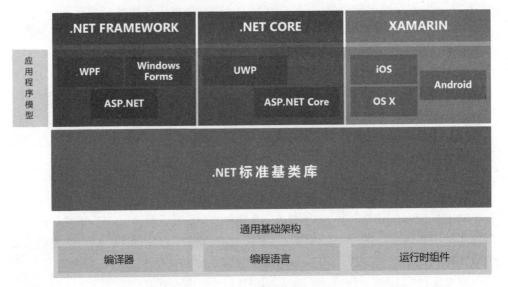

图 1.3 .NET Standard 框架图

.NET Standard 的所有版本及其支持的平台如表 1.1 所列。

表 1.1 .NET Standard 的所有版本及其支持的平台

.NET Standard 平台	版本号							
	1.0	1.1	1.2	1.3	1.4	1.5	1.6	2.0
.NET Core	1.0	1.0	1.0	1.0	1.0	1.0	1.0	2.0
.NET Framework	4.5	4.5	4.5.1	4.6	4.6.1	4.6.1	4.6.1	4.6.1
Mono	4.6	4.6	4.6	4.6	4.6	4.6	4.6	5.4
Xamarin.iOS	10.0	10.0	10.0	10.0	10.0	10.0	10.0	10.14
Xamarin.Mac	3.0	3.0	3.0	3.0	3.0	3.0	3.0	3.8
Xamarin.Android	7.0	7.0	7.0	7.0	7.0	7.0	7.0	8.0
通用 Windows 平台	10.0	10.0	10.0	10.0	10.0	10.0.16299	10.0.16299	10.0.16299
Windows	8.0	8.0	8.1	—	—	—	—	—
Windows Phone	8.1	8.1	8.1	—	—	—	—	—
Windows Phone SliverLight	8.0	—	—	—	—	—	—	—

1.2.2 .NET Core 的特点

1. 跨平台

.NET Core 是一个跨平台的.NET 实现。.NET Core 由微软在 Windows、Mac

OS 和 Linux 操作系统上提供.NET 实现,用以支持在多平台上运行.NET 应用。

与其他.NET 产品相比,.NET Core 能够适应广泛的新平台、新的工作负载和新的编译器工具。

.NET Core 中与平台无关的库在所有平台上都能够按照构建时的原样运行。.NET Core 为了支持多个操作系统,只有极少部分特定于操作系统的功能代码是单独实现,或者使用条件编译实现的。

.NET Core 中混合存在特定于平台和与平台无关的库。可以通过图 1.4 看出 CoreFx 中 90% 的代码是与平台无关的,这部分代码所实现的功能在所有平台上都是一样的。不限平台的代码可实现在所有平台上使用同一个程序集。CoreFx 中有 10% 的代码是与特定平台相关的代码。

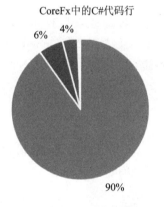
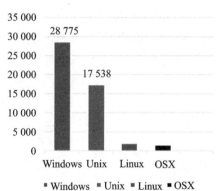

图 1.4 CoreFx:每个平台的代码行数

2. 跨语言

C♯、Visual Basic 和 F♯ 编译器以及.NET Core 工具已集成到或可以集成到多个文本编辑器和 IDE 中,包括 Visual Studio、Visual Studio Code、Sublime Text 和 Vim,使得.NET Core 的开发可以在开发者各自钟爱的编码环境和 OS 中进行,如图 1.5 所示。

3. 与其他.NET 实现的比较

有丑才有美,有低才有高,将.NET Core 与现有的.NET 实现进行比较,这可能是了解.NET Core 是什么的最简单的方法。

4. 与.NET Framework 比较

.NET 由微软于 2000 年首次发布,而后发展至今。Windows 操作系统上的.NET Framework 一直是微软的主要.NET 实现。

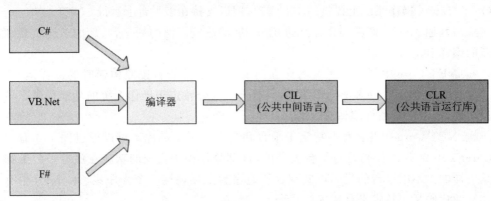

图 1.5 跨语言

.NET Core 与 .NET Framework 的主要差异在于：

① 应用模板——因为 .NET Framework 中的许多模板都是基于 Windows 操作系统的，如 WPF(基于 DirectX 生成)，所以 .NET Core 不会支持 .NET Framework 的所有应用模板。但 .NET Core 和 .NET Framework 两者都支持控制台和 ASP.NET Core 应用模板。

② API——.NET Core 是 .NET Framework 的一个子集，包含很多与 .NET Framework 相同的 API。随着时间的推移，这个子集会不断地扩大。

③ 子系统——.NET Core 实现 .NET Framework 中子系统的子级，目的是实现更简单的 .NET 实现和编程模型。例如，不支持代码访问安全性(CAS)，但支持反射。

④ 平台——.NET Framework 只支持 Windows 操作系统，而 .NET Core 除了支持 Windows 操作系统外，还支持 Mac OS 和 Linux。

⑤ 开源——.NET Core 是开源项目，而 .NET Framework 的代码是部分开源的。

5. 与 Mono 比较

Mono 是一个开源的跨平台 .NET 实现项目，于 2004 年首次发布。可以把它看作是 .NET Framework 的非官方版本。Mono 项目团队依赖于 Microsoft 发布的 .NET 标准(尤其是 ECMA 335)，以便实现兼容性。

.NET Core 与 Mono 的主要差异在于：

① 应用模板——Mono 通过 Xamarin 产品支持 .NET Framework 应用模板(例如，Windows Forms)和其他应用模板(例如，Xamarin.iOS)的子集。而 .NET Core 不支持这些内容。

② API——Mono 是 .NET Framework 的大型子集。Mono 的 API 使用与 .NET Framework 相同的程序集名称和组成要素。

③ 平台——Mono 支持很多平台和 CPU。

④ 开源——Mono 和.NET Core 两者都使用 MIT 许可证,且都属于.NET Foundation 项目。

⑤ 焦点——最近几年,Mono 的主要焦点是移动平台,而.NET Core 的焦点是云平台。

1.3 .NET Core SDK 下载安装

.NET Core 主要以两种方式发行:以包方式在 NuGet.org 上发行,以及以独立安装包形式发行。下面仅介绍后一种方式。

通常情况下,首先安装.NET Core SDK,然后就可以开始进行.NET Core 开发了,具体步骤如下。

① .NET Core 的入门页下载.NET Core 的独立安装包。在浏览器中输入网址 https://www.microsoft.com/net/download/windows,下载最新版的.NET Core SDK。

在浏览器中输入网址 https://dotnet.microsoft.com/download/dotnet-core/2.1,会发现此页有.NET Core 2.1 的各种下载版本。在图 1.6 所示页面中单击"x64"或"x86"下载本书所需要的版本。

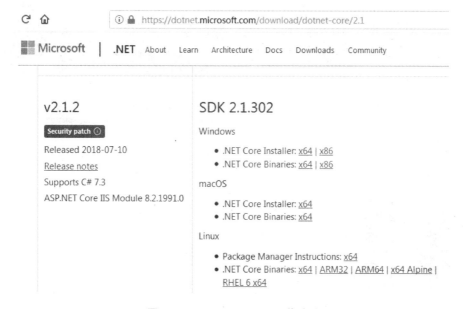

图 1.6 .NET Core SDK 下载地址

② 浏览器自动跳转到下载页面,弹出下载对话框,如图 1.7 所示。

③ 下载完成后,找到刚才下载的 dotnet-sdk-2.1.302-win-x64.exe 文件并双击,显示如图 1.8 所示页面,此时单击"Install"按钮。

④ 显示安装进度对话框,如图 1.9 所示。

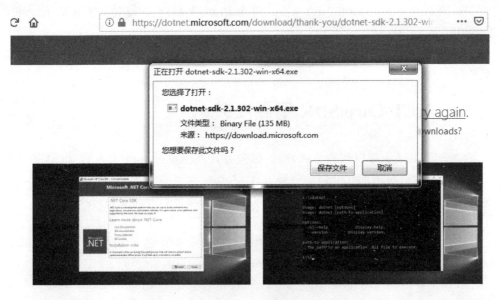

图 1.7 正在下载 .NET Core SDK 安装包

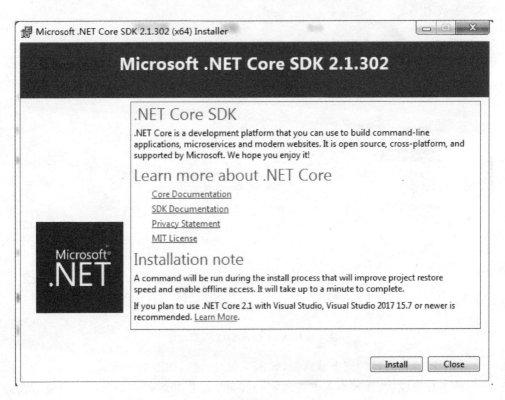

图 1.8 安装 .NET Core SDK

图 1.9　.NET Core SDK 安装进度

⑤ 安装成功后,开始开发第一个示例。在桌面菜单上选择"开始"→"运行"菜单项,在弹出的窗口中输入 cmd 查找 cmd 命令文件,如图 1.10 所示,然后双击此文件打开命令行窗口。

图 1.10　查找 cmd 命令文件

⑥ 在新建的命令行窗口中输入以下命令,如图 1.11 所示:

dotnet new console - o myApp

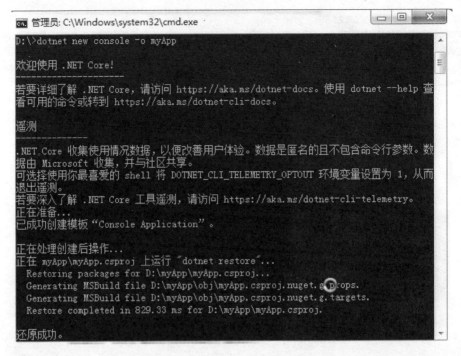

图 1.11　创建 .NET Core 应用程序

⑦ 此时，通过 dotnet 命令创建了一个新的控制台应用程序。-o 参数指示创建一个名为 myApp 的目录，并把新创建的控制台应用程序存储到 myApp 目录中。myApp 文件夹中的主文件是 program.cs，已经编写了将"Hello World!"输出到控制台的代码。

⑧ 通过以下命令进入新创建的应用程序目录中，如图 1.12 所示。

```
cd myApp
```

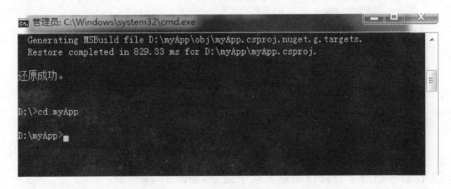

图 1.12　进入 myApp 目录

⑨ 现在运行第一个程序，在命令行窗口中输入如下命令，运行结果如图 1.13 所示。

```
dotnet run
```

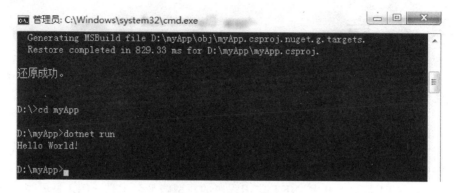

图 1.13　程序运行结果

第 2 章

dotnet 命令

dotnet 命令属于 .NET Core CLI 的一部分,是微软提供的命令行工具以供在开发程序中使用,主要用它进行代码的编译、NuGet 包的管理、程序的运行和测试等。.NET Core CLI 是用于开发 .NET 应用程序的新型跨平台工具。CLI 是更高级别的工具(如集成开发环境(IDE)、编辑器和持续集成)的基础。

在安装完 .NET Core SDK 2.1 之后,就默认安装了如表 2.1 所列的 dotnet 命令。

表 2.1 dotnet 命令

基本命令	项目命令	高级命令
new	add package	nuget delete
restore	add reference	nuget locals
build	remove package	nuget push
publish	remove reference	msbuild
run	list reference	dotnet install script
test		
vstest		
pack		
migrate		
clean		
sln		
help		
store		

2.1 dotnet 命令结构

dotnet 命令结构包含 dotnet 指令或可能的命令参数和选项。大部分 dotnet 命令都是此种结构模式,例如在第一个示例中使用命令创建了一个新控制台应用程序,在命令行窗口中从应用程序目录运行该应用程序,具体命令如下:

```
dotnet new console -o myApp
dotnet run
```

2.1.1　dotnet 命令结构介绍

dotnet 命令有如下形式：

dotnet [command] [arguments] [--additional-deps] [--additionalprobingpath] [-d|--diagnostics] [--fx-version] [-h|--help] [--info] [--list-runtimes] [--list-sdks] [--roll-forward-on-no-candidate-fx] [-v|--verbosity] [--version]

dotnet 命令具有两项职责，即运行依赖于框架的应用或执行指令。它也可以自行调用，并提供简短的使用说明。

dotnet 自行作为指令使用的唯一情况是运行依赖于框架的应用，在 dotnet 后指定应用名称，便可执行该应用程序。例如，dotnet myApp.dll。

当 dotnet 执行指令时，dotnet 会先确定要使用的 SDK 版本。如果在命令选项中未指定版本，则驱动程序使用可用的最新版本。如果要指定某个版本，而不是最新的版本，则使用--fx-version〈VERSION〉选项。确定 SDK 版本后，dotnet 执行指令。

指令[command]指特定功能的实现。例如，dotnet build 生成代码，dotnet publish 发布代码。

参数[arguments]指被指令调用的、在命令行上传递的参数。例如，在执行 dotnet publish myApp.csproj 时，myApp.csproj 参数指要发布的项目，并被传递给 publish 命令。

选项指在命令行上所输入的指令的选项，具体选项如表 2.2 所列。

表 2.2　dotnet 命令选项

选　项	说　明
--additional-deps〈PATH〉	其他 deps.json 文件的路径
--additionalprobingpath〈PATH〉	包含要进行探测的探测策略和程序集的路径
-d|--diagnostics	启用诊断输出
--fx-version〈VERSION〉	运行应用程序所使用的已安装.NET Core 运行时的版本
-h|--help	显示有关命令的简短帮助和指令列表
--info	显示.NET Core 版本信息、操作系统信息等其他信息
--list-runtimes	显示已安装的.NET Core 运行时
--list-sdks	显示已安装的.NET Core SDK
--roll-forward-on-no-candidate-fx	在没有候选共享框架的情况下前滚

续表 2.2

选 项	说 明
-v|--verbosity <LEVEL>	设置命令的详细级别。允许使用的值为 q[uiet]、m[inimal]、n[ormal]、d[etailed]和 diag[nostic]
--version	显示使用中的 .NET Core SDK 版本

2.1.2 dotnet 命令示例

dotnet 命令的几个示例如下：

① 显示 dotnet 的相关帮助，执行结果如图 2.1 所示。

dotnet -h

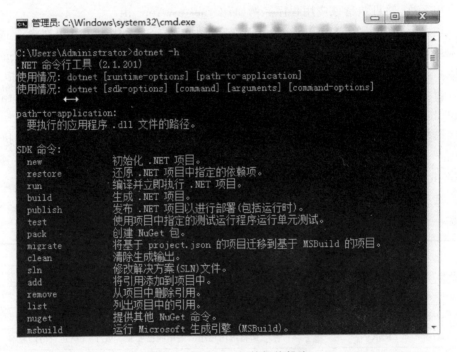

图 2.1 显示 dotnet 的相关帮助

② 在指定目录创建新的 .NET Core 控制台应用程序，执行结果如图 2.2 所示。

dotnet new console

③ 在指定目录中对项目及其依赖项进行编译，执行结果如图 2.3 所示。

dotnet build

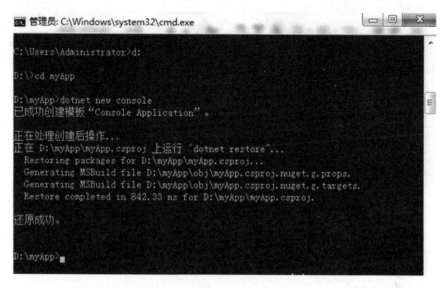

图 2.2　在指定目录创建新的 .NET Core 控制台应用程序

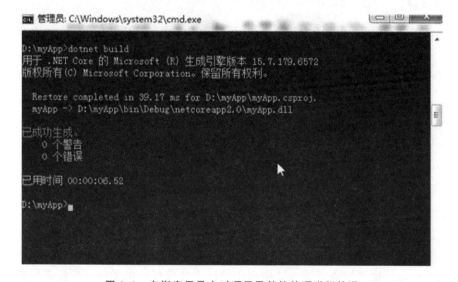

图 2.3　在指定目录中对项目及其依赖项进行编译

④ 在指定目录中运行名为 myApp.dll 的控制台应用程序,执行结果如图 2.4 所示。

dotnet myApp.dll

说明:当执行 dotnet myApp.dll 时,发现原来在当前 myApp 目录中不存在 myApp.dll 应用程序,而是在 bin\Debug\netcoreapp2.0 目录中生成,因此,可以成功执行 dotnet bin\Debug\netcoreapp2.0\myApp.dll。

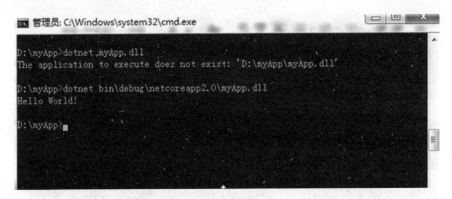

图 2.4　在指定目录中运行名为 myApp.dll 的控制台应用程序

2.2　dotnet new

2.2.1　介　绍

dotnet new 命令：根据指定的模板和选项，在硬盘上创建新的项目、配置文件或解决方案。

命令格式

dotnet new 〈TEMPLATE〉 [--force] [-i|--install] [-lang|--language][-n|--name] [--nuget-source] [-o|--output] [-u|--uninstall] [Template options]

dotnet new 〈TEMPLATE〉 [-l|--list] [--type]

dotnet new [-h|--help]

参　数

TEMPLATE

创建新项目时可以使用的模板名称。

此命令包含默认的模板列表。使用命令"dotnet new -l"获取可用模板的列表，如图 2.5 所示。也可以查看表 2.3 中列出的已安装模板。每个项目模板都可能有附加选项，如表 2.4 所列。

表 2.3　.NET Core SDK 2.1.300 预装的模板

模板名称	模板参数名称	语　言
控制台应用程序	console	[C#],F#,VB
类库	classlib	[C#],F#,VB
单元测试项目	mstest	[C#],F#,VB

续表 2.3

模板名称	模板参数名称	语　言
xUnit 测试项目	xunit	[C#],F#,VB
Razor 页	page	[C#]
MVC ViewImports	viewimports	[C#]
MVC ViewStart	viewstart	[C#]
ASP.NET Core 空项目	web	[C#],F#
ASP.NET Core Web 应用程序（Model-View-Controller）	mvc	[C#],F#
ASP.NET Core Web 应用程序	razor	[C#]
含 Angular 的 ASP.NET Core	angular	[C#]
含 React.js 的 ASP.NET Core	react	[C#]
含 React.js 和 Redux 的 ASP.NET Core	reactredux	[C#]
ASP.NET Core Web API	webapi	[C#],F#
Razor 类库	razorclasslib	[C#]
global.json 文件	globaljson	
NuGet 配置	nugetconfig	
Web 配置	webconfig	
解决方案文件	sln	

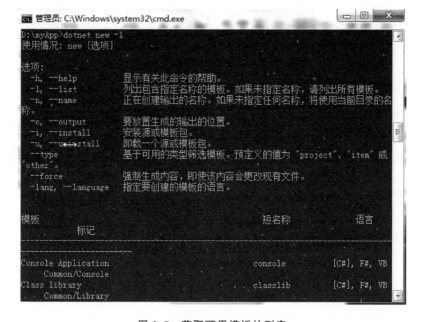

图 2.5　获取可用模板的列表

表 2.4　模板选项

模板选项	说　明	适用模板
no-restore	在项目创建期间不执行隐式还原	console，angular，react，reactredux，razorclasslib，classlib，mstest，xunit，web，webapi，mvc，razor
-f|--framework〈FRAMEWORK〉	指定目标框架。值为 netcoreapp2.0(.NET Core 类库)或 netstandard2.0(.NET Standard 类库)。默认值为 netstandard2.0	classlib
-p|--enable-pack	允许使用命令"dotnet pack"为项目打包	mstest，xunit
--sdk-version〈VERSION_NUMBER〉	指定要在 global.json 文件中使用的.NET Core SDK 版本	globaljson
--use-launch-settings	在生成的模板输出中添加 launchSettings.json	web，webapi，mvc，razor
-au|--auth〈AUTHENTICATION_TYPE〉	要使用的身份验证类型。可能的值为： • None：不进行身份验证(默认)。 • IndividualB2C：使用 Azure AD B2C 进行个人身份验证。 • SingleOrg：对一个租户进行组织身份验证。 • Windows：进行 Windows 身份验证	web，webapi，mvc，razor
	• Individual：进行个人身份验证。 • MultiOrg：对多个租户进行组织身份验证	mvc，razor
--aad-b2c-instance〈INSTANCE〉	要连接到的 Azure Active Directory B2C 实例。与 IndividualB2C 身份验证结合使用。默认值为 https://login.microsoftonline.com/tfp/	web，webapi，mvc，razor
-ssp|--susi-policy-id〈ID〉	此项目的登录和注册策略 ID。与 IndividualB2C 身份验证结合使用	web，webapi，mvc，razor

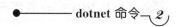

续表 2.4

模板选项	说　明	适用模板
--client-id〈ID〉	此项目的客户端 ID。与 IndividualB2C 或 SingleOrg 或 MultiOrg（mvc，razor）身份验证结合使用。 默认值为 11111111-1111-1111-11111111111111111	web，webapi，mvc，razor
--domain〈DOMAIN〉	目录租户的域。与 SingleOrg 或 IndividualB2C 身份验证结合使用。 默认值为 qualified.domain.name	web，webapi，mvc，razor
--tenant-id〈ID〉	要连接到的目录的 TenantId ID。与 SingleOrg 身份验证结合使用。 默认值为 22222222-2222-2222-2222-222222222222	web，webapi，mvc，razor
-r|--org-read-access	允许此应用程序对目录进行读取访问。仅适用于 SingleOrg 或 MultiOrg 身份验证	web，webapi，mvc，razor
--use-launch-settings	在生成的模板输出中添加 launchSettings.json	web，webapi，mvc，razor
-uld|--use-local-db	指定应使用 LocalDB，而不使用 SQLite。仅适用于 Individual 或 IndividualB2C 身份验证	web，webapi，mvc，razor
-rp|--reset-password-policy-id〈ID〉	此项目的重置密码策略 ID。与 IndividualB2C 身份验证结合使用	mvc，razor
-ep|--edit-profile-policy-id〈ID〉	此项目的编辑配置文件策略 ID。与 IndividualB2C 身份验证结合使用	mvc，razor
--callback-path〈PATH〉	重定向 URI 的应用程序基路径中的请求路径。与 SingleOrg 或 IndividualB2C 身份验证结合使用。 默认值为 /signin-oidc	mvc，razor
--use-browserlink	在项目中添加 BrowserLink	mvc，razor
-na|--namespace〈NAMESPACE_NAME〉	生成的代码的命名空间。 默认值为 MyApp.Namespace	page，viewimports
-np|--no-pagemodel	创建不含 PageModel 的页	page

命令选项

dotnet new 命令的选项说明如表 2.5 所列。

表 2.5 dotnet new 命令选项

选 项	说 明
--force	强制生成内容,即使会更改已经存在的文件
-h|--help	显示帮助信息。可针对 dotnet new 命令本身或任何模板(如 dotnet new mvc --help)调用它
-i|--install ⟨PATH|NUGET_ID⟩	从提供的 PATH 或 NUGET_ID 安装源或模板包。默认情况下,dotnet new 为该版本传递 *,表示最后一个稳定的包版本
-l|--list	列出包含指定名称的模板
-lang|--language {C#|F#|VB}	要创建的模板的语言。接受的语言因模板而异。对于某些模板无效。如果操作系统将"#"解释为特殊字符,则应在语言参数值上使用引号,如 dotnet new console -lang "F#"
-n|--name ⟨OUTPUT_NAME⟩	所创建的应用或文件的名称。如果未指定名称,则使用当前目录的名称
--nuget-source	指定在安装期间要使用的 NuGet 源
-o|--output ⟨OUTPUT_DIRECTORY⟩	用于放置生成的应用或文件的位置。默认为当前目录
--type	根据可用类型筛选模板。预定义值为"project""item"或"other"
-u|--uninstall ⟨PATH|NUGET_ID⟩	从提供的 PATH 或 NUGET_ID 卸载源或模板包。若要使用 PATH 卸载模板,则需要完全限定路径

2.2.2 示 例

dotnet new 命令的几个使用示例如下:

① 在当前目录中创建 F#控制台应用程序项目,如图 2.6 所示:

dotnet new console -lang F#

② 在指定目录中创建 .NET Standard 类库项目,如图 2.7 所示(仅适用于 .NET Core SDK 2.0 或更高版本):

dotnet new classlib -lang C# -o MyLib1

③ 在当前目录中新建一个没有设置身份验证的 ASP.NET Core C# MVC 应用程序项目,如图 2.8 所示:

dotnet new mvc -au None

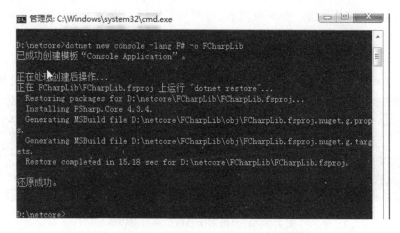

图 2.6　在当前目录中创建 F#控制台应用程序项目

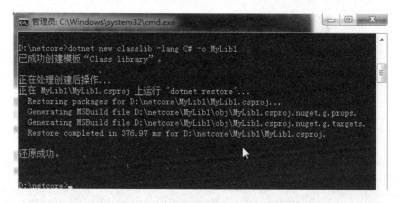

图 2.7　在指定目录中创建 .NET Standard 类库项目

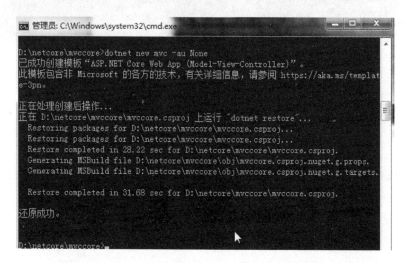

图 2.8　在当前目录中新建一个没有设置身份验证的 ASP.NET Core C# MVC 应用程序项目

④ 列出适用于 MVC 的所有模板,如图 2.9 所示:

dotnet new mvc -l

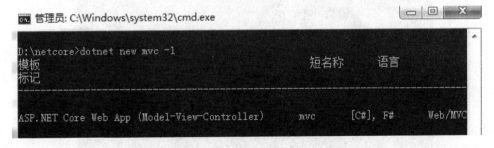

图 2.9　列出适用于 MVC 的所有模板

⑤ 在当前目录中创建 global.json,将 SDK 版本设为 2.1.0,如图 2.10 所示(仅适用于 .NET Core SDK 2.0 或更高版本):

dotnet new globaljson --sdk-version 2.1.0

```
{
  "sdk": {
    "version": "2.1.0"
  }
}
```

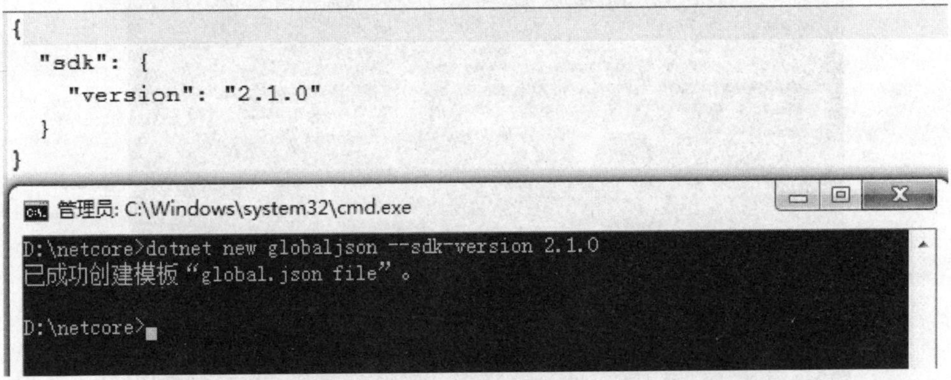

图 2.10　在当前目录中创建 global.json,将 SDK 版本设为 2.1.0

2.3　dotnet restore

dotnet restore 命令:恢复项目的依赖项和工具。

首先通过寻找当前目录下的项目文件,并使用 NuGet 来还原整个项目的依赖项,然后遍历每个目录生成项目文件,接着继续还原该项目文件中的依赖项。

从 .Net Core 2.0 开始,当发出表 2.6 中的命令时,将隐式运行 dotnet restore 命令。

表 2.6 可隐式运行 dotnet restore 命令的命令清单

序 号	命 令
1	dotnet new
2	dotnet build
3	dotnet build-server
4	dotnet run
5	dotnet test
6	dotnet pack
7	dotnet publish

有时,隐式运行 dotnet restore 命令可能不方便。例如,在 Visual Studio Team Services 的持续集成生成中(或者某些自动化构建工具中)需要显式调用 dotnet restore 命令,以控制还原发生的时间和网络使用量。若要避免隐式运行 dotnet restore 命令,可以在执行表 2.6 中的任意命令时加上 --no-restore 选项来禁用隐式还原。

为了还原依赖项,需要提供 NuGet 包所在位置的源。源通常是通过 NuGet.config 配置文件来提供的。安装 CLI 工具时提供一个默认的配置文件。可以通过在项目目录中创建自己的 NuGet.config 文件来指定其他源,也可以在命令提示符处指定每次调用的其他源。

dotnet restore 命令的行为会受 Nuget.config 文件(如果有)中某些设置的影响。例如,在 NuGet.config 中设置 globalPackagesFolder 值将变更 NuGet 包的默认存放位置,这是在 dotnet restore 命令中指定 --packages 选项的替代方法。

命令格式

dotnet restore [〈ROOT〉] [--configfile] [--disable-parallel] [--force] [--ignore-failed-sources] [--no-cache] [--no-dependencies] [--packages] [-r|--runtime] [-s|--source] [-v|--verbosity]

dotnet restore [-h|--help]

参　数

ROOT

指要还原的项目文件的可选路径。

命令选项

dotnet restore 命令的选项说明如表 2.7 所列。

表 2.7 dotnet restore 命令选项

选 项	说 明
--configfile ⟨FILE⟩	供还原操作使用的 NuGet 配置文件（NuGet.config）
--disable-parallel	禁用并行还原多个项目
--force	强制解析所有依赖项，即使上次还原已成功，也不例外。等效于删除 project.assets.json 文件
-h|--help	显示有关命令的帮助信息
--ignore-failed-sources	将包源失败视为警告
--no-cache	不缓存包和 HTTP 请求
--no-dependencies	忽略项目到项目(P2P)的引用，并仅还原根项目，不还原引用
--packages ⟨PACKAGES_DIRECTORY⟩	指定在还原操作期间放置还原包的位置。若未指定，将使用默认的 NuGet 包缓存，它可在所有操作系统上的用户主目录中的 .nuget/packages 目录中被找到
-r|--runtime ⟨RUNTIME_IDENTIFIER⟩	指定程序包还原的运行时。这用于还原 .csproj 文件中的、在 ⟨RuntimeIdentifiers⟩ 标记中未显式列出的运行时的程序包
-s|--source ⟨SOURCE⟩	指定要在还原操作期间使用的 NuGet 包源。此设置会替代 NuGet.config 文件中指定的所有源。多次指定此选项可以提供多个源
--verbosity ⟨LEVEL⟩	设置命令的详细级别。允许使用的值为 q[uiet]、m[inimal]、n[ormal]、d[etailed] 和 diag[nostic]

2.4 dotnet sln

2.4.1 介 绍

dotnet sln 命令：修改 .NET Core 解决方案文件。

使用 dotnet sln 命令可以很方便地在解决方案文件中添加、删除和列出项目。使用 dotnet sln 命令的前提是必须存在解决方案文件。此命令的选项只有一个[-h\|--help]。

命令格式

dotnet sln [⟨SOLUTION_NAME⟩] add ⟨PROJECT⟩⟨PROJECT⟩...
dotnet sln [⟨SOLUTION_NAME⟩] add ⟨GLOBBING_PATTERN⟩
dotnet sln [⟨SOLUTION_NAME⟩] remove ⟨PROJECT⟩⟨PROJECT⟩...
dotnet sln [⟨SOLUTION_NAME⟩] remove ⟨GLOBBING_PATTERN⟩

dotnet sln [<SOLUTION_NAME>] list
dotnet sln [-h|--help]

指　令

add <PROJECT> ...

add <GLOBBING_PATTERN>

将一个或多个项目添加到解决方案文件中。基于 Unix/Linux 的终端支持通配模式。

remove <PROJECT> ...

remove <GLOBBING_PATTERN>

从解决方案文件中删除一个或多个项目。基于 Unix/Linux 的终端支持通配模式。

list

列出解决方案文件中的所有项目。

参　数

SOLUTION_NAME

要使用的解决方案文件的名称。如果未指定，则此命令会搜索当前目录来获取一个解决方案文件。如果目录中有多个解决方案文件，则必须指定一个。

2.4.2　示　例

dotnet sln 命令的几个使用示例如下：

① 创建一个解决方案文件，如图 2.11 所示：

dotnet new sln

图 2.11　创建一个解决方案文件

② 将一个 C# 项目添加到解决方案中，如图 2.12 所示：

dotnet sln netcore.sln add MyLib1/MyLib1.csproj

```
Microsoft Visual Studio Solution File, Format Version 12.00
# Visual Studio 15
VisualStudioVersion = 15.0.26124.0
MinimumVisualStudioVersion = 15.0.26124.0
Project("{FAE04EC0-301F-11D3-BF4B-00C04F79EFBC}") = "MyLib1", "MyLib1\MyLib1.csproj", "{E7D0D852-94CF-4311-A7FD-C4CA54E010B7}"
EndProject
Global
    GlobalSection(SolutionConfigurationPlatforms) = preSolution
        Debug|Any CPU = Debug|Any CPU
        Debug|x64 = Debug|x64
        Debug|x86 = Debug|x86
        Release|Any CPU = Release|Any CPU
        Release|x64 = Release|x64
        Release|x86 = Release|x86
    EndGlobalSection
    GlobalSection(SolutionProperties) = preS
        HideSolutionNode = FALSE
    EndGlobalSection
    GlobalSection(ProjectConfigurationPlatfo
        {E7D0D852-94CF-4311-A7FD-C4CA54E010B
        {E7D0D852-94CF-4311-A7FD-C4CA54E010B
        {E7D0D852-94CF-4311-A7FD-C4CA54E010B
```

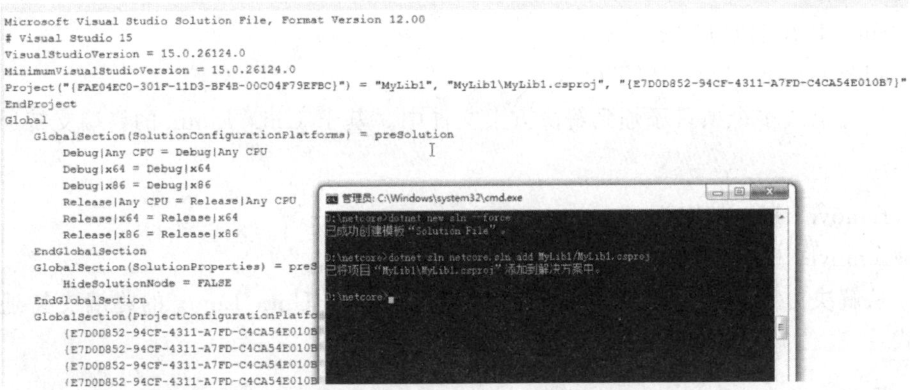

图 2.12　将一个 C# 项目添加到解决方案中

③ 从解决方案中删除一个 C# 项目，如图 2.13 所示：

dotnet sln netcore.sln remove MyLib1/MyLib1.csproj

```
Microsoft Visual Studio Solution File, Format Version 12.00
# Visual Studio 15
VisualStudioVersion = 15.0.26124.0
MinimumVisualStudioVersion = 15.0.26124.0
Global
    GlobalSection(SolutionProperties) = preSolution
        HideSolutionNode = FALSE
    EndGlobalSection
EndGlobal
```

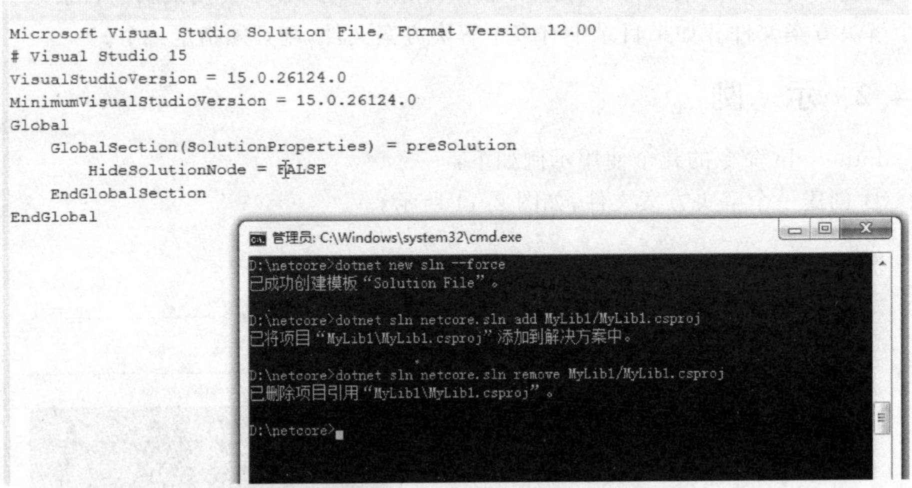

图 2.13　从解决方案中删除一个 C# 项目

④ 将多个 C# 项目添加到解决方案中：

dotnet sln netcore.sln add MyLib1/MyLib1.csproj MyLib2/MyLib2.csproj

⑤ 从解决方案中删除多个 C# 项目：

dotnet sln netcore.sln remove MyLib1/MyLib1.csproj MyLib2/MyLib2.csproj

2.5 dotnet build

2.5.1 介 绍

dotnet build 命令:将项目及其依赖项生成为一组二进制文件。

二进制文件包括中间语言(IL)文件(.dll 扩展名)和用于调试的符号文件(.pdb 扩展名)。同时会生成两个 JSON 文件,一个是记录应用程序依赖项的文件(类似 *.deps.json),另一个是指定应用程序的共享运行时及其版本的文件(类似 *.runtime-config.json)。

project.assets.json 文件记录了应用程序中的依赖项。如果没有此文件,则在应用程序进行构建时,构建工具将无法解析引用程序集,进而导致生成错误。此文件在 dotnet restore 命令执行时创建。

dotnet build 命令使用 MSBuild 生成项目,因此它支持并行生成和增量生成。

dotnet build 命令除接受自己的选项外,也接受 MSBuild 的选项,如用来设置属性的"/p"或用来定义记录器的"/l"。

命令格式

dotnet build [〈PROJECT〉] [-c|--configuration] [-f|--framework] [--force] [--no-dependencies] [--no-incremental] [--no-restore] [-o|--output] [-r|--runtime] [-v|--verbosity] [--version-suffix]

dotnet build [-h|--help]

参 数

PROJECT

要生成的项目文件。如果未指定项目文件,则 MSBuild 会在当前工作目录中搜索文件扩展名是以 proj 结尾的文件,并使用该文件。dotnet build 命令的选项说明如表 2.8 所列。

表 2.8 dotnet build 命令选项

选 项	说 明
-c|--configuration 〈Debug|Release〉	定义生成配置。默认值为 Debug
-f|--framework 〈FRAMEWORK〉	编译特定框架。必须在项目文件中定义该框架
--force	强制解析所有依赖项,即使最后一次还原已成功,也不例外。等效于删除 project.assets.json 文件
-h|--help	显示有关命令的帮助信息

续表 2.8

选 项	说 明
--no-dependencies	忽略项目到项目（P2P）的引用，并仅生成根项目
--no-incremental	禁用增量生成，并强制完全重新生成项目依赖项关系图
--no-restore	在生成期间不执行隐式还原
-o|--output ⟨OUTPUT_DIRECTORY⟩	放置生成二进制文件的目录。指定此选项时还需要定义 --framework
-r|--runtime ⟨RUNTIME_IDENTIFIER⟩	针对给定运行时发布项目
-v|--verbosity ⟨LEVEL⟩	设置命令的详细级别。允许使用的 LEVEL 值为 q[uiet]、m[inimal]、n[ormal]、d[etailed]和 diag[nostic]
--version-suffix ⟨VERSION_SUFFIX⟩	在项目文件的版本字段中定义星号（*）版本后缀。格式遵循 NuGet 的版本准则

2.5.2 示 例

dotnet build 命令的几个使用示例如下：

① 生成项目及其依赖项：

dotnet build

② 使用"发布"配置生成项目及其依赖项：

dotnet build --configuration Release

③ 针对特定运行时（本例中为 Ubuntu 16.04）生成项目及其依赖项：

dotnet build --runtime ubuntu.16.04-x64

2.6 dotnet pack

2.6.1 介 绍

dotnet pack 命令：生成项目并创建 NuGet 包。

将被打包项目的 NuGet 依赖项添加到 .nuspec 文件中，以便在安装包时可以进行正确解析。项目到项目的引用不会被打包到项目内。目前，如果具有项目到项目的依赖项，则每个项目均必须包含一个包。

默认情况下，dotnet pack 命令先构建项目。如果希望避免此行为，则传递 --no-build 选项。此选项在持续集成（CI）生成方案中通常是非常有用的。此命令的具体选项说明如表 2.9 所列。

可向 dotnet pack 命令提供 MSBuild 属性，用于打包进程。

命令格式

dotnet pack [〈PROJECT〉] [-c | --configuration] [--force] [- -include-source] [--include-symbols] [--no-build] [--no-dependencies] [--no-restore] [-o|--output] [--runtime] [-s|--serviceable] [-v|--verbosity] [--version-suffix]

dotnet pack [-h|--help]

参　数

PROJECT

要打包的项目。它可能是文件扩展名为 csproj 的文件或目录的路径。如果未指定，则默认为当前目录。

表 2.9　dotnet pack 命令选项

选　项	说　明
-c|--configuration {Debug|Release}	定义生成配置。默认值为 Debug
--force	强制解析所有依赖项，即使最后一次还原已成功，也不例外。等效于删除 project.assets.json 文件
-h|--help	显示有关命令的帮助信息
--include-source	包括 PDB 和源文件。源文件放入 NuGet 结果的 src 文件夹中
--include-symbols	生成常规包和带符号的包
--no-build	在打包前跳过生成项目。还将隐式设置 --no-restore 标记
--no-dependencies	忽略项目间的引用，仅还原根项目
--no-restore	运行此命令时不执行隐式还原
-o|--output 〈OUTPUT_DIRECTORY〉	将生成的包放置在指定目录中
--runtime 〈RUNTIME_IDENTIFIER〉	指定要为其还原的包的目标运行时
-s|--serviceable	设置包中可用的标志
--version-suffix 〈VERSION_SUFFIX〉	定义项目中 MSBuild 属性 $(VersionSuffix) 的值
-v|--verbosity 〈LEVEL〉	设置命令的详细级别。允许使用的 LEVEL 值为 q[uiet]、m[inimal]、n[ormal]、d[etailed] 和 diag[nostic]

2.6.2　示　例

dotnet pack 命令的几个使用示例如下：

① 打包当前目录中的项目，如图 2.14 所示：

dotnet pack

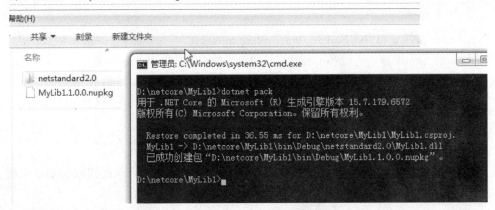

图 2.14 打包当前目录中的项目

② 打包 MyLib2 项目：

dotnet pack MyLib2/MyLib2.csproj

③ 打包当前目录中的项目，并将生成的包放置到 nupkgs 文件夹中：

dotnet pack --output nupkgs

④ 使用 PackageVersion MSBuild 属性将包版本设置为 2.1.0：

dotnet pack /p:PackageVersion=2.1.0

⑤ 打包特定目标框架的项目：

dotnet pack /p:TargetFrameworks=net45

2.7 dotnet run

2.7.1 介 绍

dotnet run 命令：无须任何显式编译或启动命令即可运行源代码。

输出文件会写到默认位置，即 bin/〈configuration〉/〈target〉处。例如，如果是使用 .NET Core 2.0 开发的应用程序并且运行 dotnet run 命令，则会在 bin/Debug/netcoreapp2.0 目录中生成应用程序文件，如果目录中已经存在相同文件，则将根据需要覆盖文件。临时文件将被置于 obj 目录中。

如果该项目指定多个框架，则在不使用 -f|--framework〈FRAMEWORK〉选项指定框架时，执行 dotnet run 命令将导致错误。

应在项目上下文，而不是生成程序集中使用 dotnet run 命令。如果尝试改为运行依赖于框架的应用程序 DLL,则必须在不使用命令的情况下使用 dotnet。例如,若要运行 myApp.dll,则使用:

```
dotnet myApp.dll
```

若要运行应用程序,则 dotnet run 命令需从 NuGet 缓存中解析共享运行时之外的应用程序依赖项。因为此命令使用缓存的依赖项,所以,不推荐在生产环境中使用 dotnet run 命令来运行应用程序。推荐使用 dotnet publish 命令创建需要发布的内容。

命令格式

dotnet run [-c|--configuration] [-f|--framework] [--force] [--launch-profile] [--no-build] [--no-dependencies] [--no-launch-profile] [--no-restore] [-p|--project] [--runtime] [-v|--verbosity] [[--] [application arguments]]

dotnet run [-h|--help]

dotnet run 命令的选项说明如表 2.10 所列。

表 2.10 dotnet run 命令选项

选 项	说 明
--	dotnet run 命令选项与正在运行的应用程序的参数的分隔符。在此分隔符后的所有参数均传递给已运行的应用程序
-c|--configuration ⟨Debug|Release⟩	定义生成配置。默认值为 Debug
-f|--framework ⟨FRAMEWORK⟩	使用指定框架生成并运行应用程序。框架必须在项目文件中指定
--force	强制解析所有依赖项,即使最后一次还原已成功,也不例外。等效于删除 project.assets.json 文件
-h|--help	显示有关命令的帮助信息
--launch-profile ⟨NAME⟩	启动应用程序时要使用的启动配置(如果有)的名称。启动配置在 launchSettings.json 文件中定义,启动配置的名称通常称为 Development、Staging 和 Production
--no-build	运行前跳过生成项目操作
--no-dependencies	忽略项目到项目的引用,并仅还原根项目,不还原引用
--no-launch-profile	不尝试使用 launchSettings.json 文件配置应用程序
--no-restore	运行此命令时不执行隐式还原
-p|--project ⟨PATH⟩	指定要运行的项目文件的路径(文件夹名称或完整路径)。如果未指定,则默认为当前目录
--runtime ⟨RUNTIME_IDENTIFIER⟩	要还原包的目标运行时
-v|--verbosity ⟨LEVEL⟩	设置命令的详细级别。允许使用的 LEVEL 值为 q[uiet]、m[inimal]、n[ormal]、d[etailed]和 diag[nostic]

2.7.2 示 例

dotnet run 命令的几个使用示例如下：

① 运行当前目录中的项目：

dotnet run

② 运行指定的项目，如图 2.15 所示：

dotnet run --project d:\myApp\myApp.csproj

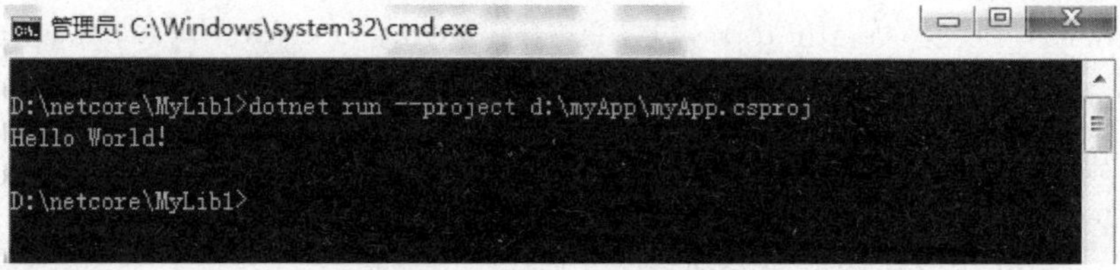

图 2.15 运行指定的项目

③ 运行当前目录中的项目（在本例中，--help 参数被传递到应用程序中，因为使用了空白的"--"选项），如图 2.16 所示：

dotnet run --configuration Release -- --help

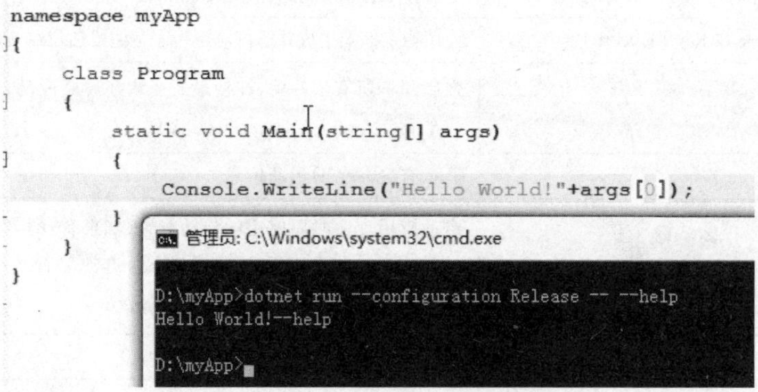

图 2.16 运行当前目录中的项目

2.8 dotnet publish

2.8.1 介 绍

dotnet publish 命令：将编译的应用程序文件及其依赖项文件打包并发布到文件

夹,以方便部署到托管系统中。

dotnet publish 命令的输出内容如图 2.17 所示。

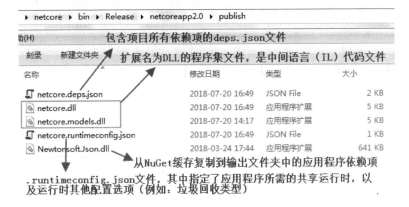

图 2.17 项目发布结构

命令格式

dotnet publish [〈PROJECT〉] [-c|--configuration] [-f|--framework] [--force] [--manifest] [--no-build] [--no-dependencies] [--no-restore] [-o|--output] [-r|--runtime] [--self-contained] [-v|--verbosity] [--version-suffix]

dotnet publish [-h|--help]

参　数

PROJECT

要发布的项目。如果未指定,则默认为当前目录。

dotnet publish 命令的选项说明如表 2.11 所列。

表 2.11　dotnet publish 命令选项

选　项	说　明
-c|--configuration {Debug|Release}	定义生成配置。默认值为 Debug
-f|--framework 〈FRAMEWORK〉	为发布应用程序而指定的目标框架。必须在项目文件中指定目标框架
--force	强制解析所有依赖项,即使最后一次还原已成功,也不例外。等效于删除 project.assets.json 文件
-h|--help	显示有关命令的帮助信息
--manifest 〈PATH_TO_MANIFEST_FILE〉	指向目标清单文件的路径,该文件包含要通过发布步骤执行的包的列表。若要指定多个清单,则请为每个清单添加一个 --manifest 选项。自 .NET Core 2.0 SDK 起,可以使用此选项

续表 2.11

选项	说明
--no-build	发布前不生成项目。还会隐式设置 --no-restore 标记
--no-dependencies	忽略项目间的引用,仅还原根项目
--no-restore	运行此命令时不执行隐式还原
-o|--output ⟨OUTPUT_DIRECTORY⟩	指定输出目录的路径。若未指定,则: • 对于依赖于框架的部署,默认目录为 ./bin/{Debug|Release}/[framework]/publish/; • 对于独立部署,默认目录为 ./bin/{Debug|Release}/[framework]/[runtime]/publish/
--self-contained	与应用程序一同发布.NET Core 运行时,免除在目标计算机上安装运行时。如果指定了运行时标识符,则默认值为 true
-r|--runtime ⟨RUNTIME_IDENTIFIER⟩	发布针对给定运行时的应用程序。这将在创建独立部署(SCD)时使用。默认为发布依赖于框架的部署(FDD)
-v|--verbosity ⟨LEVEL⟩	设置命令的详细级别。允许使用的 LEVEL 值为 q[uiet]、m[inimal]、n[ormal]、d[etailed]和 diag[nostic]
--version-suffix ⟨VERSION_SUFFIX⟩	定义项目中属性 $⟨VersionSuffix⟩$ 的值

2.8.2 示 例

dotnet publish 命令的几个使用示例如下:

① 在当前目录中发布项目:

dotnet publish

② 使用指定的项目文件发布应用程序,如图 2.18 所示:

dotnet publish d:\myApp\myApp.csproj

图 2.18 发布应用程序

2.9 dotnet add package

2.9.1 介绍

dotnet add package 命令：向项目文件添加包引用。

运行该命令后，还有一个兼容性检查，以确保包与项目中的框架兼容。如果通过了该检查，则将〈PackageReference〉元素添加到项目文件中并运行 dotnet restore 命令。

命令格式

dotnet add [〈PROJECT〉] package 〈PACKAGE_NAME〉 [-h|--help] [-f|--framework] [-n|--no-restore] [--package-directory] [-s|--source] [-v|--version]

参　数

PROJECT

指定的项目文件。如果未指定，则此命令会搜索当前目录来获取一个项目文件。

PACKAGE_NAME

要添加的包引用。

dotnet add package 命令的选项说明如表 2.12 所列。

表 2.12 dotnet add package 命令选项

选项	说明
-h|--help	显示有关命令的帮助信息
-f|--framework 〈FRAMEWORK〉	仅针对特定框架添加包引用
-n|--no-restore	在没有执行还原预览和兼容性检查的情况下添加包引用
--package-directory 〈PACKAGE_DIRECTORY〉	将包还原到指定目录
-s|--source 〈SOURCE〉	在使用还原操作期间所用的特定的 NuGet 包源
-v|--version 〈VERSION〉	要添加的包的版本

2.9.2 示例

dotnet add package 命令的几个使用示例如下：

① 将 Newtonsoft.Json NuGet 包添加到项目中，结果如图 2.19 所示：

dotnet add package Newtonsoft.Json

在执行了以上命令后，netcore.csproj 文件中将写入用于引用的包的〈PackageReference〉元素，示例代码如下：

ASP.NET Core 应用开发入门教程

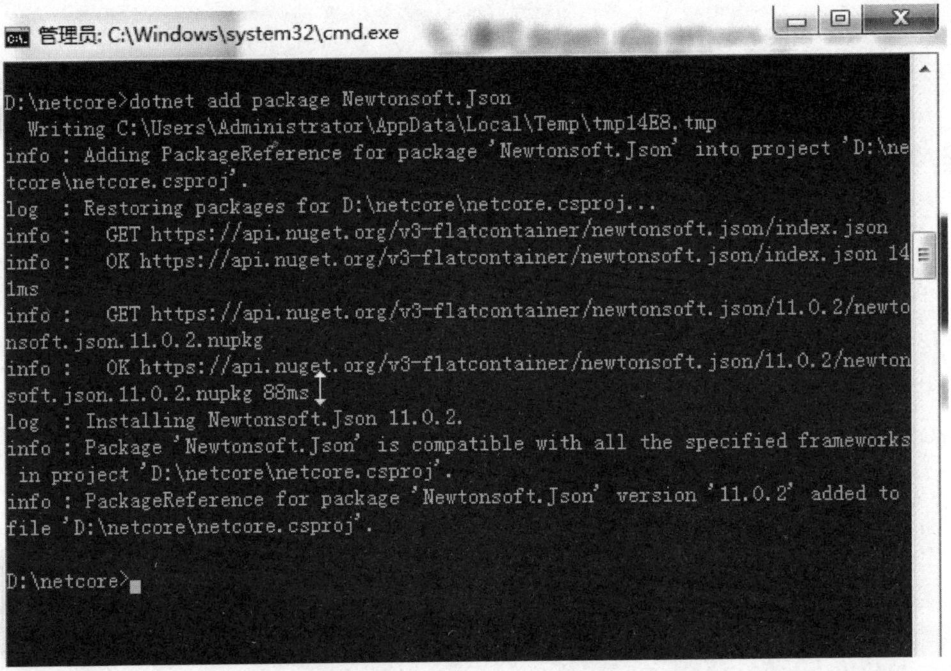

图 2.19 将包添加到项目

```
<PackageReference Include = "Newtonsoft.Json" Version = "11.0.2" />
```

② 向项目添加特定版本的包，结果如图 2.20 所示：

dotnet add package Newtonsoft.Json -v 10.0.0

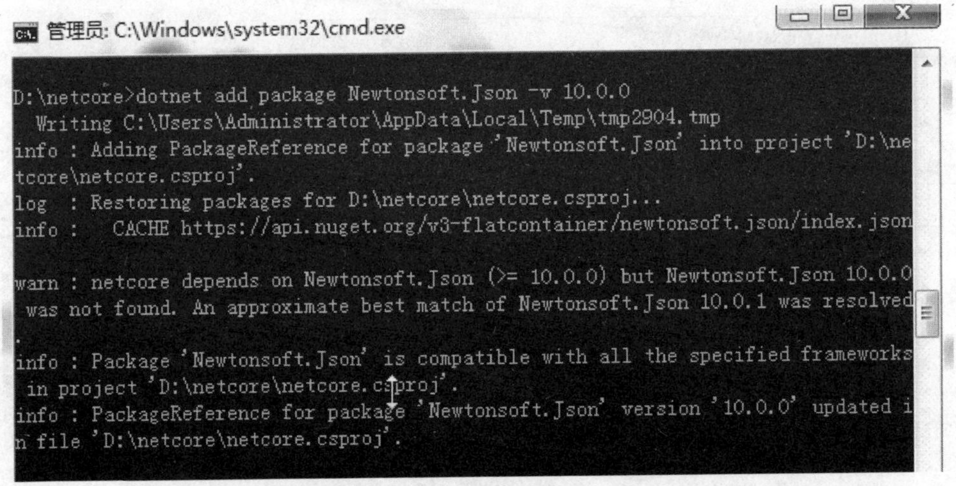

图 2.20 向项目添加特定版本的包

③ 使用特定的 NuGet 源添加包：

dotnet add package Microsoft.AspNetCore.StaticFiles -s https://dotnet.myget.org/F/dotnet-core/api/v3/index.json

2.10 dotnet add reference

2.10.1 介　绍

dotnet add reference 命令：方便地向项目添加项目引用。

运行该命令后，会将〈ProjectReference〉元素添加到项目文件中。

命令格式

dotnet add [〈PROJECT〉] reference [-f|--framework]〈PROJECT_REFERENCES〉 [-h|--help]

参　数

PROJECT

指定的项目文件。如果未指定，则此命令会搜索当前目录来获取一个项目文件。

PROJECT_REFERENCES

要添加的项目到项目（P2P）的引用。指定一个或多个项目。基于 Unix/Linux 的系统支持 glob 模式。

dotnet add reference 命令的选项说明如表 2.13 所列。

表 2.13　dotnet add reference 命令选项

选　项	说　明
-h|--help	显示有关命令的简短帮助
-f|--framework〈FRAMEWORK〉	仅针对特定框架添加项目引用

2.10.2 示　例

dotnet add reference 命令的几个使用示例如下：

① 添加项目引用，如图 2.21 所示：

dotnet add netcore.csproj reference MyLib1/MyLib1.csproj

② 向当前目录中的项目添加多个项目引用，如图 2.22 所示：

dotnet add reference lib1/lib1.csproj lib2/lib2.csproj

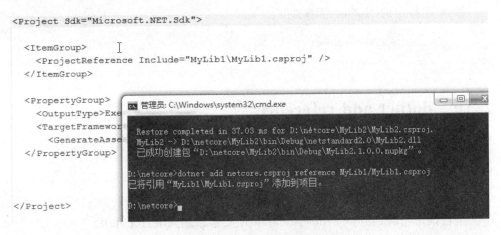

图 2.21 添加项目引用

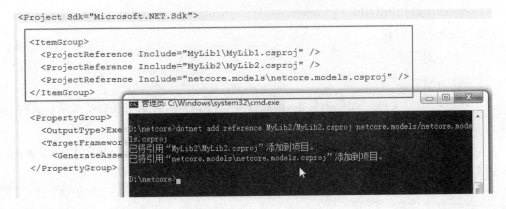

图 2.22 添加多个项目引用

2.11 dotnet 命令综合示例

下面通过一个综合示例来展示 dotnet 命令的使用方法：

① 选择"开始"→"运行"菜单项，并输入 cmd 命令，在弹出的命令行窗口中依次输入 d:和 mkdir netcore 两条命令，以指定在 D 盘下创建目录，如图 2.23 所示。

② 在命令行窗口中输入 cd netcore 命令(进入目录)，如图 2.24 所示。

③ 输入 dotnet new console -lang C♯ 命令在目录中创建一个 C♯ 控制台程序，此时会发现 dotnet restore 已经被隐式执行，如图 2.25 所示。

④ 输入 dotnet new sln 命令创建一个解决方案文件，如图 2.26 所示。

⑤ 输入 dotnet sln netcore.sln add netcore.csproj 命令把在第③步创建的项目添加到解决方案中，如图 2.27 所示。

图 2.23　在命令行中创建目录

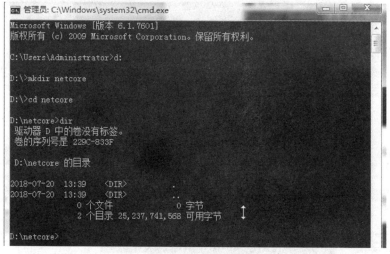

图 2.24　进入目录

图 2.25　创建应用程序

ASP.NET Core 应用开发入门教程

图 2.26 创建解决方案

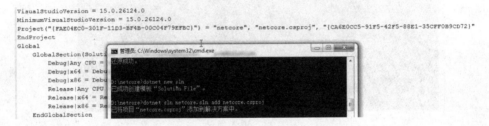

图 2.27 给解决方案添加项目

⑥ 输入 dotnet add package Newtonsoft.Json 命令，将 Newtonsoft.Json NuGet 包添加到项目中，如图 2.28 所示。

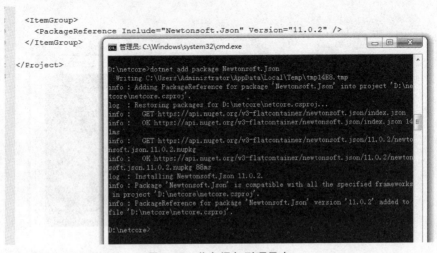

图 2.28 将包添加到项目中

⑦ 通过输入 dotnet new classlib -o netcore.models 命令来创建一个类库，如图 2.29 所示。

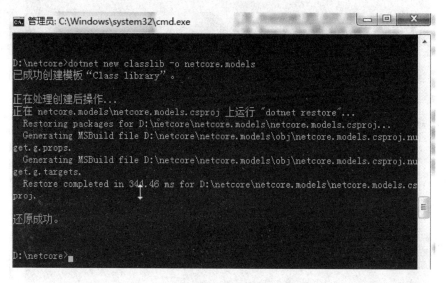

图 2.29 创建类库

⑧ 输入 dotnet add netcore.csproj reference netcore.models\netcore.models.csproj 命令，把刚才创建的类型引用到 netcore 项目中，如图 2.30 所示。

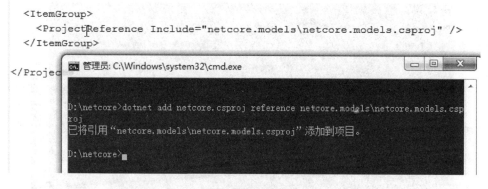

图 2.30 添加引用

⑨ 此时通过 dotnet build 命令来编译 netcore 项目时会报错，如图 2.31 所示。这个错误是由 .NET Core 自动填写程序集信息而引起的，这时只要在 netcore.csproj 文件的〈PropertyGroup〉节点中添加一行如下设置即可解决：

〈GenerateAssemblyInfo〉false〈/GenerateAssemblyInfo〉

⑩ 再次运行 dotnet build 命令则编译成功，如图 2.32 所示。

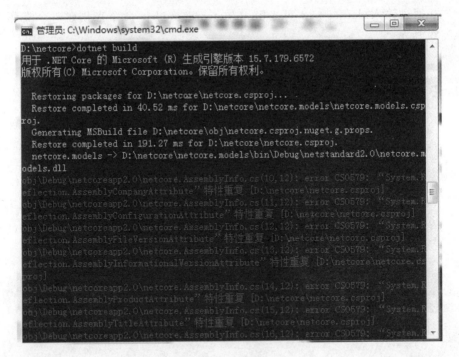

图 2.31 编译失败

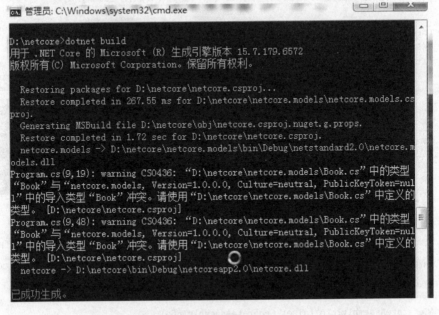

图 2.32 编译成功

⑪ 输入 dotnet run 命令运行应用程序，如图 2.33 所示。

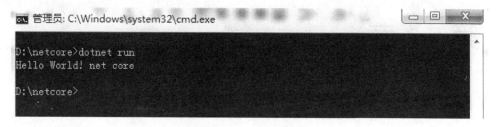

图 2.33　运行应用程序

⑫ 输入 dotnet publish 命令发布刚才编译成功的程序，如图 2.34 所示。

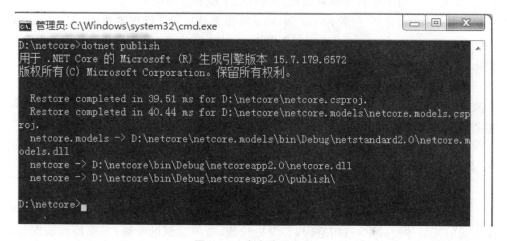

图 2.34　发布应用程序

第 3 章

Visual Studio 2017 与 NuGet

Visual Studio 2017 是适用于软件开发人员的集成开发工具,是一种交互式开发环境（IDE）,可用于查看和编辑代码,并调试、生成和发布应用。

Visual Studio 适用于 Windows 和 Mac 操作系统。Mac 操作系统的 Visual Studio 的许多功能与 Visual Studio 2017 相同,并且针对开发跨平台应用和移动应用进行了优化。

本章主要介绍 Windows 操作系统下的 Visual Studio 2017,介绍它的基本功能,并演示一些可使用 Visual Studio 完成的操作,包括创建简单项目、使用 IntelliSense 辅助编码和调试应用,以便在程序执行过程中查看变量的值。

3.1 安装 Visual Studio 2017

Visual Studio 2017 具有一种同以往不同的全新安装方式,带来全新的轻量级和模块化的安装体验,减少占用空间,实现更快速、更可定制的安装,并支持离线安装;它还减少了最小内存需求量,对系统的影响更小。

3.1.1 检查计算机安装环境

在开始安装 Visual Studio 2017 之前需要做好以下工作:

① 查看系统要求。这些要求有助于了解计算机是否支持 Visual Studio 2017。

② 应用最新的 Windows 更新。这些更新可确保计算机包含最新的安全更新程序和 Visual Studio 所需的系统组件。

③ 重新启动计算机。重新启动可确保挂起的任何安装或更新都不会影响 Visual Studio 的安装。

④ 释放空间。通过运行磁盘清理应用程序等方式,从 %SystemDrive% 中删除不需要的文件和应用程序。

3.1.2 下载 Visual Studio 2017

从 Visual Studio 的官方下载网站（https://visualstudio.microsoft.com/zhhans/downloads/）下载 Visual Studio 的引导程序文件。如图 3.1 所示,请选择所

需的 Visual Studio 2017 版本,单击对应版本的下载按钮,在弹出的对话框中单击"保存文件"按钮。如果没有找到 Visual Studio 2017 的下载按钮,则可以单击"旧版本"按钮。

图 3.1　下载 Visual Studio 引导程序文件

3.1.3　运行 Visual Studio 2017 安装程序

运行 Visual Studio 2017 安装程序的步骤是:
① 找到下载的与下面文件名类似的引导程序文件,双击执行它:
- 对于 Visual Studio Enterprise 版本,请运行 vs_enterprise.exe 文件;
- 对于 Visual Studio Professional 版本,请运行 vs_professional.exe 文件;
- 对于 Visual Studio Community 版本,请运行 vs_community.exe 文件。
② 如果收到用户账户控制通知,请单击"是"按钮。
③ 通常会要求确认 Microsoft 许可条款和 Microsoft 隐私声明,如图 3.2 所示。单击"继续"按钮。

图 3.2　隐私声明

3.1.4　选择工作负载

此时,安装程序进入到选择所需的功能集或工作负载的自定义安装阶段,这时可以选择和安装工作负载。工作负载是开发人员习惯使用的编程语言或平台所需的一些功能。具体安装方法如下:

① 在正在安装 Visual Studio 的对话框中找到所需的工作负载,如图 3.3 所示。

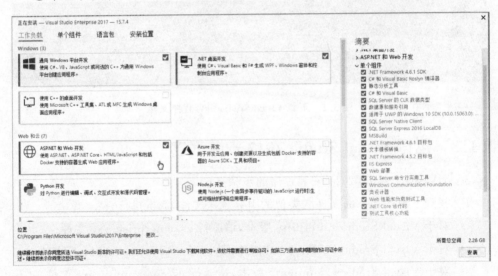

图 3.3　工作负载

② 选择所需的工作负载,如". NET 桌面开发""ASP. NET 和 Web 开发"等,然后单击"安装"按钮。

③ 接着会出现多个显示 Visual Studio 安装进度的状态屏幕。

④ 安装完新的工作负载和组件后,单击"启动"按钮。

3.1.5 逐个选择组件(可选)

如果不想使用工作负载功能来自定义 Visual Studio 2017 安装,则可以改为逐个安装组件。若要选择单个组件,则请从 Visual Studio 安装程序中单击"单个组件",在列表中选择所需项,然后按照提示进行操作,如图 3.4 所示。

图 3.4 单个组件

3.1.6 安装语言包(可选)

默认情况下,安装程序首次运行时会尝试匹配操作系统语言。但也可以自行选择语言,从 Visual Studio 2017 安装程序中单击"语言包",如图 3.5 所示,然后按照提示进行操作。

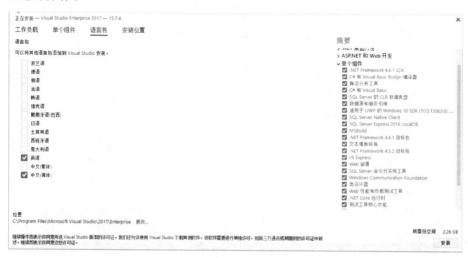

图 3.5 语言包

3.1.7 更改安装位置(可选)

Visual Studio 2017 的 15.7 版本新增的功能是：目前可减少系统驱动器上 Visual Studio 的安装量，可以选择将下载缓存、共享组件、SDK 和工具移动到不同驱动器上，并将 Visual Studio 安装在运行最快的驱动器上，如图 3.6 所示。

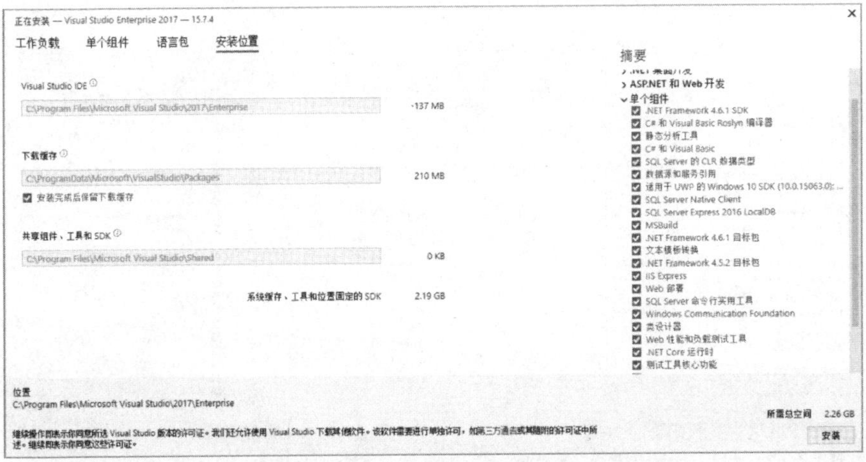

图 3.6　安装位置

3.1.8 起始页介绍

Visual Studio 2017 安装完成后，双击桌面上的 Visual Studio 2017 快捷图标启动 Visual Studio 2017，之后最先看到的是起始页，如图 3.7 所示。将起始页设想成一个"集成中心"，有助于更快地找到所需的命令和项目文件。起始页的内容如下：

图 3.7　起始页

"最近"区域显示出最近处理的项目和文件夹。在"新建项目"区域下,可以单击链接转到"新建项目"对话框中,或者在"打开"区域下,打开现有的项目或代码文件夹。窗口右侧是最新的"开发人员新闻"。

如果关闭了"起始页"并希望再次查看,则可选择"文件"→"起始页"菜单项重新打开它,如图3.8所示。

图3.8 从"文件"菜单中打开起始页

3.2 使用 Visual Studio 创建程序

1. 创建一个项目

现在来创建一个 ASP.NET Core 项目。项目类型随附了所需的全部模板文件,不需要添加任何内容。具体步骤如下:

① 选择"文件"→"新建"→"项目"菜单项,如图3.9所示。

② 还可以在"起始页"的"新建项目"区域下的搜索框中输入 asp.net 进行项目类型查询,如图3.10所示。在列表框中选择"ASP.NET 网站(Razor v3)C♯"。

③ 在"新建项目"对话框中会显示几个项目模板。模板包含给定项目类型所需的基本文件和设置。在"新建项目"对话框左侧的窗格中展开"Visual C♯",然后选择".NET Core"。在中间窗格中,选择"ASP.NET Core Web 应用程序",在"名称"

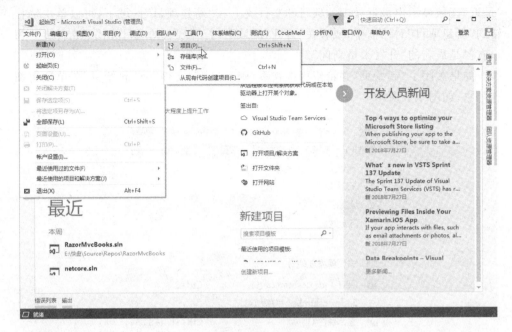

图 3.9 创建项目

文本框中输入"RazorMvcBooks",然后单击"确定"按钮,如图 3.11 所示。

2. 添加工作负载(可选)

如果未显示"ASP. NET Core Web 应用程序"项目模板,则可通过添加"ASP.NET 和 Web 开发"工作负载来获取它。可通过以下两种方式添加此工作负载。

图 3.10 新建项目

方式 1:打开"新建项目"对话框

具体步骤是:

① 单击"新建项目"对话框左窗格中的"打开 Visual Studio 安装程序"链接,如图 3.12 所示。

② 启动 Visual Studio 安装程序(见图 3.13)。选择"ASP.NET 和 Web 开发"工作负载,然后单击"修改"按钮,如图 3.14 所示。

方式 2:使用"工具"菜单栏

具体步骤是:

① 依次选择"工具"→"获取工具和功能"菜单项。

② 启动 Visual Studio 安装程序(见图 3.13)。选择"ASP.NET 和 Web 开发"工作负载,然后单击"修改"按钮,如图 3.14 所示。

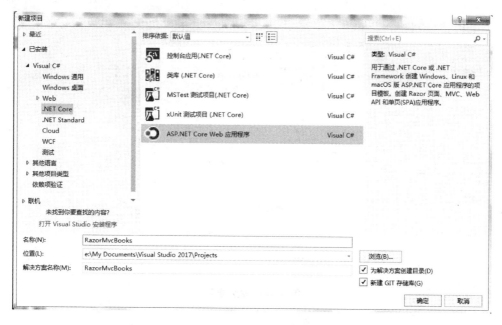

图 3.11　填写项目名称

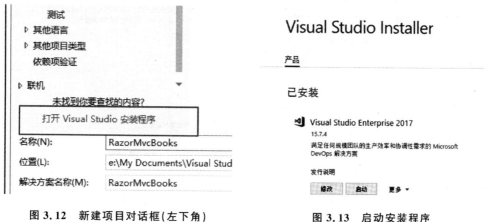

图 3.12　新建项目对话框(左下角)　　　图 3.13　启动安装程序

3. 添加项目模板

具体步骤是：

① 在图 3.11 中单击"确定"按钮，打开如图 3.15 所示"新建 ASP.NET Core Web 应用程序"对话框，然后选择"Web 应用程序"项目模板。

② 从顶部下拉菜单中选择"ASP.NET Core 2.0"，单击"确定"按钮，如图 3.15 所示。

③ 在项目创建成功之后，显示如图 3.16 所示的 Visual Studio 主界面。

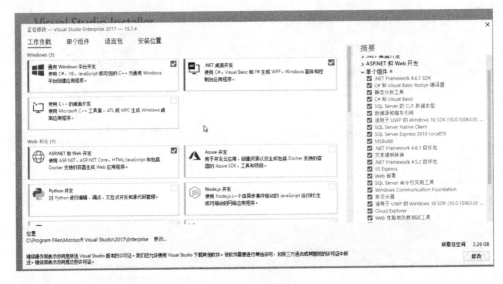

图 3.14 安装变更

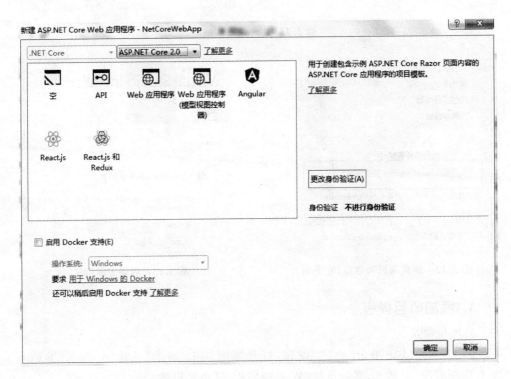

图 3.15 创建 ASP.NET Core Web 应用程序

Visual Studio 2017 与 NuGet

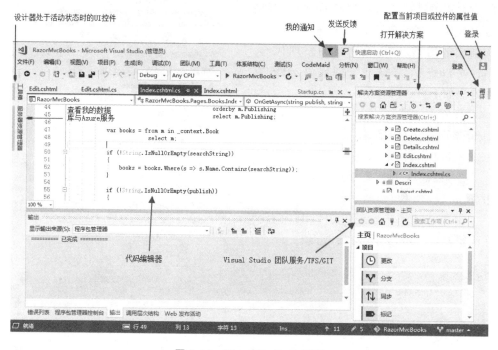

图 3.16 Visual Studio 主界面

3.3 Visual Studio 功能简介

下面介绍几个最可能用到的 Visual Studio 的关键工具。

3.3.1 菜 单

Visual Studio 2017 顶部的菜单栏将命令分组为不同的类别,如图 3.17 所示。例如,"项目"菜单包含与开发者正在处理的项目相关的命令。在"工具"菜单上,可通过选择"选项"命令来自定义 Visual Studio 2017。在"视图"菜单上,可通过选择"错误列表"命令来打开"错误列表"窗口。

图 3.17 Visual Studio 菜单栏

3.3.2 解决方案资源管理器

在 Visual Studio 2017 中选择"视图"→"解决方案资源管理器"菜单项,打开"解决方案资源管理器",可以显示项目、解决方案或代码文件夹中的文件和文件夹层次结构的图形表示,从而帮助整理代码。用户可以查看、导航和管理代码文件,操作方法如下:

① 在3.2节创建了一个解决方案,其中包含一个名为 RazorMvcBooks 的 ASP.NET Core 项目。

② 在"解决方案资源管理器"中选中 Program.cs 文件,如图 3.18 所示。

③ 双击 Program.cs 文件,其内容会在编辑器中显示,如图 3.19 所示。

图 3.18 选择文件

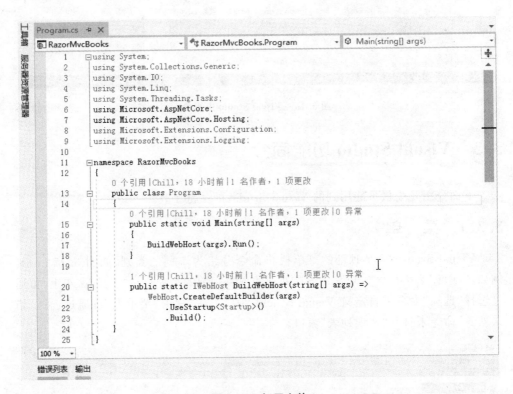

图 3.19 打开文件

④ 项目中还有一个"Pages"文件夹,其中包含布局文件和 Razor 页面文件。例如,将/Pages/Index.cshtml 文件打开,如图 3.20 所示。

图 3.20　选择 Index.cshtml 文件

⑤ 在 Index.cshtml 文件中使用了 Razor 语法,该语法基于标准标记和内联 C# 的组合来呈现 HTML,如图 3.21 所示。

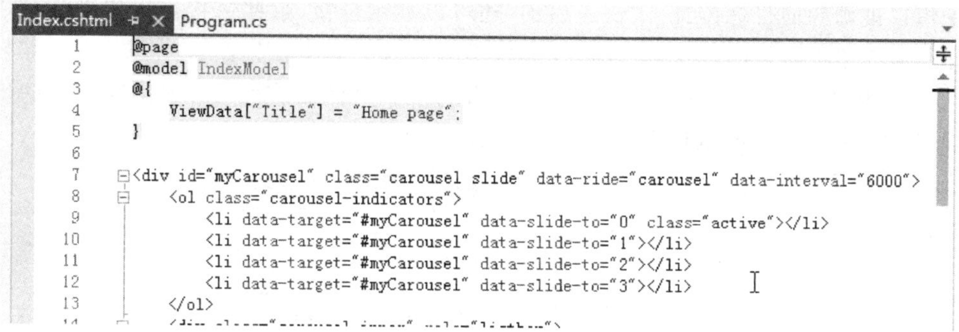

图 3.21　显示 Index.cshtml 文件

⑥ 解决方案还包含 wwwroot 文件夹,用来存放静态文件,这些文件包括 CSS、图片和 JavaScript 文件,如图 3.22 所示。

⑦ 解决方案还有各种在运行时用来管理项目、包以及应用程序的配置文件。例如,默认应用程序配置存储在 appsettings.json 中。可以通过为"开发"环境提供 appsettings.Development.json 文件之类的方式,并基于每个环境来替代部分或所有的设置,如图 3.23 所示。

图 3.22　wwwroot 文件夹

图 3.23　配置文件

3.3.3　快速启动

"快速启动"搜索框是在 Visual Studio 2017 中执行任何操作的快捷方式。使用快速启动搜索框可以在 Visual Studio 2017 中快速找到所需内容。只需输入要查找内容的相关文本，Visual Studio 2017 就会列出与查找文本相关的结果列表，这些结果可以准确导向目标位置。"快速启动"还可以显示链接，这些链接可以启动任何工作负载或单个组件的 Visual Studio 2017 安装程序。具本操作如下：

① 将 RazorMvcBooks 输入到"快速启动"框中，然后在"下拉列表"中选择"生成→生成 RazorMvcBooks"，如图 3.24 所示。

图 3.24　快速启动

② "输出"窗口显示成功生成的消息，如图 3.25 所示。

图 3.25 输出信息

③ "输出"窗口中 Visual Studio 发送通知(例如,调试和错误消息、编译器警告、发布状态消息等)的位置。在"输出"窗口中可以看到生成的信息很简单。

下面通过设置来查看更详细的输出消息:

① 选择"工具"→"选项"菜单项,如图 3.26 所示。

图 3.26 "工具"→"选项"菜单项

② 在"选项"对话框的左侧窗格中选择"项目和解决方案"→"生成并运行"选项。在"MSBuild 项目生成输出详细级别"下拉列表框中选择"普通",然后单击"确定"按钮,如图 3.27 所示。

③ 选择"生成"→"重新生成解决方案"菜单项,或者右击"解决方案资源管理器",然后选择 RazorMvcBooks,从上下文菜单中选择"重新生成"。这次"输出"窗口显示了生成过程中更详细的输出信息,如图 3.28 所示。

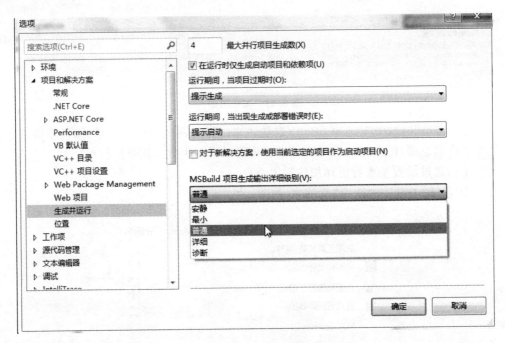

图 3.27 设置选项

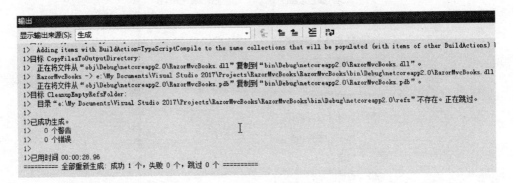

图 3.28 项目生成信息

3.3.4 编辑器

编辑器窗口会占用大部分空间,用于显示代码,同时也可通过该窗口编辑源代码和设计 UI,这里是使用 Visual Studio 2017 完成大部分编码工作的地方。编辑器中的文本已自动着色,用于指示代码的不同内容,如关键字或类型。代码中的垂直短虚线指示哪两个大括号相匹配,行号有助于以后查找代码。可以通过单击带减号的小方块来折叠或展开代码,这个代码大纲功能可以隐藏不需要的代码,最大限度地减少屏幕混乱。

编辑器的操作方法如下：

① 在"编辑器"窗口中打开名为 Program.cs 的文件，如图 3.29 所示。

图 3.29　打开 Program.cs 文件

② 当键入的代码中存在错误或潜在问题时，会在问题代码下使用波浪形下画线发出实时警报，这样就可以立即修改错误，而无需等到编译或运行时才发现错误。如果将光标悬停在波浪线上，将看到关于此错误的其他信息。左边距中也可能会出现一个灯泡，提供修改此错误的建议，如图 3.30 所示。

图 3.30　错误提示

③ 可以在文本编辑器上下文菜单中打开调用层次结构窗口，以显示调用方法、被调用方法和插入点下的方法（插入点），如图 3.31 所示。

图 3.31　查看调用层次结构

④ 在编辑器中使用"CodeLens"功能可快速了解代码内容。"CodeLens"能够查找代码引用、代码更改、链接错误、工作项、代码评审和单元测试,所有操作都在编辑器上进行,如图 3.32 所示。CodeLens 仅在 Visual Studio Enterprise 和 Visual Studio Professional 版中可用,在 Visual Studio Community 版中不可用。

图 3.32　CodeLens

⑤ 在"速览定义"窗口中,可在当前上下文中显示方法或类型的定义,且无须离开当前位置,如图 3.33 所示。

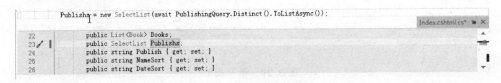

图 3.33　"速览定义"窗口

通过"转到定义"菜单项可直接进入到选中方法或对象的位置,如图 3.34 所示。

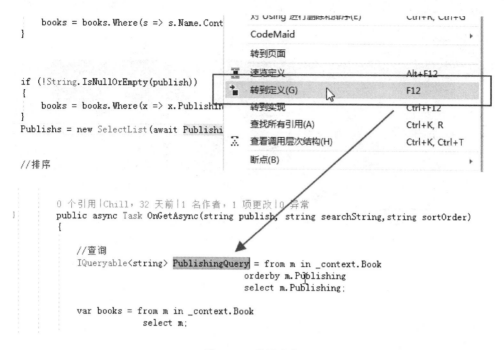

图 3.34 转到定义

3.3.5 运行和调试应用程序

操作过程如下：

① 在 Visual Studio 2017 中单击 IIS Express，可在调试模式下生成并运行应用，如图 3.35 所示。(或者按 F5 键或选择"调试"→"开始调试"菜单项。)

图 3.35 启动调试

备注：如果看到显示内容为"无法连接到 Web 服务器'IIS Express'"的错误消息，则请关闭 Visual Studio 2017；然后再右击 Visual Studio 2017，并选择"以管理员身份运行"选项打开它。接着，再次运行应用。

② Visual Studio 2017 会启动浏览器窗口，如图 3.36 所示。

③ 在浏览器中选择 About 菜单，About 页面将呈现 About.cshtml.cs 文件中设

置的文本，如图 3.37 所示。

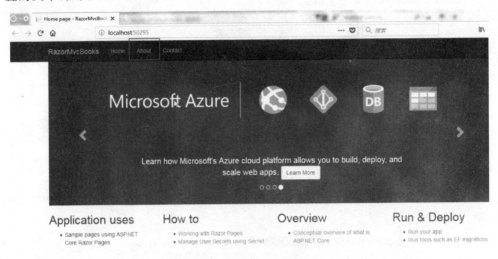

图 3.36　启动浏览器

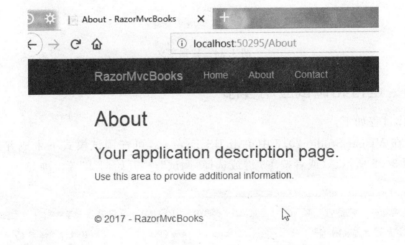

图 3.37　About 页面

3.3.6　调试代码

编写代码时，需要运行并测试该代码是否存在错误。调试时可通过 Visual Studio 2017 逐句执行代码，一次执行一条语句来逐步检查变量；也可设置仅当指定条件为真时才到达的断点。代码运行时可以监视变量的值等内容。

调试过程如下：

① 不需要关闭之前打开的浏览器。现在转换到 Visual Studio 2017 界面，打开 Pages/About.cshtml.cs 文件（如果尚未打开），如图 3.38 所示。

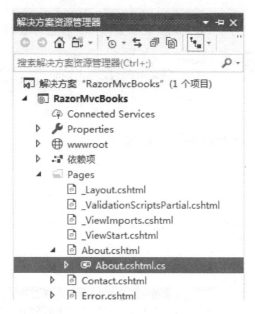

图 3.38 选择 About.cshtml.cs 代码文件

② 在 OnGet 方法的第一行设置一个断点,即让程序在该行暂停执行,然后单击编辑器的最左侧边距,还可单击代码行上的任意位置,然后按 F9 键。此时,最左侧边距中将显示一个红圈,同时代码突出显示为红色,如图 3.39 所示。

图 3.39 设置断点

③ 转换到浏览器界面,并刷新 About 页界,这时会触发 Visual Studio 2017 中的

断点。

④ 此时,将光标悬停在 Message 变量上即可查看它的值,如图 3.40 所示。

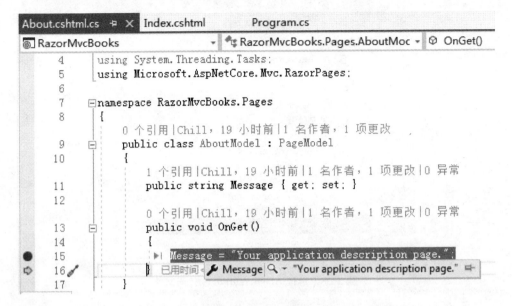

图 3.40　程序中断

⑤ 也可以右击 Message 并选择"添加监视",将变量添加到监视窗口来查看它的值,如图 3.41 所示。

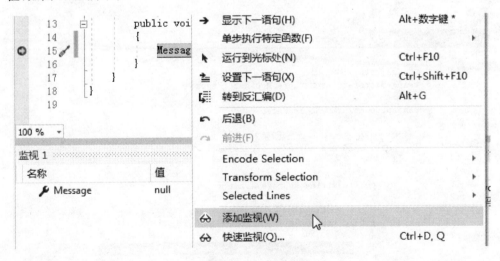

图 3.41　添加监视

注意,此时焦点返回到 Visual Studio 2017 代码编辑器,设置有断点的代码行突出显示为黄色,表示它是程序将执行的下一个代码行。

⑥ 使用与添加断点相同的方式删除应用程序断点。

⑦ 在文本"Use this area to provide additional information."行下方添加一行文字"添加一行中文"并保存文件,如图 3.42 所示。

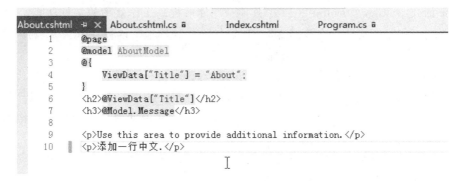

图 3.42　添加文字

⑧ 返回浏览器窗口,查看更新的文本,如图 3.43 所示。(如果看不到更新的文本,则请刷新浏览器。)

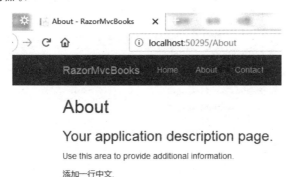

图 3.43　查看更新的文本

⑨ 选择"调试"→"停止调试"菜单项来停止调试,如图 3.44 所示。(或者按 Shift+F5 键或单击"停止调试"工具按钮。)

图 3.44　停止调试

3.3.7 使用重构和 IntelliSense

下面介绍 Visual Studio 2017 中其他一些能够提高工作效率的常见功能。

"重构"包括智能重命名变量、移动选定代码行到单独的函数、移动代码到其他位置、重新排序函数参数以及更多的操作。

"IntelliSense"又可称为智能提示,是一组常用功能的涵盖性术语。这些功能可用于在编辑器中直接显示代码的类型信息,并且可在某些情况下编写小段代码,就如同在编辑器中拥有了基本文档内联,从而节省了在单独帮助窗口中查看类型信息的时间。智能提示功能因语言的不同而异。

下面通过示例来了解如何借助重构和智能提示功能更有效地进行编码。

首先,重命名 Message 变量,步骤是:

① 双击 Message 变量将其选中。

② 输入变量名称 AboutMessage。

③ 单击左侧灯泡图标,显示可用的快速操作。选择"将"Message"重命名为"AboutMessage"",如图 3.45 所示。

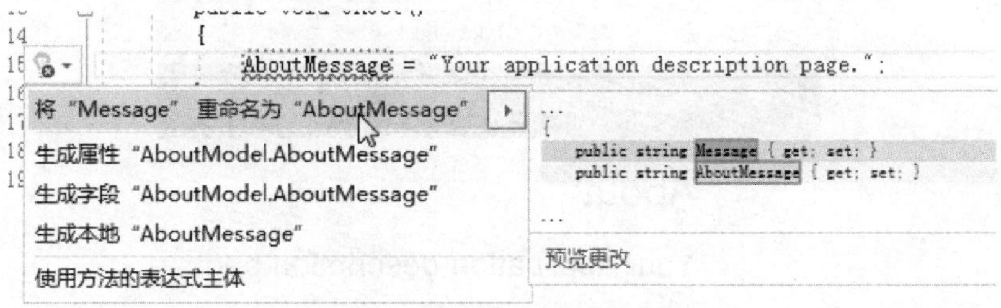

图 3.45 重命名变量

④ 该变量会在整个项目中进行重命名,本例中有三处地方,如图 3.46 所示。

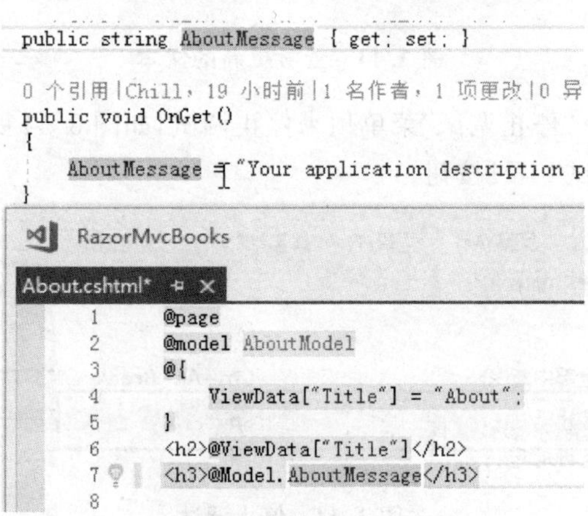

图 3.46 重命名结果

其次，介绍"智能提示"功能，步骤是：

① 在"AboutMessage = " Your application description page.""行的上方键入"DateTime dt= DateTime."。此时会弹出一个下拉列表框，显示 DateTime 类的成员。另外，当前所选成员的说明会显示在单独的框中，如图 3.47 所示。

图 3.47　IntelliSense

② 通过双击或按 Tab 键选择名为"Now"的属性。通过添加分号";"来结束该代码行。

③ 在此行代码下方键入或复制以下代码行：

string year = dt.Year.ToString();
　　AboutMessage = "Your application description page." + year + "年";

然后，再次使用重构来使代码更加简洁，步骤是：

① 单击"DateTime dt= DateTime.Now;"行中的 dt 变量。

注意，该行的边距中会显示一个小螺丝刀图标。

② 单击螺丝刀图标，查看 Visual Studio 2017 提供的建议，如图 3.48 所示。在本示例中，它显示内联临时变量重构，此时可在无须更改整体行为的情况下删除代码行。

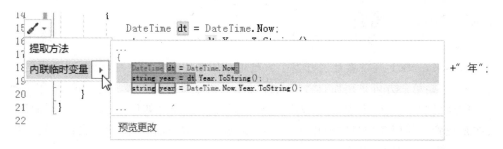

图 3.48　内联临时变量重构

③ 单击"内联临时变量"，重构代码，结果如图 3.49 所示。

最后，按 Ctrl+F5 键重新运行程序，输出内容如图 3.50 所示。

```
0 个引用 | Chill, 19 小时前 | 1 名作者，1 项更改 | 0 异常
public void OnGet()
{
    string year = DateTime.Now.Year.ToString();

    AboutMessage = "Your application description page."+year +" 年";
}
```

图 3.49　代码重构结果

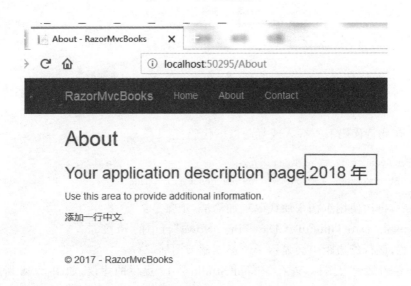

图 3.50　浏览器中的浏览结果

3.4　NuGet 简介

多年来，.NET 开发者社区开发了大量可重用的库和实用工具，以帮助完成常见的任务，这些第三方可重用库日益成为开发人员的主要工具。要想把这些库合并到自己的项目中，会面临发现、安装和维护三大挑战。首先是如何找到需要使用的库？其次是找到后如何在项目中引用它？最后是安装成功后如何跟踪项目更新？这就用到了接下来要介绍的 NuGet 工具。

NuGet 是微软支持的一种适用于任何开发平台的基本工具，用于简化添加、删除和更新 Visual Studio 项目中使用的第三方类库及其依赖项，通过这个工具，开发人员可以创建、共享和使用有用的代码。

3.4.1　包、创建者、主机和使用者之间的关系

NuGet 包是具有 .nupkg 扩展名的单个 ZIP 文件，"包"中含编译代码（Dll）和与

该代码相关的其他文件以及描述性清单（包含包版本号等信息）。开发人员将自己的代码创建成"包"，并将其发布到公用或专用主机上。"包"的使用者从主机上获取这些"包"，将它们添加到自己的项目中，并在项目代码中调用包的功能。随后，NuGet自身负责处理所有中间的详细信息。

NuGet支持公用主机Nuget.org、专用主机，甚至直接在本地文件系统以私密方式托管包。因此，可以使用NuGet包来共享组织或工作组专用的代码。在托管自己的NuGet源中提供了对相关选项的说明，通过配置选项，还可以精确控制任何给定计算机可以访问的主机，从而确保程序包是从特定源获取的；此外，还可以将NuGet包作为一种便捷的方式，将自己的代码用于除自己项目之外的任何其他项目，如图3.51所示。

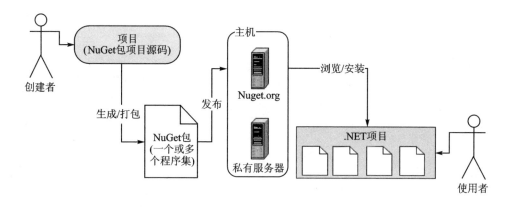

图3.51 包、创建者、主机和使用者之间的关系

3.4.2 包的兼容性

"兼容"包指"包"中所包含的程序集是由与项目所使用的框架兼容的.NET Framework生成的。开发人员可以创建针对某一个框架的"包"，也可以创建更多框架的"包"。为了最大限度地利用"包"的兼容性，开发人员开发的"包"应该使用所有.NET和.NET Core项目都可以使用的.NET Standard。对于创建者和使用者而言，这是最有效的方式，因为单个"包"适用于所有项目。

另外，如果需要.NET Standard之外的API"包"，开发人员应该创建支持不同目标框架的单独的程序集，并将所有这些程序集包含在同一个"包"中（简称多目标"包"）。当使用者安装此类"包"时，NuGet将仅提取项目需要的程序集，这样能将包在该项目生成的最终应用程序和/或程序集中的占用量降到最低。当然，多目标"包"对创建者来说更难维护。

3.4.3 NuGet 工具

除托管支持外,NuGet 还提供各种供创建者和使用者使用的工具,如表 3.1 所列。

表 3.1 NuGet 工具

工具	平台	适用方案	说明
nuget.exe CLI	全部	创建、使用	提供所有 NuGet 功能,包括一些专门适用于包创建者、仅适用于使用者和适用于两者的命令。例如,包创建者使用 nuget pack 命令通过各种程序集和相关文件创建包,包使用者使用 nuget install 命令在项目文件夹中包含包,而所有人都可使用 nuget config 命令设置 NuGet 配置变量。作为与平台无关的工具,NuGet CLI 不会与 Visual Studio 项目交互
dotnet CLI	全部	创建、使用	直接在.NET Core 工具链中提供特定的 NuGet CLI 功能。与 NuGet CLI 一样,dotnet CLI 不会与 Visual Studio 项目交互
包管理器控制台	Windows 版 Visual Studio	使用	提供用于在 Visual Studio 项目中安装和管理包的 PowerShell 命令
包管理器 UI	Windows 版 Visual Studio	使用	提供用于在 Visual Studio 项目中安装和管理包的易用 UI
管理 NuGet UI	Mac 版 Visual Studio	使用	提供用于在 Mac 版 Visual Studio 项目中安装和管理包的易用 UI
MSBuild	Windows	创建、使用	支持创建包和还原项目中直接通过 MSBuild 工具链使用的包

3.4.4 管理依赖项

复用现成的工具,不再重复造轮子,这是"包"管理系统最强大的功能之一。NuGet 的另一个重要用途就是代表项目来管理该依赖关系树或"关系图"。简单来说,仅需要关注在项目中直接使用的"包"。如果任何这些"包"本身使用其他"包"(这些"包"仍可以使用其他"包"),则 NuGet 将负责管理所有这些下层依赖项。

如图 3.52 所示是一个依赖于四个包的项目,这些包同时也依赖于许多其他包。

注意,包 B 有三个不同的使用者,并且每个使用者可能为该"包"指定不同的版本(未显示)。这是一种常见情况,特别是对于广泛使用的"包"。幸运的是,NuGet 将自动确定包 B 的哪一个版本可以满足所有使用者。随后,无论依赖项关系图多么复杂,NuGet 都将对所有包执行相同的操作。

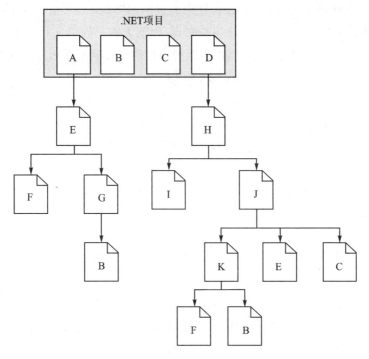

图 3.52　依赖于四个包的项目

3.4.5　跟踪引用和还原包

软件项目可以在开发人员计算机、源代码管理存储库、生成服务器等位置之间移动，因此不适合将 NuGet 包中的程序集直接集成到软件项目中，这样做不仅会使项目的每个副本浪费一些不必要的空间，还会使包中的程序集难以更新到新版本。

NuGet 创建一个项目所依赖的"包"的引用列表，这个引用列表中记录了"包"的标识符和版本号，以及顶层和下层的依赖关系。每当从一个新的计算机上打开项目时，NuGet 通过引用列表随时从公用和/或专用主机重新安装这些"包"，即"还原"这些包。当将项目提交到源代码管理存储库或以其他方式进行共享时，只需包含引用列表，而不需要包含任何"包"中的程序集文件。

如果用 Visual Studio 2017 打开项目，则它将自动还原"包"。NuGet 会查看引用列表中每一个包的信息，并下载解压这些包，进行自动恢复。

NuGet 维护的引用列表有两种包管理格式，分别称为：

① packages.config(NuGet 1.0+)：这是一种 XML 文件，用于维护项目中所有依赖项的简单列表，包括其他已安装"包"的依赖项。已安装或已还原的"包"存储在 packages 文件夹中。

② PackageReference(或"项目文件中的包引用")(NuGet 4.0+)：用于维护直接位于项目文件中的项目顶层依赖项的列表，因此无需单独的文件。关联文件 obj/project.assets.json 是动态生成的，用于维护项目的依赖项关系图以及所有下层依赖项，以确保项目所需要的包都已经安装。

任何特定项目中所用的包管理格式取决于项目类型和 NuGet(和/或 Visual Studio)的可用版本。若要确认当前使用的格式，则只需在安装第一个包后在项目根目录中查找 packages.config。如果没有该文件，则直接在项目文件中查找⟨PackageReference⟩元素。

建议使用 PackageReference。出于与旧版兼容的目的，对 packages.config 进行维护，但对其不再进行新的开发与升级。

3.5　在 Visual Studio 中安装和使用包

本节将演练如何通过 NuGet 在 2.11 节的示例的项目中添加 Newtonsoft.Json 包。安装 NuGet 程序包可以使用程序包管理器 UI 或程序包管理器控制台。在安装包时，NuGet 会将依赖项记录在项目文件或 packages.config 文件中。

Windows 上的 Visual Studio 2012 及更高版本中都包括"NuGet 包管理器"。该程序包管理器提供程序包管理器 UI 和程序包管理器控制台，通过它可以运行大部分的 NuGet 操作。

3.5.1　程序包管理器 UI

程序包管理器 UI 的使用方法如下：

① 在 Visual Studio 2017 中选择"工具"→"选项"菜单项，弹出"选项"对话框，在左侧列表框中选择"NuGet 包管理器"下的常规，根据自己需要，设置默认包管理格式，如图 3.53 所示。

② 在解决方案资源管理器中，右击"引用"，选择"管理 NuGet 程序包"，如图 3.54 所示。

③ 将"程序包源"下拉列表框选择为"nuget.org"，在"浏览"选项卡的文本框中输入"Newtonsoft.Json"，在下拉列表中选择该包，然后单击"安装"按钮，如图 3.55 所示。

④ 如果系统提示查看更改，则单击"确定"按钮，如图 3.56 所示。

⑤ 接受任何许可证提示，如图 3.57 所示。

Visual Studio 2017 与 NuGet 3

图 3.53　程序包格式设置

图 3.54　管理 NuGet 程序包

图 3.55　浏览包

ASP.NET Core 应用开发入门教程

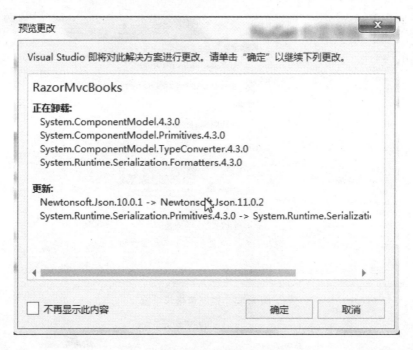

图 3.56　安装包

图 3.57　接受许可

3.5.2　程序包管理器控制台

程序包管理器控制台的使用方法如下：

① 选择"工具"→"NuGet 包管理器"→"程序包管理器控制台"菜单项，将控制台

打开后,检查"默认项目"下拉列表框中是否显示了在程序包中要安装的项目。如果在解决方案中有一个项目,则它已被选中,如图 3.58 所示。

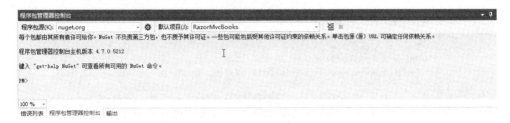

图 3.58　程序包管理器控制台

② 输入 Install-Package Newtonsoft.json 命令,控制台窗口会显示该命令的输出,如图 3.59 所示。错误通常指示程序包与项目的目标框架不兼容。

图 3.59　在程序包管理器控制台上安装包

3.5.3　在应用中使用 Newtonsoft.Json API

使用 Newtonsoft.Json API 的方法是:

① 通过 3.5.1 小节和 3.5.2 小节的两种方式中的一种,已经安装了 Newtonsoft.Json 包。下面使用项目中的 Newtonsoft.Json 包调用 JsonConvert.SerializeObject 方法,将对象转换为字符串,在页面上显示出来。

② 打开 About.cshtml.cs 文件,然后在 OnGet 方法中插入以下代码:

```
Account account = new Account
{
    Name = "John Doe",
    Phone = "13300223344",
    CreateTime = new DateTime(2018, 7, 20, 0, 0, 0, DateTimeKind.Utc),
};
string json = JsonConvert.SerializeObject(account, Formatting.Indented);
Message = "Your application description page." + json;
```

③ 尽管已将 Newtonsoft.Json 包添加到项目中,但还是无法使用,所以在"Json-

Convert"下方仍会出现红色波浪线,如图3.60所示。

```
public void OnGet()
{
    Account account = new Account
    {
        Name = "John Doe",
        Phone = "13300223344",
        CreateTime = new DateTime(2018, 7, 20, 0, 0, 0, DateTimeKind.Utc),
    };
    string json = JsonConvert.SerializeObject(account, Formatting.Indented);

    Message = "Your application description page."+json;
}
```

图3.60　错误提示

④ 在文件头部添加 using Newtonsoft.Json,如图3.61所示。

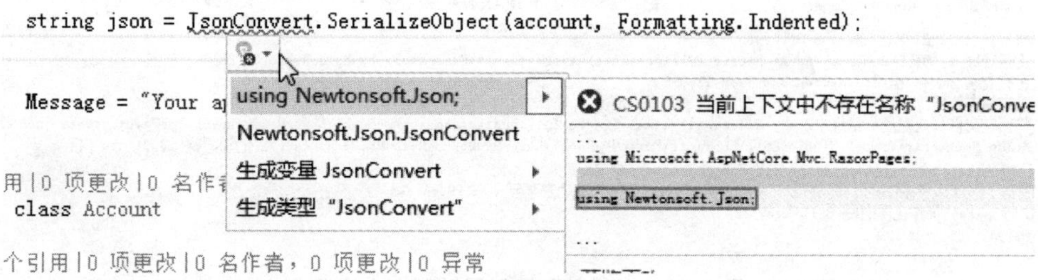

图3.61　智能提示

⑤ 按F5键或选择"调试"→"开始调试"菜单项。

⑥ 在浏览器中单击"About"标签,查看JSON文本的内容,如图3.62所示。

图3.62　浏览About页面

第 4 章

ASP.NET Core 简介

ASP.NET Core 是一个开源跨平台框架,用于生成基于 Internet 连接的应用程序,例如 Web 应用和服务。ASP.NET Core 应用可以在 .NET Core 或 .NET Framework 上运行。可以在 Windows、Mac 和 Linux 上跨平台开发和运行 ASP.NET Core 应用。

使用 ASP.NET Core 可以:
① 创建 Web 应用程序和服务、IoT 应用和移动后端。
② 在 Windows、macOS 和 Linux 上使用喜爱的开发工具。
③ 部署到云或本地。
④ 在.NET Core 或 .NET Framework 上运行。

4.1 为何使用 ASP.NET Core

ASP.NET Core 是重新设计的 ASP.NET 4.x,更改了体系结构,形成了更精简的模块化框架。

ASP.NET Core 具有如下优点:
① 可生成 Web UI 和 Web API 的统一场景。
② 具有新式客户端框架和开发流程。
③ 具有基于环境的云就绪配置系统。
④ 内置依赖项注入。
⑤ 具有轻型的高性能模块化 HTTP 请求管道。
⑥ 能够在 IIS、Nginx、Apache、Docker 上进行托管或在自己的进程中进行自托管。
⑦ 当面向.NET Core 时,可以使用并行应用版本来控制。
⑧ 是简化新式 Web 开发的工具。
⑨ 能够在 Windows、macOS 和 Linux 上开发和运行。
⑩ 是开源代码和以社区为中心。

ASP.NET Core 是完全作为 NuGet 包的一部分提供的。借助 NuGet 包,可以将应用优化为只包含必需的依赖项。实际上,面向.NET Core 的 ASP.NET Core 2.x

应用只需要使用一个 NuGet 包,这样做可以提升安全性、减少维护和提高性能。

ASP.NET Core 可以面向.NET Core 或.NET Framework。面向.NET Framework 的 ASP.NET Core 应用无法跨平台,它们仅在 Windows 上运行。通常,ASP.NET Core 由.NET Standard 库组成。使用.NET Standard 2.0 编写的应用可在 NET Standard 2.0 支持的任何位置运行。

ASP.NET Core 与常用客户端框架和库(包括 Angular、React 和 Bootstrap)无缝集成。

4.2 ASP.NET Core 启动的秘密

对于 ASP.NET Core 应用程序来说,要记住非常重要的一点是:其本质上是一个独立的控制台应用,而并非必须在 IIS 内部托管且并不需要由 IIS 来启动运行(而这正是 ASP.NET Core 跨平台的基石)。ASP.NET Core 应用程序拥有一个内置的 Self-Hosted(自托管)的 Web Server(Web 服务器),用来处理外部请求。

不管是托管还是自托管,都离不开 Host (宿主)。在 ASP.NET Core 应用中,通过配置并启动一个 Host 来完成应用程序的启动和其生命周期的管理(见图 4.1),而 Host 的主要职责就是对 Web Server 的配置和 Pipeline(请求处理管道)的构建。

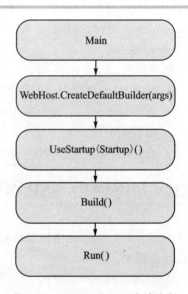

图 4.1 ASP.NET Core 启动流程

4.2.1 ASP.NET Core 启动流程

图 4.1 描述了一个总体的启动流程,从图中知道,ASP.NET Core 应用程序的启动主要包含以下 4 步:

① CreateDefaultBuilder():创建 IWebHostBuilder。
② UseStartup():指定启动类,为后续服务的注册及中间件的注册提供入口。
③ Build():IWebHostBuilder 负责创建 IWebHost。
④ Run():启动 IWebHost。

因此,ASP.NET Core 应用的启动,本质上是启动作为宿主的 WebHost 对象,其主要涉及两个关键对象 IWebHostBuilder 和 IWebHost,它们的内部实现是 ASP.NET Core 应用的核心所在。下面结合源码并梳理调用堆栈来一探究竟!

4.2.2 宿主构造器：WebHostBuilder

在启动 IWebHost 宿主之前，需要完成对 IWebHost 的创建和配置。应用程序在 Main 方法之后通过调用 CreateDefaultBuilder() 方法创建并配置 WebHostBuilder。CreateDefaultBuilder() 方法完成的主要功能如图 4.2 所示。

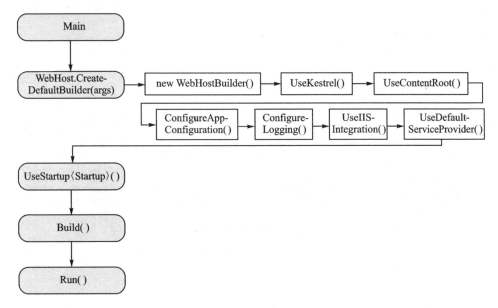

图 4.2　CreateDefaultBuilder() 方法

从图 4.2 中可以看出，CreateDefaultBuilder() 方法主要做了七件大事：

① new WebHostBuilder()：构造了 WebHostBuilder 对象。

② UseKestrel()：使用 Kestrel 作为 Web 服务器。

③ UseContentRoot()：指定了 Web 宿主使用的内容根目录（content root），比如 Views，默认为当前应用程序根目录。

④ ConfigureAppConfiguration()：设置当前应用程序的配置，主要是读取 appsettings.json 配置文件和开发环境中配置的 UserSecrets，以及添加环境变量和命令行参数。

⑤ ConfigureLogging()：读取配置文件中的 Logging 节点，配置日志系统。

⑥ UseIISIntegration()：使用 IISIntegration 中间件。

⑦ UseDefaultServiceProvider()：设置默认的依赖注入容器。

WebHostBuilder 类中存在一个重要的集合：

private readonly List<Action<WebHostBuilderContext, IServiceCollection>> _configureServicesDelegates;

通过 ConfigureServices() 方法将需要的 Action 加进来。

创建完 WebHostBuilder 后,通过调用 UseStartup()来指定启动类,为后续服务的注册及中间件的注册提供入口。

4.2.3 UseStartup〈Startup〉()

在 CreateDefaultBuilder()之后调用 UseStartup〈Startup〉(),指定 Startup 为启动类。Startup 类用于注册服务和应用的请求管道。UseStartup()方法完成的主要功能如图 4.3 所示。

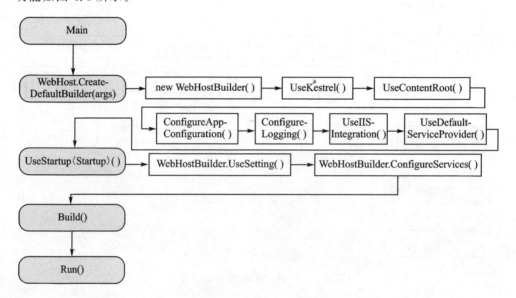

图 4.3 UseStartup()方法

ASP.NET Core 框架本身提供了一个 DI(依赖注入)系统,并且可以非常灵活地去扩展,很容易地切换为其他的 DI 框架(如 Autofac、Ninject 等)。在 ASP.NET Core 中,所有的实例都是通过这个 DI 系统来获取的,并要求应用程序也使用 DI 系统,以便能够开发出更具弹性、更易维护和测试的应用程序。总之在 ASP.NET Core 中,一切皆注入。关于"依赖注入"这里就不再多说了。

从图 4.3 可以看出,UseStartup()方法主要做了两件大事:

① UseSetting():用来配置 WebHost 中的 IConfiguration 对象。WebHost 中的 Configuration 只限于在 WebHost 中使用,并且不能配置它的数据源,它只会读取以"ASPNETCORE_"开头的环境变量。虽然不能配置它的数据源,但是它提供了一个 UseSetting()方法,还提供了一个设置_config 的机会,即构造了 WebHostBuilder 对象。

② ConfigureServices():用来注册应用服务。

ASP.NET Core 应用使用 Startup 类,而且类名按约定命名为 Startup。

通过 WebHostBuilder.UseStartup〈TStartup〉方法指定 Startup 类,代码如下:

```
    public static IWebHostBuilder UseStartup(this IWebHostBuilder hostBuilder, Type 
startupType)
    {
        var startupAssemblyName = startupType.GetTypeInfo().Assembly.GetName().Name;

        return hostBuilder
            .UseSetting(WebHostDefaults.ApplicationKey, startupAssemblyName)
            .ConfigureServices(services =>
            {
                if(typeof(IStartup).GetTypeInfo().IsAssignableFrom(startupType.GetTypeInfo()))
                {
                    services.AddSingleton(typeof(IStartup), startupType);
                }
                else
                {
                    services.AddSingleton(typeof(IStartup), sp =>
                    {
                        var hostingEnvironment = sp.GetRequiredService<IHostingEnvironment>();
                        return new ConventionBasedStartup(StartupLoader.LoadMethods(sp, 
startupType, hostingEnvironment.EnvironmentName));
                    });
                }
            });
    }
```

首先获取 Startup 类对应的 AssemblyName，并调用 UseSetting 方法将其设置为 WebHostDefaults.ApplicationKey 的值（"applicationName"）。UseSetting 是一个用于设置 Key-Value 的方法，一些常用的配置均会通过此方法写入_config 中。

然后调用 WebHostBuilder 的 ConfigureServices 方法，将一个 Action 写入 WebHostBuilder 的 configureServicesDelegates 中。这个 Action 的意思是：如果这个被指定的类 StartupType 是一个实现了 IStartup 的类，那么将其通过 AddSingleton 注册到 Services 的 ServiceCollection 中；如果不是，那么将其"转换"成 ConventionBasedStartup，并在实现了 IStartup 的类后再进行注册。这里涉及一个 StartupLoader 的 LoadMethods() 方法，会通过字符串方式查找类似带有"ConfigureServices"或"Configure{environmentName}Services"文字的方法。

注意：这里只是将一个 Action 写入了 configureServicesDelegates 中，而不是已经执行了对 IStartup 的注册，因为这个 Action 尚未执行，Services 也还不存在。

Startup 类必须要有 Configure 方法，用于每次进行 HTTP 请求的处理管道配置和一些系统配置，比如后面要讲的中间件（Middleware）就是在 Configure 方法中配置的。而 ConfigureServices 方法在 Configure 方法之前被调用，是一个可选的方法，并可在 ConfigureServices 中依赖注入接口或一些全局的框架,比如 EntityFramework 或 MVC 等。

当应用启动时,Startup 类的执行顺序是:构造→ConfigureServices→Configure,代码如下:

```
public class Startup
{
    // Use this method to add services to the container.
    public void ConfigureServices(IServiceCollection services)
    {
        ...
    }

    // Use this method to configure the HTTP request pipeline.
    public void Configure(IApplicationBuilder app){
        ...
    }
}
```

Startup 类构造函数接受由主机定义的依赖关系。在 Startup 类中注入依赖关系的常见用途为注入:
- IHostingEnvironment 以按环境配置服务。
- IConfiguration 以在启动过程中配置应用。

注入 IHostingEnvironment 的替代方法是使用基于约定的方法。应用可以为不同的环境单独定义 Startup 类(例如,StartupDevelopment),相应 Startup 类会在运行时得到选择。优先考虑名称后缀与当前环境相匹配的类。如果应用在开发环境中运行并包含 Startup 类和 StartupDevelopment 类,则使用 StartupDevelopment 类。

4.2.4　WebHostBuilder.Build()

在 ASP.Net Core 中定义的 IWebHost 用来表示 Web 应用的宿主,并提供了一个默认的 Web 宿主实现 WebHost。宿主的创建是通过调用 IWebHostBuilder 的 Build()方法来完成的,该方法完成的主要功能如图 4.4 所示。

Build()方法的核心主要在于对 WebHost 的创建,这又可以划分为三个部分:
① 构建依赖注入容器,注册初始通用服务,语句为"BuildCommonServices();"。
② 实例化 WebHost,语句为"var host = new WebHost(...);"。
③ 初始化 WebHost,也就是构建由中间件组成的请求处理管道,语句为"host.Initialize();"。

Build()方法的代码如下:

ASP.NET Core 简介

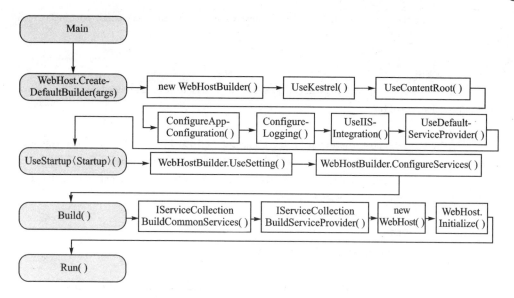

图 4.4　Build( )方法

```
public IWebHost Build()
{
    var hostingServices = BuildCommonServices(out var hostingStartupErrors);
    var applicationServices = hostingServices.Clone();
    var hostingServiceProvider = GetProviderFromFactory(hostingServices);

    AddApplicationServices(applicationServices, hostingServiceProvider);

    var host = new WebHost( applicationServices, hostingServiceProvider, _options,
_config, hostingStartupErrors);

    host.Initialize();
    return host;
}
```

1. 注册初始通用服务

BuildCommonServices 最为重要，它将构造第一个 ServiceCollection，这里称为 hostingServices，并包含环境变量、配置信息、日志和注册服务等。

BuildCommonService 方法主要做三件事：

① 查找 HostingStartupAttribute 属性以应用其他程序集中的启动配置。

② 注册通用服务。

③ 若配置了启动程序集，则发现并以 IStartup 类型注入到 IoC 容器中。

创建并配置好的 WebHostBuilder 可以通过 Build 方法创建 WebHost 了。首先创建 BuildCommonServices，代码如下：

```csharp
        private IServiceCollection BuildCommonServices(out AggregateException hostingStartupErrors)
        {
            ...省略
            foreach (string hostingStartupAssembly in (IEnumerable<string>) this._options.HostingStartupAssemblies)
            {
                foreach (HostingStartupAttribute customAttribute in Assembly.Load(new AssemblyName(hostingStartupAssembly)).GetCustomAttributes<HostingStartupAttribute>())
                    ((IHostingStartup) Activator.CreateInstance(customAttribute.HostingStartupType)).Configure((IWebHostBuilder) this);
            }
            ...省略
            if (!string.IsNullOrEmpty(_options.StartupAssembly))
            {
                var startupType = StartupLoader.FindStartupType(_options.StartupAssembly, _hostingEnvironment.EnvironmentName);

                if (typeof(IStartup).GetTypeInfo().IsAssignableFrom(startupType.GetTypeInfo()))
                {
                    services.AddSingleton(typeof(IStartup), startupType);
                }
                else
                {
                    services.AddSingleton(typeof(IStartup), sp =>
                    {
                        var hostingEnvironment = sp.GetRequiredService<IHostingEnvironment>();
                        var methods = StartupLoader.LoadMethods(sp, startupType, hostingEnvironment.EnvironmentName);
                        return new ConventionBasedStartup(methods);
                    });
                }
                ...省略
                foreach (var configureServices in _configureServicesDelegates)
                {
                    configureServices(_context, services);
                }
                return services;
            }
        }
```

在以上方法中创建了 ServiceCollection services，然后通过各种 Add 方法注册了很多内容进去，接着 foreach 之前暂存在 configureServicesDelegates 中的各个 Action 被传入 services 中逐一执行，使之前需要注册的内容被注册到 services 中，其中就包括 Startup，注意这里仅是进行了注册，而未执行 Startup 的方法。

2. 创建 IWebHost

处理好的 services 被 BuildCommonServices 返回后赋值给 hostingServices，然后 hostingServices 经过 Clone()生成 applicationServices，再经过 GetProviderFromFactory(hostingServices)生成 IServiceProvider hostingServiceProvider。经过一系列的处理后，就可以创建 WebHost 了。创建 IWebHost 的代码如下：

```
public IWebHost Build()
{
    //省略部分代码
    var host = new WebHost(
                applicationServices,
                hostingServiceProvider,
                _options,
                _config,
                hostingStartupErrors);
    try
    {
        host.Initialize();
        return host;
    }
}
```

将生成的 applicationServices 和 hostingServiceProvider 作为参数传递给新生成的 WebHost，接着执行 WebHost 的 Initialize()进行初始化。

4.2.5 WebHost.Initialize()

host.Initialize()方法用于初始化 WebHost，步骤是：

① 把在 StartUp.ConfigureServices()方法中注册的服务加到应用程序服务集合(applicationServiceCollection)中。

② 构造 StartUp 服务类。

Initialize()方法完成的主要功能如图 4.5 所示。

WebHost 的 Initialize()方法完成的主要工作如下列代码所示：

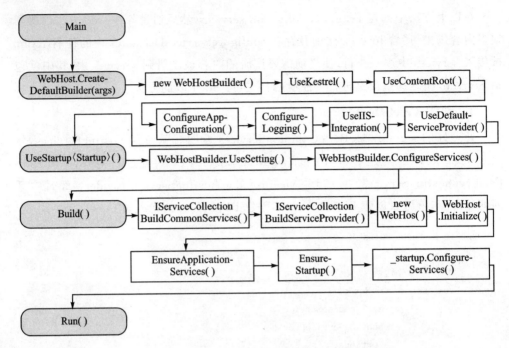

图 4.5 Initialize( )方法

```
public void Initialize()
{
    EnsureApplicationServices();
    ...省略
}
private void EnsureApplicationServices()
{
    if (_applicationServices == null)
    {
        EnsureStartup();
        _applicationServices = _startup.ConfigureServices(_applicationServiceCollection);
    }
}
private void EnsureStartup()
{
    if (_startup != null)
    {
        return;
    }
    _startup = _hostingServiceProvider.GetService<IStartup>();
    ...省略
}
```

调用 EnsureApplicationServices() 方法，实际上就是调用 StartUp. ConfigureServices() 方法，后者大家肯定熟悉，就是把那些在 ConfigureServices() 中注册的服务放到_applicationServices 字段里。

通过_hostingServiceProvider. GetService<IStartup>() 可以获取到_startup，然后调用_startup 的 ConfigureServices 方法，这就是用于依赖注入的 startup 类的 ConfigureServices 方法。

所以，_applicationServices 是根据_applicationServiceCollection 加上在_startup 中注册的内容之后重新生成的 IServiceProvider 的服务对象。

4.2.6 WebHost.Run()

WebHost 创建完毕，最后一步就是运行起来，WebHost 的 Run() 会内部调用 StartAsync() 方法，然后再在控制台输出一些信息，并且传入一个 CancellationToken 允许随时中断程序，代码如下：

```
public static void Run(this IWebHost host)
{
    host.RunAsync().GetAwaiter().GetResult();
}
private static async Task RunAsync(this IWebHost host, CancellationToken token, string shutdownMessage)
{
    using (host)
    {
        await host.StartAsync(token);

        var hostingEnvironment = host.Services.GetService<IHostingEnvironment>();
        var options = host.Services.GetRequiredService<WebHostOptions>();

        ...省略

        await host.WaitForTokenShutdownAsync(token);
    }
}
```

4.2.7 构建请求处理管道

请求处理管道的构建(见图 4.6)包含三个主要部分：
① 构建管道；
② 注册 Startup 中绑定的服务；
③ 配置 IServer。

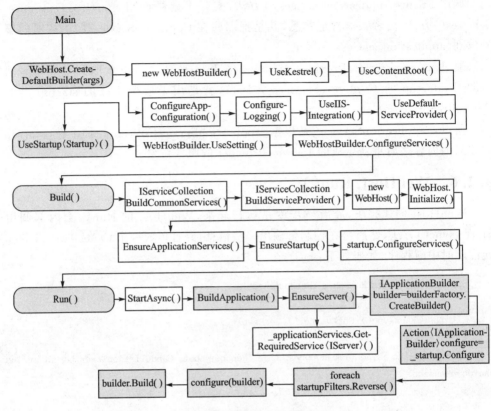

图 4.6 构建请求处理管道

结合图 4.6，构建请求处理管道的步骤是：

① 执行 BuildApplication() 方法用于请求管道的构建，这主要是利用 IApplicationBuilder 来创建的，代码如下：

```
private RequestDelegate BuildApplication()
{
    try
    {
        _applicationServicesException?.Throw();
        EnsureServer();

        var builderFactory = _applicationServices.GetRequiredService<IApplicationBuilderFactory>();
        var builder = builderFactory.CreateBuilder(Server.Features);
        builder.ApplicationServices = _applicationServices;

        var startupFilters = _applicationServices.GetService<IEnumerable<IStartup-
```

```
Filter>>();
            Action<IApplicationBuilder> configure = _startup.Configure;
            foreach (var filter in startupFilters.Reverse())
            {
                configure = filter.Configure(configure);
            }
            configure(builder);

            return builder.Build();
        }
        catch (Exception ex)
        {
            ...省略
        }
    }
```

以上代码主要完成如下工作：

首先，通过 _applicationServices.GetRequiredService<IApplicationBuilder-Factory>() 获取 ApplicationBuilderFactory，并将监听地址列表作为参数传入它的 CreateBuilder() 方法，创建 IApplicationBuilder，并把注册的服务转移给它，也就是把 _applicationServices 赋值给它的 ApplicationServices 属性。

其次，通过 _applicationServices.GetService<IEnumerable<IStartupFilter>>() 获取定义的 IStartupFilter，先使用循环语句 foreach 循环处理 IStartupFilter，并与 _startup 的 Configure 方法一起构造一个服务委托链，然后在 Build() 中对 IApplicationBuilder 进行处理生成请求处理管道。

② 调用 EnsureServer() 方法确保 Server 存在，并监听正确的端口，代码如下：

```
    private void EnsureServer()
    {
        if (Server == null)
        {
            Server = _applicationServices.GetRequiredService<IServer>();
            var serverAddressesFeature = Server.Features?.Get<IServerAddressesFeature>();
            var addresses = serverAddressesFeature?.Addresses;

            if (addresses != null && !addresses.IsReadOnly && addresses.Count == 0)
            {
                var urls = _config[WebHostDefaults.ServerUrlsKey] ?? _config[DeprecatedServerUrlsKey];
                if (!string.IsNullOrEmpty(urls))
```

```
                    {
                            serverAddressesFeature.PreferHostingUrls = WebHostUtilities.Parse-
Bool(_config, WebHostDefaults.PreferHostingUrlsKey);

                            foreach (var value in urls.Split(new[] {';'}, StringSplitOptions.Remove-
EmptyEntries))
                            {
                                addresses.Add(value);
                            }
                        }
                    }
                }
```

以上方法通过 GetRequiredService<IServer>() 获取 Server 并配置监听地址,其作用就是配合 Server 去创建上下文,大概的过程是:

① 服务器把请求的信息放入 IFeatureCollection 变量里;

② 利用上面的信息构造上下文;

③ 通过调用 ProcessRequestAsync() 方法来处理请求,此时请求进入处理管道 (Pipeline)。

④ 触发服务委托链,相当于调用里面的 UseXXX() 方法。

4.2.8 启动 WebHost

WebHost 的启动过程如图 4.7 所示,主要分为两步:

① 启动 Server 以监听请求;

② 启动 HostedService。

启动 WebHost 的代码如下:

```
public virtual async Task StartAsync(CancellationToken cancellationToken = default)
{
    HostingEventSource.Log.HostStart();

    _logger = _applicationServices.GetRequiredService<ILogger<WebHost>>();

    _logger.Starting();
    var application = BuildApplication();
    _applicationLifetime = _applicationServices.GetRequiredService<IApplication-
Lifetime>() as ApplicationLifetime;
```

ASP.NET Core 简介 4

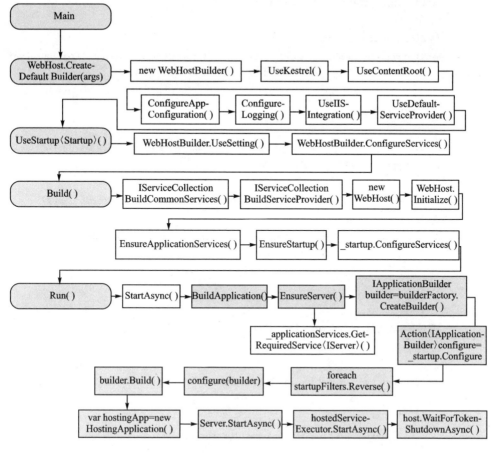

图 4.7 启动 WebHost

```
_hostedServiceExecutor = _applicationServices.GetRequiredService<HostedService-
Executor>();

var diagnosticSource = _applicationServices.GetRequiredService<Diagnostic
Listener>();

var httpContextFactory = _applicationServices.GetRequiredService<IHttpContext-
Factory>();

var hostingApp = new HostingApplication(application, _logger, diagnosticSource,
httpContextFactory);

await Server.StartAsync(hostingApp, cancellationToken).ConfigureAwait(false);

// Fire IApplicationLifetime.Started
```

```
        _applicationLifetime?.NotifyStarted();

        // Fire IHostedService.Start
        await _hostedServiceExecutor. StartAsync ( cancellationToken ). ConfigureAwait
(false);

        _logger.Started();

        // Log the fact that we did load hosting startup assemblies.
        if (_logger.IsEnabled(LogLevel.Debug))
        {
            foreach (var assembly in _options.GetFinalHostingStartupAssemblies())
            {
                _logger.LogDebug("Loaded hosting startup assembly {assemblyName}", assembly);
            }
        }

        if (_hostingStartupErrors != null)
        {
            foreach (var exception in _hostingStartupErrors.InnerExceptions)
            {
                _logger.HostingStartupAssemblyError(exception);
            }
        }
    }
```

4.2.9 启动 Server

Server 本身并不清楚 HttpContext 的细节,因此它需要接收一个 IHttpApplication 类型的参数。从命名来看,IHttpApplication 代表一个 HTTP 应用程序,用来负责 HttpContext 的创建,定义代码如下:

```
    public interface IHttpApplication<TContext>
    {
        TContext CreateContext(IFeatureCollection contextFeatures);
        Task ProcessRequestAsync(TContext context);
        void DisposeContext(TContext context, Exception exception);
    }
```

以上代码主要定义了三个方法，CreateContext 方法用来创建请求上下文，ProcessRequestAsync 方法用来处理请求，DisposeContext 方法用来释放上下文。而请求上下文用来携带请求和返回响应的核心参数，其贯穿于整个请求处理管道之中。ASP.NET Core 中提供了默认的实现 HostingApplication，其构造函数接收一个 RequestDelegate _application（也就是链接中间件形成的处理管道），通过调用 ProcessRequestAsync 方法来调用上面创建的 RequestDelegate 委托，以完成对 HttpContext 的处理请求，代码如下：

```
public class HostingApplication : IHttpApplication<HostingApplication.Context>
{
    private readonly RequestDelegate _application;

    public Task ProcessRequestAsync(Context context)
    {
        return _application(context.HttpContext);
    }
}
```

最后启动 Server，代码如下：

```
 var httpContextFactory = _applicationServices.GetRequiredService<IHttpContextFactory>();
 var hostingApp = new HostingApplication(_application, _logger, diagnosticSource, httpContextFactory);
 await Server.StartAsync(hostingApp, cancellationToken).ConfigureAwait(false);
```

Server 会绑定一个监听端口，注册 HTTP 连接事件，最终通过上面的 hostingApp 来切入到应用程序中，以完成整个请求的处理。

所以，首先传入上面创建的 _application 和 httpContextFactory 来生成 HostingApplication，并将 HostingApplication 传入 Server 的 StartAsync() 中，当 Server 监听到请求之后，后面的工作由 HostingApplication 来完成。

4.2.10 启动 IHostedService

IHostedService 接口用来定义后台任务，通过实现该接口并注册到 IoC 容器中，IHostedService 接口的实现对象会随着 ASP.NET Core 程序的启动而启动，终止而终止。

WebHost 会调用 HostedServiceExecutor 的 StartAsync 方法，从而完成对 HostedService 的启动，代码如下：

```
_applicationLifetime = _applicationServices.GetRequiredService<IApplicationLifetime>() as ApplicationLifetime;
```

```
_hostedServiceExecutor = _applicationServices.GetRequiredService<HostedService-
Executor>();
```

```
await _hostedServiceExecutor.StartAsync(cancellationToken).ConfigureAwait(false);
```

hostedServiceExecutor.StartAsync()方法用来开启一个后台运行的服务，一些需要后台运行的操作比如定期刷新缓存等可以放到这里。

4.3 ASP.NET Core 中间件

4.3.1 什么是中间件

已经知道，任何一个 Web 框架都把 HTTP 请求封装成一个管道，每一次的请求都经过管道的一系列操作，最终到达程序代码中。那么中间件就是在应用程序管道中的一个组件，用来拦截请求过程进行其他一些处理和响应。中间件可以有很多个，每一个中间件都可以对管道中的请求进行拦截，它可以决定是否将请求转移给下一个中间件。

ASP.NET Core 提供了 IApplicationBuilder 接口，通过该接口把中间件注册到 ASP.NET Core 的管道请求中去，中间件是一个典型的 AOP 应用。

可使用 Run、Map 和 Use 扩展方法来配置请求委托。可将一个单独的请求委托并行指定为匿名方法（称为并行中间件），或在可重用的类中对其进行定义。这些可重用的类和并行匿名方法即为中间件或中间件组件。请求管道中的每个中间件组件负责调用管道中的下一个组件，或在适当情况下使链发生短路。

4.3.2 中间件的运行方式

ASP.NET Core 请求管道包含一系列相继调用的请求委托，图 4.8 是一个微软官方的中间件管道请求图，可以看到，每一个中间件都可以在请求之前和之后进行操作。请求处理完成之后传递给下一个请求。

每个委托均可在下一个委托前后执行操作。此外，委托还可以决定不将请求传递给下一个委托，这称为请求管道的短路。通常需要这种短路操作，因为这样做可以避免不必要的工作。例如，静态文件中间件可以返回一个静态文件的请求并使管道的其余部分短路。需要在管道早期调用异常处理委托，以便它们可以捕获在管道后期阶段发生的异常。

最简单的 ASP.NET Core 应用示例是建立一个请求委托，并处理所有请求，其中不包括实际请求管道，而是仅针对每个 HTTP 请求调用一个匿名方法来响应，代码如下：

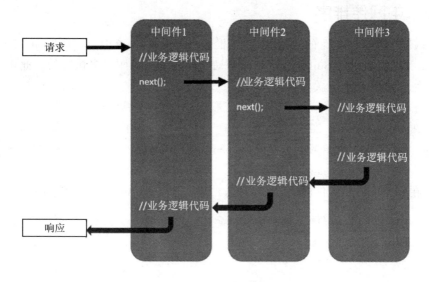

图 4.8 中间件管道请求图

```
using Microsoft.AspNetCore.Builder;
using Microsoft.AspNetCore.Hosting;
using Microsoft.AspNetCore.Http;

public class Startup
{
    public void Configure(IApplicationBuilder app)
    {
        app.Run(async context =>
        {
            await context.Response.WriteAsync("1  Hello.\r\n");
        });
        app.Run(async context =>
        {
            await context.Response.WriteAsync("2  Hello Welcome.");
        });
    }
}
```

通过浏览器访问发现,确实在第一个 app.Run 终止了管道,如图 4.9 所示。

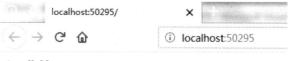

图 4.9 管道终止

4.3.3 中间件排序

默认情况下,中间件的执行顺序是根据 Startup.cs 文件中,在"public void Configure(IApplicationBuilder app){}"方法中注册的先后顺序来执行的。此排序对于安全性、性能和功能至关重要。

下面通过一个示例来了解中间件执行顺序的重要性,具体操作如下:

① 在 Visual Studio 2017 中选择"工具"→"NuGet 包管理器"→"管理解决方案的 NuGet 程序包"菜单项,安装 Microsoft.AspNetCore.Session 和 Microsoft.AspNetCore.Http.Extensions 两个包。

② 在 Startup.cs 文件的 Configure 方法中注册 Session 服务,代码如下:

```
public void Configure(IApplicationBuilder app, IHostingEnvironment env)
{
    if(env.IsDevelopment())
    {
        app.UseBrowserLink();
        app.UseDeveloperExceptionPage();
    }
    else
    {
        app.UseExceptionHandler("/Error");
    }
    app.UseStaticFiles();
    app.UseMvc();
    app.UseCookiePolicy();
    app.UseSession();
}
```

③ 在 Visual Studio 2017 中打开 About.cshtml.cs 文件,输入以下代码:

```
using Microsoft.AspNetCore.Http;

public class AboutModel: PageModel
{
    private const stringSessionKeyName = "_Name";
    private const stringSessionKeyAge = "_Age";
    public string Message { get; set; }
    public string Name { get; set; }
    public string Age { get; set; }
    public void OnGet()
```

```
        {
            if(string.IsNullOrEmpty(HttpContext.Session.GetString(SessionKeyName)))
            {
                HttpContext.Session.SetString(SessionKeyName, "彭祖");
                HttpContext.Session.SetInt32(SessionKeyAge, 800);
            }
            var name = HttpContext.Session.GetString(SessionKeyName);
            var age = HttpContext.Session.GetInt32(SessionKeyAge);
            Name = name;
            Age = age.ToString();
        }
    }
```

④ 在 Visual Studio 2017 中按 F5 键运行应用程序,在浏览器中浏览 About 页面。此时在 Visual Studio 2017 中会出现一个错误信息,如图 4.10 所示。

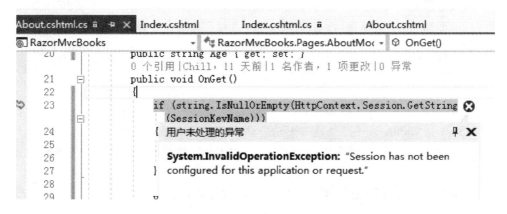

图 4.10　错误信息

⑤ 这是由于中间件的顺序不对,ASP.NET Core 的中间件的顺序是很重要的。现在修改一下 Startup.cs 文件中的代码,也就是把"app.UseCookiePolicy();"和"app.UseSession();"移到 app.UseMvc 上面,代码如下:

```
public void Configure(IApplicationBuilder app, IHostingEnvironment env)
{
    if(env.IsDevelopment())
    {
        app.UseBrowserLink();
        app.UseDeveloperExceptionPage();
    }
    else
    {
        app.UseExceptionHandler("/Error");
```

```
        }
        app.UseStaticFiles();
        app.UseCookiePolicy();
        app.UseSession();
        app.UseMvc();
    }
}
```

⑥ 在 Visual Studio 2017 中再次执行以上代码,就不会报错了。

4.3.4　Use、Run 和 Map 方法

有 3 个方法可以在管道中注册"中间件",它们是:

① app.Use()。此方法是 IApplicationBuilder 接口原生提供的,一般情况下注册时都用它。它可使管道短路(即不调用 next 请求委托)。

② app.Run()。这是一个扩展方法,它需要一个 RequestDelegate 委托,里面包含了 HTTP 的上下文信息,没有 next 参数,因为它总是在管道最后一步执行。

③ app.Map()。这也是一个扩展方法,类似于 MVC 的路由,用于一些特殊请求路径的处理。如果请求路径以给定路径开头,则执行分支,如:www.example.com/token 等。

上面的 Run 和 Map 内部也调用了 Use,算是对 IApplicationBuilder 接口的扩充。app.UseMiddleware<>()是一个功能强大的扩展方法,用于注册中间件。为什么说它功能强大呢?是因为它不但提供了注册中间件的功能,还提供了依赖注入(DI)的功能。

下面通过一个示例来了解这 3 种注册方式,代码如下:

```
using System;
using System.Collections.Generic;
using System.IO;
using System.Linq;
using System.Threading.Tasks;
using Microsoft.AspNetCore.Builder;
using Microsoft.AspNetCore.Hosting;
using Microsoft.AspNetCore.Http;
using Microsoft.Extensions.Configuration;
using Microsoft.Extensions.DependencyInjection;
using Microsoft.Extensions.FileProviders;
using RazorMvcBooks.Models;

namespace RazorMvcBooks
{
```

```csharp
public class Startup
{
    public Startup(IConfiguration configuration)
    {
        Configuration = configuration;
    }
    private static void MapTest1(IApplicationBuilder app)
    {
        app.Run(async context =>
        {
            await context.Response.WriteAsync("Map 测试 1");
        });
    }
    private static void MapTest2(IApplicationBuilder app)
    {
        app.Run(async context =>
        {
            await context.Response.WriteAsync("Map 测试 2");
        });
    }
    private static void MapTestBranch(IApplicationBuilder app)
    {
        app.Run(async context =>
        {
            var branchVer = context.Request.Query["branch"];
            await context.Response.WriteAsync($"Branch used = {branchVer} Map 分支测试");
        });
    }
    public IConfiguration Configuration { get; }
    //This method gets called by the runtime. Use this method to add services to the container.
    public void ConfigureServices(IServiceCollection services)
    {
        //services.AddDistributedMemoryCache();
        services.AddSession(options =>
        {
            // Set a short timeout for easy testing.
            options.IdleTimeout = TimeSpan.FromSeconds(10);
            options.Cookie.HttpOnly = true;
        });
        services.AddMvc();
```

```
        }
        //This method gets called by the runtime. Use this method to configure the HTTP
        //request pipeline.
        public void Configure(IApplicationBuilder app, IHostingEnvironment env)
        {
            if (env.IsDevelopment())
            {
                app.UseBrowserLink();
                app.UseDeveloperExceptionPage();
            }
            else
            {
                app.UseExceptionHandler("/Error");
            }
            app.Map("/map1", MapTest1);
            app.Map("/map2", MapTest2);
            app.MapWhen(context => context.Request.Query.ContainsKey("branch"),
MapTestBranch);

            app.Run(async context =>
            {
                await context.Response.WriteAsync(" Hello from non-Map delegate.");
            });
            app.UseStaticFiles();
            app.UseMvc();
        }
    }
}
```

在浏览器中浏览之后,表 4.1 使用前面的代码显示来自 http://localhost:50295 的请求和响应。

表 4.1 请求和响应

请 求	响 应
localhost:50295/	localhost:50295/ localhost:50295 Hello from non-Map delegate. \<p\>

续表 4.1

请　求	响　应
localhost:50295/map1	localhost:50295/map1 Map 测试 1
localhost:50295/map2	localhost:50295/map2 Map 测试 2
localhost:50295/map3	localhost:50296/map3 Hello from non-Map delegate.
localhost:50295/?branch=12	localhost:50295/?branch=12 Branch used = 12 Map 分支测试

在使用 Map 时，匹配的路径段将从 HttpRequest.Path 中删除，并将每个请求追加到 HttpRequest.PathBase 中。

MapWhen 根据给定谓词的结果创建请求管道分支。Func⟨HttpContext，bool⟩类型的任何谓词均可用于将请求映射到管道的新分支。在上面的示例中，谓词用于检测字符串变量 branch 是否存在。

可使用 app.Use 将多个请求委托链接在一起。next 参数表示管道中的下一个委托。通常可在下一个委托前后执行操作，代码如下：

```
using System;
using System.Collections.Generic;
using System.IO;
using System.Linq;
using System.Threading.Tasks;
using Microsoft.AspNetCore.Builder;
using Microsoft.AspNetCore.Hosting;
using Microsoft.AspNetCore.Http;
using Microsoft.Extensions.Configuration;
using Microsoft.Extensions.DependencyInjection;
using Microsoft.Extensions.FileProviders;
```

ASP. NET Core 应用开发入门教程

```
namespace RazorMvcBooks
{
    public class Startup
    {
        public Startup(IConfiguration configuration)
        {
            Configuration = configuration;
        }
        public IConfiguration Configuration { get; }
        // This method gets called by the runtime. Use this method to add services to the
        // container.
        public void ConfigureServices(IServiceCollection services)
        {
            // services.AddDistributedMemoryCache();
            services.AddSession(options =>
            {
                // Set a short timeout for easy testing.
                options.IdleTimeout = TimeSpan.FromSeconds(10);
                options.Cookie.HttpOnly = true;
            });
            services.AddMvc();
        }
        // This method gets called by the runtime. Use this method to configure the HTTP
        // request pipeline.
        public void Configure(IApplicationBuilder app, IHostingEnvironment env)
        {
            if (env.IsDevelopment())
            {
                app.UseBrowserLink();
                app.UseDeveloperExceptionPage();
            }
            else
            {
                app.UseExceptionHandler("/Error");
            }
            app.Use(async (context, next) =>
            {
                await context.Response.WriteAsync("1  Hello from Use start.");
                // Call the next delegate/middleware in the pipeline
                await next.Invoke();
                await context.Response.WriteAsync("2  Hello from Use End.");
            });
```

```
    app.Run(async context =>
    {
        await context.Response.WriteAsync("3   Hello from non-Map delegate. ");
    });
    app.UseStaticFiles();
    app.UseMvc();
        }
    }
}
```

通过浏览器访问，发现 app.Use 在显示"1   Hello from Use start."之后，通过调用 next.Invoke 方法，执行了 app.Run.方法，显示"3   Hello from non-Map delegate."然后又返回执行了 app.Use 中的第二步显示"2   Hello from Use End"，如图 4.11 所示。

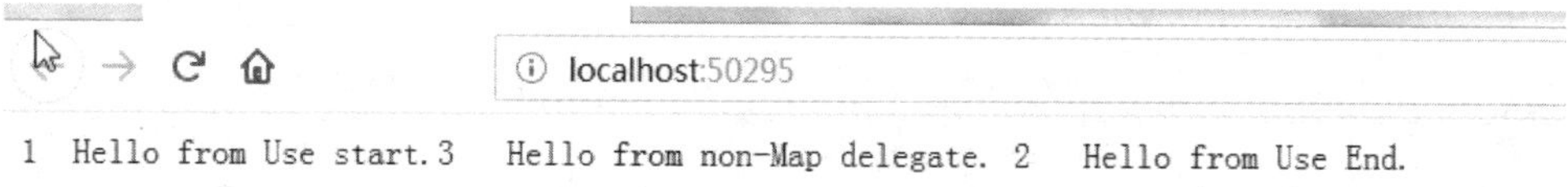

1 Hello from Use start.3   Hello from non-Map delegate. 2   Hello from Use End.

图 4.11 浏览结果

警告：

在向客户端发送响应后，请勿调用 next.Invoke。在响应启动后，针对 HttpResponse 的更改将引发异常。例如，设置标头和状态代码等的更改将引发异常。调用 next 后写入响应正文将发生以下情况：

① 可能导致违反协议，例如，写入的长度超过规定的 content-length。

② 可能损坏正文格式，例如，向 CSS 文件中写入 HTML 页脚。

HttpResponse.HasStarted 是一个有用的提示，指示是否已发送标头和/或已写入正文。

4.3.5 内置中间件

ASP.NET Core 的内置中间件组件和用于添加这些中间件的顺序的说明如表 4.2 所列。

表 4.2 中间件

中间件	描　　述	顺　　序
身份验证	提供身份验证支持	在需要 HttpContext.User 之前。OAuth 回叫的终端
CORS	配置跨域资源共享	在使用 CORS 的组件之前

续表 4.2

中间件	描述	顺序
诊断	配置诊断	在生成错误的组件之前
转接头	将代理标头转发到当前请求	在使用更新字段(如:架构、主机、客户端 IP、方法)的组件之前
HTTP 方法重写	允许传入 POST 请求重写方法	在使用已更新方法的组件之前
HTTPS 重定向	将所有 HTTP 请求重定向到 HTTPS (ASP. NET Core 2.1 或更高版本)	在使用 URL 的组件之前
HTTP 严格传输安全性 (HSTS)	添加特殊响应标头的安全增强中间件 (ASP. NET Core 2.1 或更高版本)	在发送响应之前,修改请求的组件(例如转接头、URL 重写)之后
响应缓存	提供对缓存响应的支持	在需要缓存的组件之前
响应压缩	提供对压缩响应的支持	在需要压缩的组件之前
请求本地化	提供本地化支持	在对本地化敏感的组件之前
路由	定义和约束请求路由	用于匹配路由的终端
会话	提供对管理用户会话的支持	在需要会话的组件之前
静态文件	为静态文件和目录浏览提供支持	如果请求与文件匹配,则为终端
URL 重写	提供对重写 URL 和重定向请求的支持	在使用 URL 的组件之前
WebSockets	启用 WebSockets 协议	在接受 WebSocket 请求所需的组件之前

4.4　ASP. NET Core 中的静态文件

　　静态文件(如 HTML、CSS、图像和 JavaScript)是 ASP. NET Core 应用直接提供给客户端的资源。ASP. NET Core 应用需要进行一些配置才能提供这些文件。

　　一个 ASP. NET Core 项目的静态文件一般是放在项目目录下的 wwwroot 文件夹中,文件目录如图 4.12 所示。

图 4.12　wwwroot 文件夹

4.4.1 如何将静态文件注入到项目中

为了提供静态文件，必须配置中间件以向管道添加静态文件，这是通过在 Startup.Configure 方法中调用 UseStaticFiles 扩展方法来实现的。

静态文件存储在项目的 Web 根目录中，默认目录是〈content_root〉/wwwroot，但可通过 UseWebRoot 方法更改目录。

将静态文件注入到项目中的具体操作是：

① 用 Visual Studio 2017 打开 RazorMvcBooks 项目。

② 在 Visual Studio 2017 中用 NuGet 将以下包添加到项目中。如果以.NET Framework 为目标，则将 Microsoft.AspNetCore.StaticFiles 包添加到项目中。如果以 .NET Core 为目标，则将 Microsoft.AspNetCore.All 包添加到项目中，如图 4.13 所示。

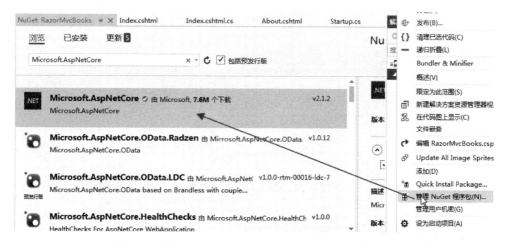

图 4.13 将 Microsoft.AspNetCore.All 包添加到项目中

③ 打开 Startup.cs 文件，并在 Configure 方法中添加 UseStaticFiles 方法调用，该方法就使用了默认的静态文件夹 wwwroot 目录，代码如下：

```
public void Configure(IApplicationBuilder app, IHostingEnvironment env)
{
    if (env.IsDevelopment())
    {
        app.UseBrowserLink();
        app.UseDeveloperExceptionPage();
    }
    else
    {
        app.UseExceptionHandler("/Error");
    }
```

```
app.UseStaticFiles();
app.UseMvc();
}
```

④ 通过 Web 根目录的相关路径访问静态文件。Web 应用程序的项目模板包含 wwwroot 文件夹中的多个文件夹。用于访问 images 子文件夹中的文件的 URL 格式为 http://〈server_address〉/images/〈image_file_name〉,例如,http://localhost:50295/images/fen1.jpg,如图 4.14 所示。

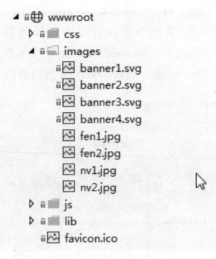

图 4.14 图片文件的存放目录

⑤ 在 Visual Studio 2017 中按 F5 键运行应用程序。在浏览器中输入以上地址,浏览静态文件图片 fen1.jpg,结果如图 4.15 所示。

图 4.15 浏览图片

⑥ 在 Visual Studio 2017 中打开 About.cshtml 文件,输入以下代码:

〈img src = "~/images/fen2.jpg" alt = "ASP.NET Image" class = "img-responsive" /〉

⑦ 在浏览器中浏览 About 页面即可显示图片,结果如图 4.16 所示。

图 4.16　在 About 中显示图片

4.4.2　自定义静态文件夹

除了由默认模板提供的静态文件夹之外,也可以创建一个自己的文件夹来提供静态文件,把一些主要的静态文件放到 wwwroot 根目录以外。创建操作如下:

① 在 Visual Studio 2017 中创建一个自己的静态文件夹 MyStaticFiles,并在 MyStaticFiles 文件夹中创建一个 images 文件夹,在 images 文件夹中放入两张图片,如图 4.17 所示。

图 4.17　自定义静态文件夹

② 打开 Startup.cs 文件,并在 Configure 方法中添加 UseStaticFiles 方法的有参调用,把自己创建的静态文件夹公开,代码如下:

```
public void Configure(IApplicationBuilder app, IHostingEnvironment env)
{
    if (env.IsDevelopment())
    {
        app.UseBrowserLink();
        app.UseDeveloperExceptionPage();
    }
    else
    {
        app.UseExceptionHandler("/Error");
    }
    app.UseStaticFiles();
    app.UseStaticFiles(new StaticFileOptions
    {
        FileProvider = new PhysicalFileProvider(Path.Combine(Directory.GetCurrentDirectory(), "MyStaticFiles")),
        RequestPath = "/StaticFiles"
    });
    app.UseMvc();
}
```

③ 上面的代码表示 MyStaticFiles 目录层次结构通过 StaticFiles URL 段公开。例如"~/StaticFiles/images/jin2.jpg"中使用 StaticFiles 作为 URL 地址中的一部分,而不是直接使用实际的文件夹名称 MyStaticFiles。代码中的 FileProvider 是物理路径,RequestPath 是对物理路径的重写。可以请求 http://⟨server_address⟩/StaticFiles/images/jin1.jpg 地址进行浏览,结果如图 4.18 所示。

图 4.18　浏览自定义静态文件夹

④ 在 Visual Studio 2017 中打开 About.cshtml 文件，输入以下代码：

〈img src = "~/StaticFiles/images/jin2.jpg" alt = "ASP.NET Image" class = "img-responsive" /〉

⑤ 在 Visual Studio 2017 中按 F5 键运行应用程序，在浏览器中输入链接地址，将看到如图 4.19 所示的效果。

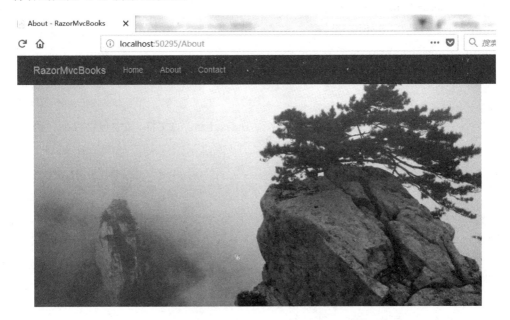

图 4.19　在 About 中显示自定义静态文件夹中的图片

4.4.3　添加默认文件支持

为了让 Web 应用程序提供默认页面，用户不必完全限定 URL，而是使用 UseDefaultFiles 扩展方法，此方法实际上不提供文件的 URL 重写程序。

除了 UseStaticFiles 和 UseDefaultFiles 扩展方法外，还有 UseFileServer 方法，该方法组合了以上两种方法的功能以及 UseDirectoryBrowser 扩展方法。添加默认文件支持的操作是：

① 在 Visual Studio 2017 中打开 Startup.cs 文件，将 Configure 方法中的静态文件中间件从 app.UseStaticFiles() 更改为 app.UseFileServer()，代码如下：

```
public void Configure(IApplicationBuilder app, IHostingEnvironment env, ILoggerFactory loggerFactory)
{
    loggerFactory.AddConsole();
    if(env.IsDevelopment())
```

```
        {
            app.UseDeveloperExceptionPage();
        }
        //添加静态文件支持管道
        app.UseStaticFiles();
        //添加静态文件和默认文件的支持,可代替 app.UseStaticFiles();
        app.UseFileServer();
        app.Run(async (context) =>
        {
            await context.Response.WriteAsync("Hello World!");
        });
    }
```

② 在 wwwroot 文件夹中创建一个名为 Index.html 的文件,其中包含以下代码:

```
<!DOCTYPE html>
<html>
<head>
<meta charset = "utf-8" />
<title>Hello static world!</title>
</head>
<body>
<h1>Hello from ASP.NET Core!</h1>
<h1>我是默认页面!</h1>
</body>
</html>
```

③ 运行应用程序,则应该显示默认页面 Index.html 中的内容。

4.4.4 设置 HTTP 响应标头

StaticFileOptions 对象可用于设置 HTTP 响应标头。除配置从 Web 根目录提供的静态文件外,以下操作还设置了 Cache-Control 标头:

① 在 Visual Studio 2017 中使用 NuGet 安装 Microsoft.AspNetCore.Http 包。

② 打开 Startup.cs 文件,在文件的最上方添加以下代码:

```
using Microsoft.AspNetCore.Http;
```

③ 在 Startup.cs 文件的 Configure 方法中,在 UseStaticFiles 中设置响应缓存,缓存时间为 10 分钟,代码如下:

```
public void Configure(IApplicationBuilder app, IHostingEnvironment env)
{
    if (env.IsDevelopment())
```

```
        {
            app.UseBrowserLink();
            app.UseDeveloperExceptionPage();
        }
        else
        {
            app.UseExceptionHandler("/Error");
        }
        app.UseStaticFiles();
        app.UseStaticFiles(new StaticFileOptions
        {
            FileProvider = new PhysicalFileProvider(Path.Combine(Directory.GetCurrent-
Directory(), "MyStaticFiles")),
            RequestPath = "/StaticFiles",
            OnPrepareResponse = ctx =>
            {
                //设置响应缓存,缓存时间为 10 分钟
                ctx.Context.Response.Headers.Append("Cache-Control", "public,max-age = 600");
            }
        });
        app.UseMvc();
    }
```

④ 在 Visual Studio 2017 中按 F5 键运行应用程序,然后在浏览器中访问 About 页面。如图 4.20 所示,如果没有添加响应缓存,则无论刷新多少次,都如图中的 1 所示,没有缓存响应;如果添加了响应缓存,则第一次浏览时如图中的 1 所示,再多次刷新后,则如图中的 2 所示,缓存响应为 1,此时图片 images 的大小为 0。

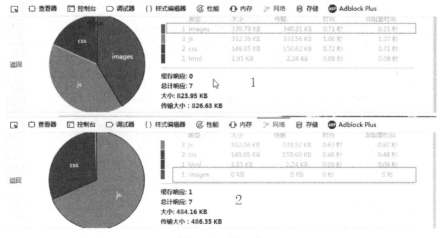

图 4.20 缓 存

4.4.5 启用目录浏览

通过目录浏览,Web 应用的用户可查看目录列表和指定目录中的文件。出于安全考虑,默认情况下不允许浏览目录的文件和文件夹。启用目录浏览的操作如下:

① 在 Visual Studio 2017 中打开 Startup.cs 文件,在 ConfigureServices 方法中调用 services.AddDirectoryBrowser,代码如下:

```
public void ConfigureServices(IServiceCollection services)
{
    services.AddDirectoryBrowser();
    services.AddMvc();
}
```

② 在 Startup.cs 文件的 Configure 方法中添加如下代码:

```
public void Configure(IApplicationBuilder app, IHostingEnvironment env)
{
    if (env.IsDevelopment())
    {
        app.UseBrowserLink();
        app.UseDeveloperExceptionPage();
    }
    else
    {
        app.UseExceptionHandler("/Error");
    }
    app.UseStaticFiles();
    app.UseStaticFiles(new StaticFileOptions
    {
        FileProvider = new PhysicalFileProvider(Path.Combine(Directory.GetCurrentDirectory(), "MyStaticFiles")),
        RequestPath = "/StaticFiles"
    });
    app.UseDirectoryBrowser(new DirectoryBrowserOptions
    {
        FileProvider = new PhysicalFileProvider(Path.Combine(Directory.GetCurrentDirectory(), "MyStaticFiles", "images")),
        RequestPath = "/StaticFiles"
    });
    app.UseMvc();
}
```

上述代码允许使用 URL http://⟨server_address⟩/StaticFiles 浏览 MyStaticFiles/

images 文件夹中的目录,并链接到每个文件和文件夹上。

③ 在 Visual Studio 2017 中按 F5 键运行应用程序。在浏览器中输入 http://localhost:50295/StaticFiles/即可看到/MyStaticFiles/images 目录下的文件了,效果如图 4.21 所示。一般来说还是使用默认目录较好。

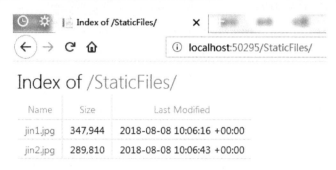

图 4.21 浏览目录下的文件

在启用了 UseDirectoryBrowser 和 UseStaticFiles 之后,其指定的目录及其子目录均可被公开访问,因此使用不当可能会泄露机密。将适合公开的文件存储在专用目录中,如〈content_root〉/wwwroot。将这些文件与 MVC 视图、Razor 页面(仅限 2.x)和配置文件等分开。

使用 UseDirectoryBrowser 和 UseStaticFiles 公开的内容的 URL 受大小写和基础文件系统字符限制的影响,例如,Windows 不区分大小写,而 macOS 和 Linux 却要区分。

在生产环境中建议禁用目录浏览。

4.5 ASP. NET Core 中的配置

通常来讲,对配置的设置会因环境(如开发、测试和生产等)的不同而应用不同的设置。.NET Core 提供了一套用于设置的 API,为程序提供了运行时从文件、命令行参数、环境变量等读取配置的方法。配置都是键值对的形式,并且支持嵌套,.NET Core 还内建了从配置反序列化为 POCO(一种具有属性的简单的.NET 类)对象的支持。

目前支持以下配置解析程序:

① 文件格式(INI,JSON,XML);

② 命令行参数;

③ 环境变量;

④ 内存中的.NET 对象;

⑤ 未加密的机密管理器存储(User Secrets);

⑥ 加密的用户存储(Azure Key Vault)。

如果现有的解析程序不能满足用户的使用场景,则还允许自定义解析程序,比如从数据库中读取。

4.5.1 配置相关的包

在 Visual Studio 2017 中选择"工具"→"NuGet 包管理器"→"管理解决方案的 NuGet 程序包"菜单项打开包管理器,在搜索框中输入"Microsoft. Extensions. Configuration",则所有与配置相关的包都会列举出来,如图 4.22 所示。

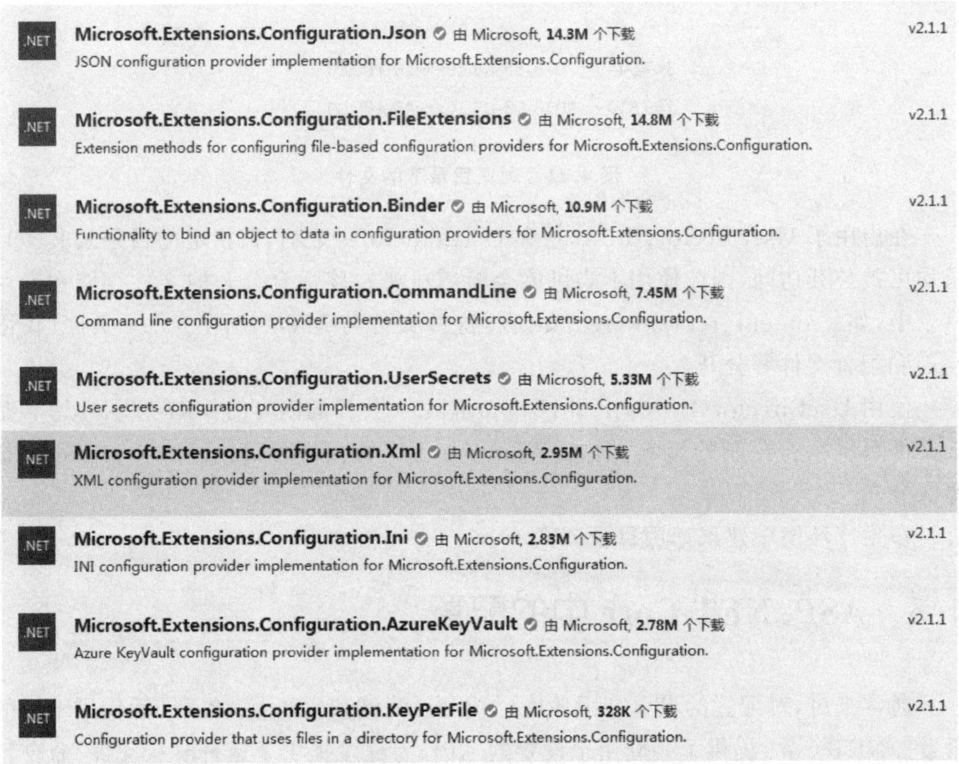

图 4.22 配置相关包

从包的名称基本上可以看出它的用途,比如:

- Microsoft. Extensions. Configuration. Json 是 Json 配置的解析程序;
- Microsoft. Extensions. Configuration. CommandLine 是命令行参数配置的解析程序;
- Microsoft. Extensions. Configuration. UserSecrets 用于在. NET Core 程序中使用 User Secrets 存储敏感数据。

4.5.2 文件配置

文件配置(以 Json 为例)的操作如下:

① 在 Visual Studio 2017 中选择"工具"→"NuGet 包管理器"→"管理解决方案的 NuGet 程序包"菜单项,安装 Microsoft.Extensions.Configuration.Json 包。

② 在 Visual Studio 2017 中打开 appsetttings.json 文件,添加如下代码:

```
{
    "Logging": {
        "IncludeScopes": false,
        "LogLevel": {
            "Default": "Warning"
        }
    },
    "DclTypes": [
        {
            "Code": 1,
            "Name": "申报"
        },
        {
            "Code": 2,
            "Name": "备案"
        },
        {
            "Code": 3,
            "Name": "注销"
        }
    ]
}
```

③ 在 Visual Studio 2017 中打开 About.cshtml.cs 文件,调用 AddJsonFile 把 Json 配置的解析程序添加到 ConfigurationBuilder 中,代码如下:

```
using System;
using System.IO;
using Microsoft.AspNetCore.Mvc.RazorPages;
using Microsoft.Extensions.Configuration;

namespace RazorMvcBooks.Pages
{
    public class AboutModel: PageModel
    {
```

```
        public static IConfigurationRoot Configuration { get; set; }
        public string Message { get; set; }
        public string Name { get; set; }
        public string Age { get; set; }
        public void OnGet()
        {
            var builder = new ConfigurationBuilder()
                .SetBasePath(Directory.GetCurrentDirectory())
                .AddJsonFile("appsettings.json");
            Configuration = builder.Build();
        }
    }
}
```

说明:

① SetBasePath 是指定从哪个目录开始查找 appsettings.json。如果 appsettings.json 在 configs 目录中，那么调用 AddJsonFile 应该指定的路径为"configs/appsettings.json"。

② 如果使用 Xml 或 Ini 作为配置文件，那么首先要安装相应的包，然后再调用相应的扩展方法 AddXmlFile("appsettings.xml")或 AddIniFile("appsettings.ini")。

1. 读取 JSON 配置

具体操作是:

① 配置包含名称/值对的分层列表，其中节点由冒号（:）分隔。若要检索某个值，则使用相应项的键访问 Configuration 索引器。

② 若在 JSON 格式的配置源中使用数组，则在由冒号分隔的字符串中使用数组的下标索引，0 表示第一项。获取上述 DclTypes 数组中第一项的名称的示例是 Configuration["DclTypes:0:Name"]。

③ 在 Visual Studio 2017 中打开 About.cshtml.cs 文件，把以下读取配置的代码写入 OnGet 方法中:

```
public void OnGet()
{
    var builder = new ConfigurationBuilder()
        .SetBasePath(Directory.GetCurrentDirectory())
        .AddJsonFile("appsettings.json");
    Configuration = builder.Build();
    Message = Configuration["Logging:IncludeScopes"];    //结果: False
    Name = Configuration["DclTypes:0:Name"];             //结果: 申报
    Code = Configuration["DclTypes:1:Code"];             //结果: 2
}
```

④ 在 Visual Studio 2017 中按 F5 键运行应用程序。在浏览器中访问 About 页面，结果如图 4.23 所示。

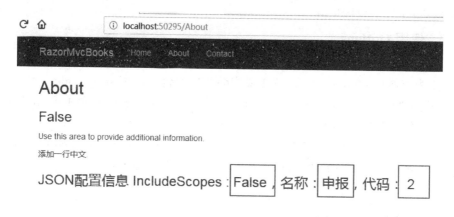

图 4.23 读取 JSON 配置

2. 使用 GetValue⟨T⟩

具体操作是：

① GetValue⟨T⟩是一个扩展方法，要想使用这个方法，首先需要安装相应的包。在 Visual Studio 2017 中选择"工具"→"NuGet 包管理器"→"管理解决方案的 NuGet 程序包"菜单项，在 NuGet 包管理界面安装 Microsoft.Extensions.Configuration.Binder 包。

② 在 Visual Studio 2017 中打开 About.cshtml.cs 文件，把以下读取配置的代码写入 OnGet 方法中：

```
public void OnGet()
{
    var builder = new ConfigurationBuilder()
        .SetBasePath(Directory.GetCurrentDirectory())
        .AddJsonFile("appsettings.json");
    Configuration = builder.Build();
    Message = Configuration.GetValue<string>("Logging:IncludeScopes");    //结果:False
    Name = Configuration.GetValue<string>("DclTypes:0:Name");             //结果:申报
    Code = Configuration.GetValue<string>("DclTypes:1:Code", "2");        //结果:2
}
```

③ 在 Visual Studio 2017 中按 F5 键运行应用程序。在浏览器中访问 About 页面。

GetValue 方法的泛型形式有两个重载，一个是 GetValue("key")，另一个可以指定默认值，即 GetValue("key",defaultValue)。如果 key 的配置不存在，且第一种情况的结果为 default(T)，则第二种情况的结果为指定的默认值。GetValue⟨T⟩适用于简单方案，并不绑定到整个部分。GetValue⟨T⟩从转换为特定类型的 GetSection(key).Value 中获取标量值。

3. 使用 Options

具体操作是：

① 为了使用 Options 方式，首先要安装相应的包。在 Visual Studio 2017 中选择"工具"→"NuGet 包管理器"→"管理解决方案的 NuGet 程序包"菜单项，在 NuGet 包管理界面安装 Microsoft.Extensions.Options.ConfigurationExtensions 包。

② 在 Visual Studio 2017 中创建和定义与配置 appsettings.json 对应的 POCO 的类。通过绑定 POCO 对象，可以检索整个对象图，代码如下：

```
using System;
using System.Collections.Generic;
using System.Linq;
using System.Threading.Tasks;

namespace RazorMvcBooks.Models
{
    public class AppsettingsOptions
    {
        public LoggingOptions Logging { get; set; }
        public List<DclType> GrantTypes { get; set; }
        public class DclType
        {
            public string Name { get; set; }
            public int Code { get; set; }
        }
    }
}

using System;
using System.Collections.Generic;
using System.Linq;
using System.Threading.Tasks;

namespace RazorMvcBooks.Models
{
    public class LoggingOptions
    {
        public bool IncludeScopes { get; set; }
        public LogLevelOptions LogLevel { get; set; }
```

```
        }
    }
using System;
using System.Collections.Generic;
using System.Linq;
using System.Threading.Tasks;

namespace RazorMvcBooks.Models
{
    public class LogLevelOptions
    {
        public string Default { get; set; }
        public string System { get; set; }
        public string Microsoft { get; set; }
    }
}
```

③ 在 Visual Studio 2017 中打开 About.cshtml.cs 文件,把以下读取配置的代码写入 OnGet 方法中:

```
public void OnGet()
{
    var builder = new ConfigurationBuilder()
        .SetBasePath(Directory.GetCurrentDirectory())
        .AddJsonFile("appsettings.json");

    Appsettings = new ServiceCollection().AddOptions()
        .Configure<AppsettingsOptions>(builder.Build())
        .BuildServiceProvider()
        .GetService<IOptions<AppsettingsOptions>>()
        .Value;
    Message = Appsettings.Logging.IncludeScopes.ToString();
    Name = Appsettings.DclTypes[0].Name;
    Code = Appsettings.DclTypes[1].Code.ToString();
}
```

④ 在 Visual Studio 2017 中按 F5 键运行应用程序。在浏览器中访问 About 页面。

4. 使用 Get<T>和 Bind

具体操作是:

① 使用上面示例中的 POCO 类。

② 在 Visual Studio 2017 中打开 About.cshtml.cs 文件,把以下读取配置的代

码写入 OnGet 方法中：

```
public void OnGet()
{
    var builder = new ConfigurationBuilder()
        .SetBasePath(Directory.GetCurrentDirectory())
        .AddJsonFile("appsettings.json");
    Configuration = builder.Build();

    Appsettings = Configuration.Get<AppsettingsOptions>();
    var loggingOptions = Configuration.GetSection("Logging").Get<LoggingOptions>();
    //Appsettings = Configuration.Bind<AppsettingsOptions>();
    //var loggingOptions = Configuration.GetSection("Logging").Bind<Logging-
    //Options>();
    Message = loggingOptions.IncludeScopes.ToString();
    Name = Appsettings.DclTypes[0].Name;
    Code = Appsettings.DclTypes[1].Code.ToString();
}
```

③ 在 Visual Studio 2017 中按 F5 键运行应用程序。在浏览器中访问 About 页面。

5. 文件变化自动重新加载配置

如果配置文件发生了变化，那么程序中的配置项能否跟着同步变化呢？答案是肯定的，这就要用到 reloadOnChange 选项。该选项属性的含义是当被监控的配置文件的内容发生改变时是否需要重新加载配置。使用 reloadOnChange 也很简单，下列代码中的 AddJsonFile 有个重载，只要指定 reloadOnChange:true 即可。

```
var builder = new ConfigurationBuilder()
    .SetBasePath(Directory.GetCurrentDirectory())
    .AddJsonFile("appsettings.json", optional: false, reloadOnChange: true);
```

4.5.3　XML 配置

XML 配置的读写方式与 JSON 格式的读写差不多。若想在 XML 格式的配置源中使用数组，则应向每个元素提供一个 name 索引，并使用该索引访问以下值：

```
<DclTypes>
<DclType Name="申报">
<Code>1</Code>
</DclType>
<DclType Name="注销">
<Code>2</Code>
```

```
</DclType>
</DclTypes>
Console.Write($"{Configuration["DclType:申报:Code"]}");
// Output: 1
```

4.5.4 按环境配置

在环境中设置的任意配置值将替换先前两个提供程序中设置的配置值。

对于在环境变量中指定的分层配置值,冒号(:)可能不适用于所有平台,而所有平台均支持采用双下划线(__)。

当与配置 API 交互时,冒号(:)适用于所有平台。

按环境配置的具体操作是:

① 在 Visual Studio 2017 中选择"工具"→"NuGet 包管理器"→"管理解决方案的 NuGet 程序包"菜单项,安装 Microsoft.Extensions.Configuration.EnvironmentVariables 包。

② 在 Visual Studio 2017 中打开 About.cshtml.cs 文件,并添加如下代码:

```
public void OnGet()
{
    var builder = new ConfigurationBuilder()
        .AddEnvironmentVariables();
    Configuration = builder.Build();
    Message = string.Empty;
    var evn = Environment.GetEnvironmentVariables();
    foreach (var key in evn.Keys)
    {
        Message += evn[key].ToString();
    }
}
```

③ 通过 Environment.GetEnvironmentVariables()方法获取所有的环境变量,如图 4.24 所示。这个方法返回一个 Directory 集合。

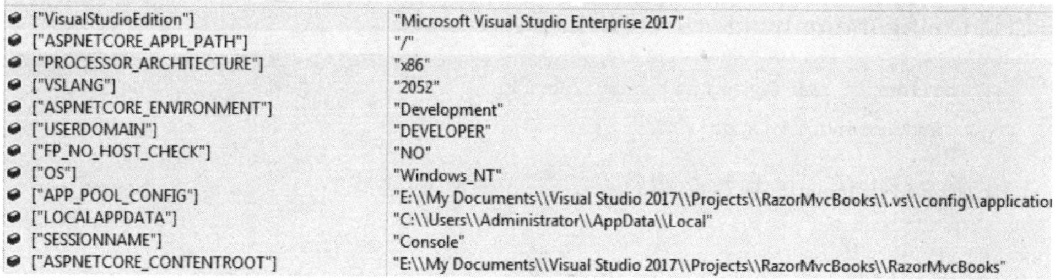

["VisualStudioEdition"]	"Microsoft Visual Studio Enterprise 2017"
["ASPNETCORE_APPL_PATH"]	"/"
["PROCESSOR_ARCHITECTURE"]	"x86"
["VSLANG"]	"2052"
["ASPNETCORE_ENVIRONMENT"]	"Development"
["USERDOMAIN"]	"DEVELOPER"
["FP_NO_HOST_CHECK"]	"NO"
["OS"]	"Windows_NT"
["APP_POOL_CONFIG"]	"E:\\My Documents\\Visual Studio 2017\\Projects\\RazorMvcBooks\\.vs\\config\\application
["LOCALAPPDATA"]	"C:\\Users\\Administrator\\AppData\\Local"
["SESSIONNAME"]	"Console"
["ASPNETCORE_CONTENTROOT"]	"E:\\My Documents\\Visual Studio 2017\\Projects\\RazorMvcBooks\\RazorMvcBooks"

图 4.24 环境变量集合

④ 也可以根据名称来获取环境变量。例如：

Environment.GetEnvironmentVariable("ASPNETCORE_ENVIRONMENT");

4.5.5 在 Razor 页面中访问配置

要想在 Razor 页面中访问配置文件的配置值，需要先打开 About.cshtml 文件，然后用 using 把 Microsoft.Extensions.Configuration 命名空间添加到页面中，并将 IConfiguration 注入页面或视图中，代码如下：

```
@page
@using Microsoft.Extensions.Configuration
@inject IConfiguration Configuration

@model AboutModel
@{
    ViewData["Title"] = "About";
}
<h2>@ViewData["Title"]</h2>
<h3>@Model.Message</h3>
<p>Use this area to provide additional information.</p>
<p>添加一行中文。</p>
<h3>JSON 配置信息　IncludeScopes：@Configuration["Logging:IncludeScopes"]，名称：@Configuration["DclTypes:0:Name"]，代码：@Configuration["DclTypes:1:Code"]</h3>
```

4.5.6 其他配置方式

对于内存中的配置，要调用 AddInMemoryCollection()，其余与 JSON 的配置一样，而且效果也一样。

对于命令行参数的配置，首先要安装 Microsoft.Extensions.Configuration.CommandLine 包，然后调用 AddCommandLine() 扩展方法将命令行配置解析程序添加到 ConfigurationBuilder 中，代码如下：

```
var builder = new ConfigurationBuilder()
    .AddCommandLine(args);
```

在命令行上传递的参数必须符合表 4.3 所列的格式。

表 4.3 参数格式

参数个数	键前缀	示 例
单一参数	无前缀	key1＝value1
	单划线（-）	key2＝value2
	双划线（--）	--key3＝value3
	正斜杠（/）	/key4＝value4
多个参数	单划线（-）	-key1 value1
	双划线（--）	--key2 value2
	正斜杠（/）	/key3 value3

注：①单个参数的值可以为 NULL，多个参数的值不能为 NULL。

②如果提供了重复的键，则使用最后一个键值对。

如果两个解析程序都有相同的配置，那么添加解析程序的顺序就非常重要了，因为后加入的会覆盖前面的。

另外，建议将环境变量的配置解析程序放到最后。

4.6 ASP.NET Core 中的日志记录

记录各种级别的日志是所有应用不可或缺的功能。关于日志记录的实现，已经拥有太多第三方框架可供选择，比如 Log4Net、NLog、Loggr 和 Serilog 等，当然还可以选择微软原生的诊断框架（Microsoft.Diagnostics.Trace/Debug/TraceSource）来实现对日志的记录。

作为一个面对无数选择的开发人员，很可能先选择一个熟悉的日志框架，稍后可能不得不切换到另一个日志框架。因此，开发人员很可能想要编写自己的日志记录 API 包装器，以调用其或其公司选择的任何特定日志框架。类似地，开发人员可能在自己的应用程序中使用一个特定的日志记录框架，但却发现自己要利用的库之一正在使用另一个日志记录框架，使得自己不得不编写一个侦听器，用于将消息从一个日志记录框架传递到另一个日志记录框架。

ASP.NET Core 支持适用于各种日志记录提供程序的日志记录 API。通过内置提供程序，可向一个或多个目标发送日志，还可插入第三方记录框架。ASP.NET Core 不提供异步记录器方法，因为日志记录的速度非常快，使用异步的代价不值得。如果发现自己的实际情况与上述不同，则应考虑更改记录方式。如果数据存储速度较慢，则先将日志消息写入快速存储，稍后再将其转移至低速存储，例如，记录到由另一进程读取和暂留以减缓存储的消息队列中。

ASP.NET Core 日志模块中提供了用于控制台（Microsoft.Extensions.Logging.Console）、调试（Microsoft.Extensions.Logging.Debug）、事件日志

（Microsoft.Extensions.Logging.EventLog）和 TraceSource（Microsoft.Estensions.Logging.TraceSource）等功能的程序。

4.6.1 日志模型三要素

正如上面内容写到的那样，当日志输出的目标多起来之后，写日志就会变得很麻烦。仔细想一下，日志输出这个行为是不变的，变的只是不同的输出目标（控制台、文本文件、数据库等），所以可以把日志记录这个行为抽象出来，使日志记录器仅包含多个可输出的目标，当调用 Log 方法写日志时，由 Log 方法依次调用 Logger 中的 XxxLogger，把日志写到具体的目标上。写日志的过程如图 4.25 所示。

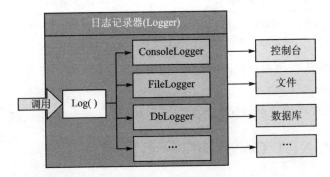

图 4.25　写日志

日志记录主要涉及三个核心对象，分别是 Logger、LoggerFactory 和 LoggerProvider，这三个对象同时也是 .NET Core 日志模型中的核心对象，并通过相应的接口（ILogger、ILoggerFactory 和 ILoggerProvider）来表示，如图 4.26 所示。

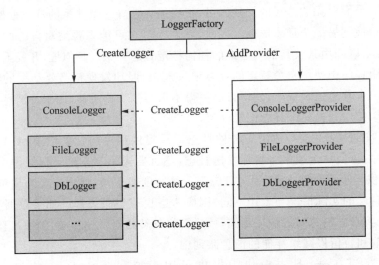

图 4.26　日志记录核心对象

LoggerProvider 和 LoggerFactory 创建的其实是不同的 Logger。Logger-Provider 创建的 Logger 提供的是真正的日志写入功能，即它的作用就是将日志消息写到对应的目的地（比如文件、数据库等）。LoggerFactory 创建一个"组合式"的 Logger，换句话说，这个 Logger 实际上是对一组 Logger 的封装，它自身并不提供真正的日志写入功能，而是委托这组内部封装的 Logger 来写日志。

一个 LoggerFactory 对象上可以添加多个 LoggerProvider 对象。在进行日志编程时，会利用 LoggerFactory 对象创建 Logger 来写日志，而这个 Logger 对象内部封装的 Logger 则通过注册到 LoggerFactory 上的这些 LoggerProvider 来提供。

4.6.2 日志记录级别

当应用程序添加一条日志记录时，必须指定日志的级别。日志级别用于控制应用程序输出日志的详细程度，以及把不同类型的日志传送给不同的日志记录器。例如，开发人员可能希望把调试消息放在本地文件中，而把错误消息记录到计算机的事件日志或数据库中。

ASP.NET Core 详尽定义了六个日志级别，并按照重要性或严重程度排序，如表 4.4 所列。

表 4.4　日志级别

日志级别	描　　述
Trace	用于记录最详细的日志消息，通常仅用于开发阶段调试问题。这些消息可能包含敏感的应用程序数据，因此不应该用于生产环境。默认时禁用。例如：Credentials：{ "User"："someuser"，"Password"："P@ssword"}
Debug	这种消息在开发阶段短期内比较有用。它们包含一些可能会对调试有所助益，但没有长期价值的信息。默认情况下这是最详细的日志。例如：Entering method Configure with flag set to true
Information	这种消息被用于跟踪应用程序的一般流程。与 Verbose 级别的消息相反，这些日志应该有一定的长期价值。例如：Request received for path /foo
Warning	当应用程序出现错误或其他不会导致程序停止的流程异常或意外事件时使用警告级别，以供日后调查。在一个通用的地方处理警告级别的异常。例如：Login failed for IP 127.0.0.1 或 FileNotFoundException for file foo.txt
Error	当应用程序由于某些故障而停止工作时，需要记录错误日志。这些消息应该指明当前活动或操作（如当前的 HTTP 请求），而不是应用程序范围的故障。例如：Cannot insert record due to duplicate key violation
Critical	当应用程序或系统崩溃、遇到灾难性故障，需要立即被关注时，应当记录关键级别的日志。例如：数据丢失、磁盘空间不够等

ASP.NET Core 应用开发入门教程

Logging 包为每个 LogLevel 值提供 Helper 扩展方法,允许开发人员调用,例如,LogInformation,而不是更多详尽的 Log(LogLevel.Information,…)方法。每个 LogLevel 的特定扩展方法有多个重载,允许开发人员传递表 4.5 中的一些或所有参数。

表 4.5 LogLevel 参数

参数名称	描述
string data	需要记录的日志消息
EventId eventId	使用数字类型的 ID 来标记日志,这样可以将一系列事件彼此相互关联。被记录的事件 ID 应该是静态的,是特定于指定类型时间的。比如,可能会把添加商品到购物车的事件 ID 标记为 1000,然后把结单的事件 ID 标记为 1001,以便能智能过滤并处理这些日志记录
string format	日志消息的格式字符串
object[] args	用于格式化的一组对象
Exception error	用于记录异常实例

注:EventId 类型可以隐式转换为 int,所以,可以传递一个 int 参数。

4.6.3 将日志写入不同的目的地

为了在 ASP.NET 应用程序中配置日志,须在 Startup 类的 ConfigureServices 方法中注册日志服务。

下面通过一个简单实例来演示如何将具有不同等级的日志写入两个不同的目的地,其中一个是直接将格式化的日志消息输出到当前控制台,另一个则是将日志写入 Debug 输出窗口(相当于直接调用 Debug.WriteLine 方法),针对这两个日志目的地的 Logger,可以分别通过 ConsoleLoggerProvider 和 DebugLoggerProvider 两种不同的 LoggerProvider 来提供。具体操作如下:

① 在 Visual Studio 2017 中选择"工具"→"NuGet 包管理器"→"管理解决方案的 NuGet 程序包"菜单项,安装"Microsoft.Extensions.Logging.Console"和"Microsoft.Extensions.Logging.Debug"NuGet 包。在安装这两个包时,NuGet 会一起安装"Microsoft.Extensions.Logging"包,其中默认使用的 LoggerFactory 和由其创建的 Logger 定义在"Microsoft.Extensions.Logging"NuGet 包中。而 ConsoleLoggerProvider 和 DebugLoggerProvider 这两个 LoggerProvider 分别定义在"Microsoft.Extensions.Logging.Console"和"Microsoft.Extensions.Logging.Debug"包中。

② 在 Visual Studio 2017 中打开 Startup.cs 文件,在 ConfigureServices 方法中注册和配置日志服务。可以直接调用针对 ILoggerFactory 接口的扩展方法 AddConsole 和 AddDebug 分别完成针对 ConsoleLoggerProvider 和 DebugLoggerProvider 的注册,代码如下:

```csharp
using System;
using System.Collections.Generic;
using System.IO;
using System.Linq;
using System.Threading.Tasks;
using Microsoft.AspNetCore.Builder;
using Microsoft.AspNetCore.Hosting;
using Microsoft.AspNetCore.Http;
using Microsoft.Extensions.Configuration;
using Microsoft.Extensions.DependencyInjection;
using Microsoft.Extensions.FileProviders;
using Microsoft.Extensions.Logging;

namespace RazorMvcBooks
{
    public class Startup
    {
        public Startup(IConfiguration configuration)
        {
            Configuration = configuration;
        }
        public IConfiguration Configuration { get; }
        //This method gets called by the runtime. Use this method to add services to the
        //container.
        public void ConfigureServices(IServiceCollection services)
        {
            // services.AddDistributedMemoryCache();
            services.AddSession(options =>
            {
                // Set a short timeout for easy testing.
                options.IdleTimeout = TimeSpan.FromSeconds(10);
                options.Cookie.HttpOnly = true;
            });
            services.AddLogging(builder =>
            {
                builder
                    .AddConfiguration(Configuration.GetSection("Logging"))
                    .AddDebug()
                    .AddConsole();
            });
            services.AddMvc();
        }
```

```csharp
//This method gets called by the runtime. Use this method to configure the HTTP
//request pipeline.
public void Configure(IApplicationBuilder app, IHostingEnvironment env)
{
    if (env.IsDevelopment())
    {
        app.UseBrowserLink();
        app.UseDeveloperExceptionPage();
    }
    else
    {
        app.UseExceptionHandler("/Error");
    }
    app.UseStaticFiles();
    app.UseMvc();
}
}
```

③ ASP.NET Core 会基于依赖注入容器以参数的形式自动提供 ILogger 实例，开发人员可以把 ILogger 作为参数添加到构造函数中。日志记录通过如下程序来完成。

```csharp
using System;
using System.IO;
using Microsoft.AspNetCore.Mvc.RazorPages;
using Microsoft.Extensions.DependencyInjection;
using Microsoft.Extensions.Logging;
using Microsoft.Extensions.Options;

namespace RazorMvcBooks.Pages
{
    public class AboutModel : PageModel
    {
        private readonly ILogger m_logger;
        public AboutModel(ILogger<AboutModel> logger)
        {
            m_logger = logger;
        }
        public string Message { get; set; }
        public void OnGet()
        {
```

```
                int eventId = 10002;
                m_logger.LogInformation(eventId, "请把.NET Core 的版本升级到({version})", "2.0.0");
                m_logger.LogWarning(eventId, "用户数量接近上限({maximum})", 200);
                m_logger.LogError(eventId, "用户登录失败(用户名:{User},业务系统:{System})", "张三", "OA");
            }
        }
    }
```

以上程序中通过依赖注入获取了一个 Logger 对象,然后先后调用 LogInformation、LogWarning 和 LogError 这三个扩展方法来记录三条日志消息,这三个方法的命名决定了日志所采用的等级(Information、Warning 和 Error)。同时,在调用这三个方法时指定了一个表示日志记录事件 ID 的整数(10002),以及具有占位符("{version}""{maximum}""{User}"和"{System}")的消息模板和替换这些占位符的参数列表。

④ 在 Visual Studio 2017 中以项目形式启动应用程序,如图 4.27 所示。

⑤ 在浏览器中浏览 About 页面后,三条日志消息会直接按照如图 4.28 所示的形式打印到控制台上。从图中可以看出,格式化的日志消息不仅仅包含了指定的消息内容,日志的等级、类型和事件 ID 同样包含其中。不仅如此,表示日志等级的文字还会采用不同的前景色和背景色来显示。

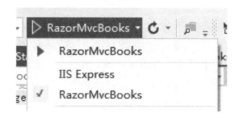

图 4.27 启动应用程序

图 4.28 不同颜色的日志

⑥ 由于还注册了另一个 DebugLoggerProvider 对象，而该对象创建的 Logger 会直接调用 Debug.WriteLine 方法写入格式化的日志消息，所以当以 Debug 模式编译并执行应用程序时，Visual Studio 2017 的输出窗口会以如图 4.29 所示的形式呈现日志消息。

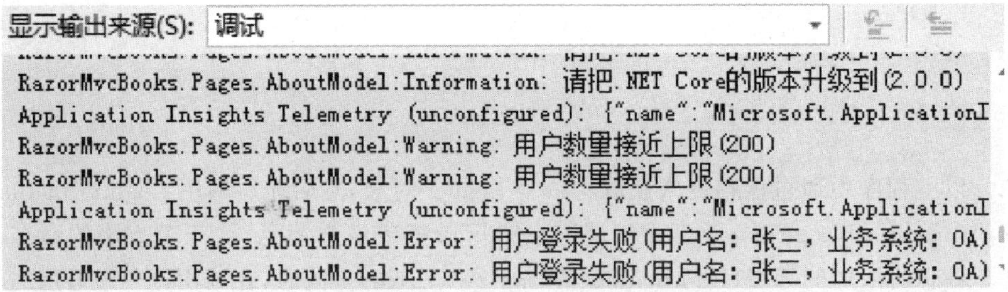

图 4.29　输出窗口中的日志信息

4.6.4　添加筛选功能

在 4.6.3 小节的示例中会看到，在第⑥步中，每次通过浏览器发出一个 Web 请求都会产生超过一条的日志记录，这是因为大多数浏览器在尝试加载一个页面时会发出多个请求（如请求 CSS 文件、请求网站的图标文件等）。需要注意一点，在控制台记录器上显示的先是日志级别（如图 4.48 中的 info），然后是类别（[Catchall Endpoint]），最后是日志消息。

在实际的应用程序中，一般希望基于应用程序级别来添加日志，而不是基于框架级别或事件。如 4.6.3 小节的示例中，就希望看到自己所希望看到的各种操作的日志信息。如下所示的代码片段就体现了这样的编程方式。

```
public void ConfigureServices(IServiceCollection services)
{
    // services.AddDistributedMemoryCache();
    services.AddSession(options =>
    {
        // Set a short timeout for easy testing.
        options.IdleTimeout = TimeSpan.FromSeconds(10);
        options.Cookie.HttpOnly = true;
    });
    services.AddLogging(builder =>
    {
        builder
            .AddConfiguration(Configuration.GetSection("Logging"))
            .AddFilter("Microsoft", LogLevel.Warning)
            .AddDebug()
```

```
            .AddConsole();
        });
        services.AddMvc();
}
```

在 Visual Studio 2017 中以项目形式启动应用程序。在浏览器中浏览 About 页面后,三条日志消息会直接按照如图 4.30 所示的形式打印到控制台上,而基于框架级别的一些日志消息则没有显示。

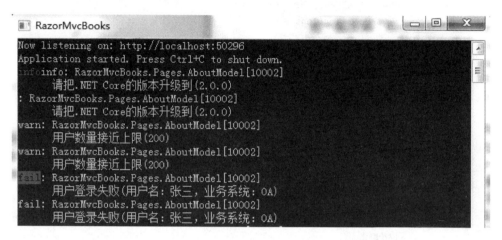

图 4.30　筛选后的日志信息

4.6.5　根据等级过滤日志消息

由于在同一个 LoggerFactory 上可以注册多个 LoggerProvider,所以当利用 LoggerFactory 创建出相应的 Logger 来写入某条日志消息时,这条消息实际上会被分发给由 LoggerProvider 提供的所有 Logger。其实,在很多情况下并不希望每个 Logger 都去写入分发给它的每条日志消息,而是希望 Logger 能够"智能"地忽略不应该由它写入的日志消息。每条日志消息都具有一个等级,日志等级是普遍采用的日志过滤策略。日志等级通过具有如下定义的枚举 LogLevel 来表示,枚举项的值决定了等级的高低,值越大,等级越高;等级越高,越需要记录。

```
public enum LogLevel
{
    Trace           = 0,
    Debug           = 1,
    Information     = 2,
    Warning         = 3,
    Error           = 4,
    Critical        = 5,
```

```
            None                = 6
}
```

可以指定全局过滤器作用于所有的日志提供者,如下段程序所示,其中忽略了低于 Error 级别的日志。

```
public void ConfigureServices(IServiceCollection services)
{
    // services.AddDistributedMemoryCache();
    services.AddSession(options =>
    {
        // Set a short timeout for easy testing.
        options.IdleTimeout = TimeSpan.FromSeconds(10);
        options.Cookie.HttpOnly = true;
    });
    services.AddLogging(builder =>
    {
        builder
            .AddConfiguration(Configuration.GetSection("Logging"))
            .AddFilter("Microsoft", LogLevel.Warning)
            .AddFilter(string.Empty, LogLevel.Error)
            .AddDebug()
            .AddConsole();
    });
    services.AddMvc();
}
```

在 Visual Studio 2017 中以项目形式启动应用程序。在浏览器中浏览 About 页面后,在控制台窗口中只显示 Error 级别的日志消息,如图 4.31 所示。

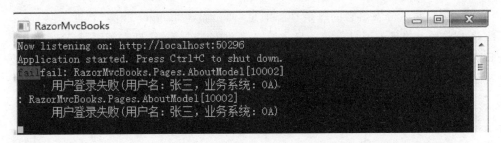

图 4.31　Error 级别日志信息

在 Visual Studio 2017 的输出窗口中也会忽略掉 Error 级别以下的日志信息,如图 4.32 所示。

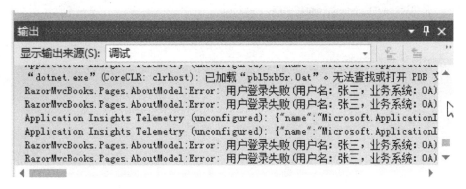

图 4.32　输出窗口中的 Error 级别日志信息

4.6.6　设置文件配置

开发人员也可以通过设置配置文件来对日志信息进行配置。具体操作是：

① 在 Visual Studio 2017 中打开 Startup.cs 文件，把注册日志服务的相关代码删除。

② 在 Visual Studio 2017 中打开 appsettings.json 文件，修改 Logging 部分，以供日志记录程序加载使用。以下代码显示了典型 appsettings.json 文件的内容：

```
{
    "Logging": {
        "IncludeScopes": false,
        "LogLevel": {
            "Default": "Warning",
            "Microsoft": "Warning"
        }
    }
}
```

LogLevel 项表示日志级别，其中 Default 项表示默认日志，其值表示记录默认日志的最低日志级别；设置 IncludeScopes 的日志项指定了是否为相关日志启用日志作用域。

③ 在 Visual Studio 2017 中打开 Program.cs 文件，然后注册日志服务。

```
using System;
using System.Collections.Generic;
using System.IO;
using System.Linq;
using System.Threading.Tasks;
using Microsoft.AspNetCore;
using Microsoft.AspNetCore.Hosting;
```

```
using Microsoft.Extensions.Configuration;
using Microsoft.Extensions.Logging;

namespace RazorMvcBooks
{
    public class Program
    {
        public static void Main(string[] args)
        {
            BuildWebHost(args).Run();
        }
        public static IWebHost BuildWebHost(string[] args) =>
            WebHost.CreateDefaultBuilder(args)
                .ConfigureAppConfiguration((hostingContext, config) =>
                {
                    var env = hostingContext.HostingEnvironment;
                    config.AddJsonFile("appsettings.json", optional: true, reloadOnChange: true);
                })
                .ConfigureLogging((hostingContext, logging) =>
                {
                    logging.AddConfiguration(hostingContext.Configuration.GetSection("Logging"));
                    logging.AddConsole();
                    logging.AddDebug();
                })
                .UseStartup<Startup>()
                .Build();
    }
}
```

④ 在 Visual Studio 2017 中以项目形式启动应用程序。在浏览器中浏览 About 页面后，在控制台窗口中只显示 Warning 级别以上的日志消息，如图 4.33 所示。

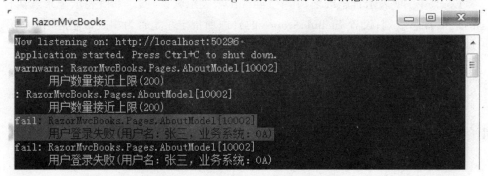

图 4.33 Warning 级别以上的日志消息

⑤ 在 Visual Studio 2017 的输出窗口中也会忽略掉 Warning 级别以下的日志信息,如图 4.34 所示。

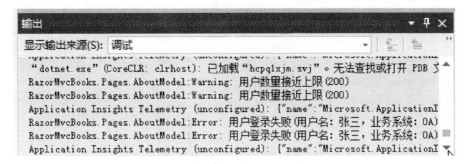

图 4.34 输出窗口中 Warning 级别以上的日志消息

4.6.7 作用域

在应用程序记录日志信息的过程中,可以将一组逻辑操作"作用域"打包为一组。作用域也是一种 IDisposable 类型,通过调用 ILogger.BeginScope〈TState〉方法来返回,它自创建起持续到释放为止。内建的 TraceSource 日志记录器会返回一个作用域实例来响应启动与停止跟踪操作。任何诸如事务 ID 这样的日志状态从刚创建时起便关联到作用域了。

作用域不是必需的,而且需要谨慎使用。它们适用于具有比较明显的开始和结束操作,比如在一个事务中调用多个资源。

通常,提供程序支持"作用域"的概念,以便可以记录下代码是如何遍历调用链的。具体操作是:

① 在 Visual Studio 2017 中打开 Program.cs 文件,把控制台显示日志的作用域打开。把原来的

```
logging.AddConsole();
```

替换为

```
logging.AddConsole(options => options.IncludeScopes = true);
```

② 在 Visual Studio 2017 中打开 About.cshtml.cs 文件,添加以下作用域代码:

```
using System;
using System.IO;
using Microsoft.AspNetCore.Mvc.RazorPages;
using Microsoft.Extensions.Configuration;
using Microsoft.Extensions.DependencyInjection;
using Microsoft.Extensions.Logging;
using Microsoft.Extensions.Options;
```

```
using RazorMvcBooks.Models;

namespace RazorMvcBooks.Pages
{
    public class AboutModel: PageModel
    {
        private readonly ILogger m_logger;
        public AboutModel(ILogger<AboutModel> logger)
        {
            m_logger = logger;
        }
        public string Message { get; set; }
        public void OnGet()
        {
            using (m_logger.BeginScope(" -- 日志信息作用域,显示日志 -- "))
            {
                int eventId = 10002;
                m_logger.LogInformation(eventId, "请把.NET Core 的版本升级到({version})", "2.0.0");
                m_logger.LogWarning(eventId, "用户数量接近上限({maximum})", 200);
                m_logger.LogError(eventId, "用户登录失败(用户名:{User},业务系统:{System})", "张三", "OA");
            }
        }
    }
}
```

③ 在 Visual Studio 2017 中以项目形式启动应用程序。在浏览器中浏览 About 页面后,在控制台窗口中显示包含作用域的日志记录,如图 4.35 所示。

图 4.35 包含作用域的日志记录

4.6.8 日志记录建议

当想在 ASP.NET Core 应用程序中实现日志时,可以参考以下有用的建议:

① 正确使用 LogLevel,这将使不同重要级别的日志消息路由到相关的输出目标上。

② 记录的日志信息要能立即识别问题所在,剔除不必要的冗余信息。

③ 保证日志内容简单明了,直指重要信息。

④ 尽管日志记录器被禁用后将不记录日志,但也请在日志方法的周围增加控制代码,以防止多余的方法调用和日志设置的开销,特别是在循环和对性能要求比较高的方法中。

⑤ 使用独有的前缀命名日志记录器以确保能快速过滤或禁用。

⑥ 使用作用域时保持谨慎,明晰动作开始和结束的界限(比如框架提供的 MVC Action 的范围),避免相互嵌套。

⑦ 应用程序日志代码应关注应用程序的业务。通过提高日志的详细程度级别来记录与框架相关的问题,而不是日志记录器自己。

4.7 在 ASP.NET Core 中使用多个环境

ASP.NET Core 基于使用环境变量的运行时环境来配置应用行为。

ASP.NET Core 提供了许多功能和约定来支持应用程序在多个环境中的行为。环境变量用来指示应用程序正在运行的环境,允许应用程序进行适当的配置。

4.7.1 环　境

ASP.NET Core 引用了一个特定的环境变量 ASPNETCORE_ENVIRONMENT 来描述应用程序当前运行的环境。ASPNETCORE_ENVIRONMENT 可设置为任意值,框架支持三个值:开发(Development)、预演(Staging)和生产(Production)。如果未设置 ASPNETCORE_ENVIRONMENT,则默认为 Production。

当前的环境设置可以通过编程方式从应用程序中被检测到;除此之外,也可以基于当前的应用程序环境在视图里使用环境标记助手来包含某些部分。适当的环境变量允许对应用程序的调试、测试或生产使用进行适当的优化。

注意:在 Windows 和 macOS 上,环境变量和值不区分大小写。无论把变量设置为 Development 或 development 或 DEVELOPMENT,其结果都是相同的。默认情况下,Linux 的环境变量和值要区分大小写。

4.7.2 在运行时确定环境

IHostingEnvironment 服务为工作环境提供了核心抽象。该服务由 ASP.NET 宿主层提供,并且能够通过 Dependency Injection 注入到启动逻辑中。在 Visual

Studio 2017 中的 ASP.NET Core 网站模板使用这种方式来加载特定的环境配置文件,并且自定义应用程序的错误处理设置。在这两种情况下,这种行为是由通过参照当前指定的环境来调用 IHostingEnvironment 的实例上的 EnvironmentName 或将 IsEnvironment 传递到适当的方法上来实现的。例如,可以使用如下代码在配置方法中设置特定环境的错误处理。

下面来看一个在 Startup 中的环境变更设置,这段代码主要实现以下功能:

① 当将 ASPNETCORE_ENVIRONMENT 设置为 Development 时,调用 UseDeveloperExceptionPage 和 UseBrowserLink,开启使用浏览器链接(BrowserLink)功能,显示特定的开发错误页面和特定的数据库错误页面。

② 当 ASPNETCORE_ENVIRONMENT 的值为 Production、Starting、Starting_2 中的三个值之一时,调用 UseExceptionHandler,配置一个标准的错误处理页面来显示响应中的任何未处理异常。

代码如下:

```
public void Configure(IApplicationBuilder app, IHostingEnvironment env)
{
    if (env.IsDevelopment())
    {
        app.UseBrowserLink();
        app.UseDeveloperExceptionPage();
    }
    if (env.IsProduction() || env.IsStaging() || env.IsEnvironment("Staging_2"))
    {
        app.UseExceptionHandler("/Error");
    }
    app.UseStaticFiles();
    app.UseMvc();
}
```

③ 在 Visual Studio 2017 的 About.cshtml 文件中使用环境标记助手,并使用 IHostingEnvironment.EnvironmentName 的值来包含或排除元素中的标记。

可能需要在运行时确定需要向客户端发送哪些内容。例如,在开发环境中通常提供非最小化的脚本和样式表,这样更容易调试。在生产和测试环境中,一般应从 CND 提供最小化的版本,当然也可以使用 environment 标签做到这一点。如果当前环境与使用 names 特性指定的环境相匹配,则 environment 标签将只提供标签范围内的内容。代码如下:

```
@page
@inject Microsoft.AspNetCore.Hosting.IHostingEnvironment hostingEnv
@model AboutModel
```

```
@{
    ViewData["Title"] = "About";
}
<h2>@ViewData["Title"]</h2>
<h3>@Model.Message</h3>

<p>ASPNETCORE_ENVIRONMENT = @hostingEnv.EnvironmentName</p>

<environment names = "Development">
<link rel = "stylesheet" href = "~/lib/bootstrap/dist/css/bootstrap.css" />
<link rel = "stylesheet" href = "~/css/site.css" />
</environment>
<environment names = "Staging,Production">
<link rel = "stylesheet" href = "https://ajax.aspnetcdn.com/ajax/bootstrap/3.3.6/css/bootstrap.min.css" asp-fallback-href = "~/lib/bootstrap/dist/css/bootstrap.min.css" asp-fallback-test-class = " sr-only" asp-fallback-test-property = " position" asp-fallback-test-value = "absolute" />
<link rel = "stylesheet" href = "~/css/site.min.css" asp-append-version = "true" />
</environment>
```

4.7.3 开发环境

Development 是在开发应用程序时所使用的环境设置,当使用 Visual Studio 2017 时,该设置可以在项目的调试配置文件中指定,比如 IIS Express。

当使用 Visual Studio 2017 创建一个 ASP.NET Core 项目时会创建一个默认设置,该默认设置可以被修改,修改后会保存在 Properties 文件夹的 launchSettings.json 文件中,该文件就是应用程序的配置文件。例如,本地计算机开发环境可以在项目的 Properties\launchSettings.json 文件中设置。以下显示了 launchSettings.json 配置文件的代码:

```
{
    "iisSettings": {
        "windowsAuthentication": false,
        "anonymousAuthentication": true,
        "iisExpress": {
            "applicationUrl": "http://localhost:50295/",
            "sslPort": 0
        }
    },
    "profiles": {
        "IIS Express": {
        "commandName": "IISExpress",
        "launchBrowser": true,
        "environmentVariables": {
            "ASPNETCORE_ENVIRONMENT": "Development"
```

ASP. NET Core 应用开发入门教程

```
        }
    },
    "RazorMvcBooks": {
        "commandName": "Project",
        "launchBrowser": true,
        "environmentVariables": {
            "ASPNETCORE_ENVIRONMENT": "Development"
        },
        "applicationUrl": "http://localhost:50296/"
    }
}
```

注意：

① 对项目配置文件或 launchSettings.json 所做的更改，在使用的 Web 服务器重启之前可能不会直接生效(尤其是，kestrel 在将要检测它的环境变化之前必须重启)。也可以为应用程序的不同配置文件创建多个不同的启动配置，包括它们需要的其他环境变量。

② launchSettings.json 中的 applicationUrl 属性可指定服务器 URL 的列表。在列表中的 URL 之间使用分号隔开。

警告：环境变量存储在 launchSettings.json 中是不安全的，并且将作为应用程序源代码仓库的一部分。在 launchSettings.json 中不应存储机密数据。

Visual Studio 2017 提供了 UI 编辑界面，可以修改 launchSettings.json 文件。可以选择"菜单"→"项目"→"RazorMvcBooks 属性"菜单项，或者在解决方案资源管理器中选中项目名称(如本例 RazorMvcBooks)并右击，在弹出的快捷菜单中选中"属性"，在弹出的项目属性对话框中选中"调试"选项卡，此时，可由 Visual Studio 2017 提供的编辑界面对 launchSettings.json 文件进行编辑，如图 4.36 所示。

图 4.36 配置调试环境

在界面中进行编辑后,单击"保存"按钮。在 Web 服务器重新启动之前,对项目配置文件所做的更改可能不会生效。必须重新启动 Kestrel 才能检测到对其环境所做的更改。在 Visual Studio 2017 中重新启动应用程序,这次选择 RazorMvcBooks 进行启动,如图 4.37 所示。

图 4.37 重新启动应用程序

在使用 Visual Studio 2017 启动应用程序时,会先读取 launchSettings.json。launchSettings.json 中的 environmentVariables 设置会替代环境变量。图 4.38 的输出显示了承载环境。

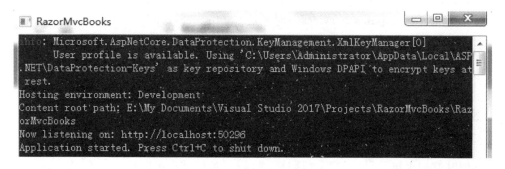

图 4.38 环境信息

4.7.4 生产环境

Production 环境是应用程序在生产中运行的被终端用户使用的环境。这个环境应该被配置为最大限度地提高安全性、性能和应用程序的健壮性。生产环境的一些不同于开发环境的通用设置包括:

① 启用缓存;
② 确保所有客户端资源被打包、压缩和尽可能从 CDN 提供;
③ 关闭诊断错误页面;
④ 启用友好的错误页面;
⑤ 启用生产日志和监控(例如 Application Insights)。

这并不是一个完整的列表,最好避免在应用程序的各个部分进行散乱的环境检查;相反,推荐尽可能在应用程序的 Startup 类中进行这样的检查。

4.7.5 基于环境的 Startup 类和方法

ASP.NET Core 支持一种基于约定协议并根据当前的环境配置进行应用程序的启动。根据应用程序处于哪一种环境,可以使用编程的方式来控制应用程序的行为,以及允许创建和管理自己的约定。

当 ASP.NET Core 应用程序启动时,Startup 类用来引导应用程序和加载其配置设置等。然而,如果一个类的命名存在 Startup{EnvironmentName}(例如 Startup-Development),并且 Hosting:Environment 环境变量与其名称相匹配,则使用该 Startup 类。

除了使用一个基于当前环境的完全独立的启动类外,也可以在 Startup 类中对应用程序如何配置做出调整。Configure() 和 ConfigureServices() 方法类似 Startup 类,以 Configure[EnvironmentName]() 和 Configure[EnvironmentName]Services() 的形式支持特定环境的版本。当设置为开发环境时,如果定义了一个 Configure-Development() 方法,则调用该方法而不是 Configure()。同样,在相同的环境里将调用 ConfigureDevelopmentServices() 而不是 ConfigureServices()。

4.8 Session 详解

4.8.1 什么是 Session

Session,译为"会话",其本来的含义是指有始有终的一系列动作/消息,比如,打电话时从拿起手机拨号到挂断电话这中间的一系列过程可以称为一个 Session。在 Web 领域是指一个浏览器窗口从打开到关闭的这一个周期,比如,从登录到选购商品,再到结账登出这样一个网上购物的过程。

HTTP 协议是一种无状态协议,不能保存信息,即在每次服务器端接收到客户端的请求时,都是一个全新的请求,服务器并不知道客户端的历史请求记录,好比一个非会员的普通顾客与一个普通超市之间的关系一样。

Session 的主要目的就是为了弥补 HTTP 的无状态特性。简单地说,就是服务器可以利用 Session 实现"面向连接"与"保持状态"的功能。

"面向连接"指的是通信双方在通信之前要先建立一个通信的渠道,比如打电话,直到对方接了电话后通信才能开始,与此相对的是写信,在把信发出去时并不能确认对方的地址是否正确,通信渠道不一定能建立,但对发信人来说,通信已经开始了。

"保持状态"则是指通信的一方能够把一系列的消息关联起来,使得消息之间能够互相依赖,比如一个服务员能够认出再次光临的老顾客,并且记得上次这个顾客坐在哪个位置。

说到 Session 就不能不说到 Cookie。通常情况下,Session 是保存在服务器的内存中,而 Cookie 则是保存在客户端。

当然也可以将 Session 从内存中拿到其他地方存储。具体来说,Cookie 机制采用的是在客户端保持状态的方案,而 Session 机制采用的是在服务器端保持状态的方案。同时也可以看到,由于采用服务器端保持状态的方案在客户端也需要保存一个标识,所以 Session 机制可以借助于 Cookie 机制来保存 Session 标识;当然,实际上也有其他不采用 Cookie 的方案。

做一个形象的比喻,保存 Session 的服务器内存就像一个银行,与之对应的浏览器 Cookie 就像银行卡,而 Session Id 就是银行卡号。浏览器中不存储任何重要信息,仅用一把"钥匙"就可以打开这个银行。

4.8.2 理解 Session 机制

当应用程序需要为某个客户端的请求创建一个 Session 时,服务器首先检查该客户端的请求里是否已包含了一个 Session 标识 ——称为 Session Id,如果已包含一个 Session Id,则说明以前曾为此客户端创建过 Session,服务器就按照 Session Id 把这个 Session 检索出来后使用(如果检索不到,可能会新建一个);如果客户端请求不包含 Session Id,则为此客户端创建一个 Session,并生成一个与此 Session 相关联的 Session Id,Session Id 的值是一个不可能重复的唯一值,并且保存在一个类似于字典的集合中,Session Id 将被在本次响应中返回给客户端保存。

共有四种方式在客户端保存 Session Id:

第一种,这是最简单、最常用的方式,它依赖于 Cookie 的实现。实现步骤是:

① 浏览器对服务器上的页面(如 http://localhost:5000/Index)发出了 HTTP GET 请求,该客户端的浏览器以前从未访问过该站点。

② Web 服务器通过 Web 应用程序返回了要显示的 HTML 作为响应。此外,Index 页面还返回了带有唯一标识(Session Id)的 Cookie,以此来跟踪该浏览器。

③ 在下一个请求中,之前设置的 Cookie 会返回给服务器。

④ 之前设置的保存在 Cookie 中的唯一标识(Session Id),现在可以在任意类型的服务器端状态机制中作为键来使用。状态可以是内存中的散列表,也可以是 SQL 数据库,还可以是缓存中的类似于字典的集合。

第二种,重写 URL 中的 QueryString。由于 Cookie 能被人为地禁止,因此,采用这种方式在 Cookie 被禁止时仍然能够把 Session Id 传递回服务器。这种实现方式是通过在请求的 URL 后面添加后缀 SESSIONID=xxx 将 SESSIONID 传递给客户端。这种方式相对复杂,不能传递敏感信息和大量数据信息,且容易被用户修改。

第三种,隐藏的表单域。在 FORM 表单中设置了一个隐藏域,并将 SESSIONID 放置其中一并发送给服务器端。

第四种,HTML 5 中的 WebStorage。它很像 Cookie 的强化版,只是存储为键值对,它不会将数据自动发送到服务器,而需要客户端主动把数据附加到请求中。

当然，第一种方式是目前的主流，但是浏览器必须要支持 Cookie，如果浏览器禁用 Cookie，那么就可以考虑其他三种实现方式！

在谈论 Session 机制的时候，常常听到这样一种误解，"只要关闭浏览器，Session 就消失了。"可以参照现实中会员卡的例子，除非顾客主动对店家提出销卡，否则店家绝对不会轻易删除顾客的资料。对 Session 来说也是一样，除非程序通知服务器删除一个 Session，否则服务器会一直保留，应用程序一般都是在用户登出时发出指令去删除 Session。浏览器从来不会在关闭之前主动通知服务器它将要关闭，因此服务器根本不会有机会知道浏览器已关闭。之所以会有这种错觉，是因为大部分 Session 机制都使用会话 Cookie 来保存 Session Id，而关闭浏览器后，这个 Session Id 就消失了，再次连接服务器时也无法找到原来的 Session。如果服务器设置的 Cookie 被保存到硬盘上，或者使用某种手段改写浏览器发出的 HTTP 请求头，把原来的 Session Id 发送给服务器，则再次打开浏览器时仍能找到原来的 Session。

恰恰是由于关闭浏览器不会导致 Session 被删除，才迫使服务器为 Session 设置了一个失效时间，当距离客户端上一次使用 Session 的时间超过这个失效时间时，服务器就认为客户端已停止了活动，之后才会把 Session 删除以节省存储空间。

没有任何一种技术是完美无瑕的，Session 也不例外。笔者认为，Session 有如下缺点：

① 如果 Session 的实现是依赖于 Cookie 的话，那么浏览器必须要支持 Cookie，否则不能实现。

② 过多的浏览器请求会在服务器端创建很多的 Session 信息，这样会对浏览器造成一定的内存压力，而且不易于维护。

4.8.3 ASP.NET Core 中的 Session

一般情况下，Session 都是存储在 Web 服务器的内存中，当服务器进程被停止或重启时，内存中的 Session 也会被清空。虽然也可以将每个 Session 加入到单独服务器上的特定应用实例中，但更好的方法是使用 Redis 或 SQL Server 分布式缓存来保存 Session，这样，Session 的信息被复制到各个不同的服务器的内存中，即使某个服务器进程停止工作，仍然能从其他服务器中取得 Session。

Session 使用应用程序维护的存储来保存客户端所有请求的数据。Session 数据由缓存支持并被视为临时数据，而站点则在没有会话数据的情况下继续运行。

ASP.NET Core 通过向客户端提供包含 Session Id 的 Cookie 来维护会话状态，该 Session Id 与每个请求一起发送给应用程序，应用程序使用 Session Id 获取会话数据。

Microsoft.AspNetCore.Session 就是在用户浏览 Web 应用时用来存储用户数据的 ASP.NET Core 方案，是 .NET Core 平台管理 Session 的中间件。Microsoft.AspNetCore.Session 提供了一系列的类来实现 Session 功能，包括 Session 持久化的设置和 Session 失效时间的设置，以及 Cookie 的设置。

Session 具有以下行为：

① 由于会话 Cookie 是针对于浏览器的，因此不能跨浏览器共享会话。

② 浏览器会话结束时删除会话 Cookie。

③ 如果收到过期的会话 Cookie，则创建使用相同会话 Cookie 的新会话。

④ 不会保留空会话，会话中必须设置了至少一个值以保存所有请求的会话。当会话未保留时，为每个新的请求生成新会话 ID。

⑤ 应用程序在上次请求后保留会话的时间有限。应用程序可以设置会话超时，或者使用 20 分钟的默认值。

⑥ 调用 ISession.Clear 方法可以删除或者在会话过期时删除会话数据。

⑦ 没有默认机制的应用代码可以告知客户端浏览器已关闭，或者客户端上的会话 Cookie 被删除或已过期。

4.8.4　ASP.NET Core 中如何使用 Session

使用 Session 的步骤是：

① 在 Visual Studio 2017 中选择"工具"→"NuGet 包管理器"→"管理解决方案的 NuGet 程序包"菜单项，安装 Microsoft.AspNetCore.Session 包。

② 在 Startup.cs 中注册 Session 服务，其中必须包含：

ⓐ 任一 IDistributedCache 缓存，该缓存用于实现会话后备存储。

ⓑ 调用 ConfigureServices 中的 AddSession。

ⓒ 调用 Configure 中的 UseSession。

具体代码如下：

```
using System;
using System.Collections.Generic;
using System.Linq;
using System.Threading.Tasks;
using Microsoft.AspNetCore.Builder;
using Microsoft.AspNetCore.Hosting;
using Microsoft.Extensions.Configuration;
using Microsoft.Extensions.DependencyInjection;

namespace RazorMvcBooks
{
    public class Startup
    {
        public Startup(IConfiguration configuration)
        {
            Configuration = configuration;
        }
        public IConfiguration Configuration { get; }
```

```
// This method gets called by the runtime. Use this method to add services to the container.
        public void ConfigureServices(IServiceCollection services)
        {
            // services.AddDistributedMemoryCache();
            services.AddSession(options =>
            {
                // Set a short timeout for easy testing.
                options.IdleTimeout = TimeSpan.FromSeconds(10);
                options.Cookie.HttpOnly = true;
            });
            services.AddMvc();
        }
        //This method gets called by the runtime. Use this method to configure the HTTP
        //request pipeline.
        public void Configure(IApplicationBuilder app, IHostingEnvironment env)
        {
            if (env.IsDevelopment())
            {
                app.UseBrowserLink();
                app.UseDeveloperExceptionPage();
            }
            else
            {
                app.UseExceptionHandler("/Error");
            }
            app.UseStaticFiles();
            app.UseMvc();
            app.UseCookiePolicy();
            app.UseSession();
        }
    }
}
```

通过以上代码,将 Session 以组件化的形式添加到项目中。在控制器类中通过 HttpContext.Session 就可以对 Session 进行读写了。

③ 在 Visual Studio 2017 中打开 About.cshtml.cs 文件,输入以下代码:

```
public class AboutModel : PageModel
{
    private const string SessionKeyName = "_Name";
    private const string SessionKeyAge = "_Age";
    public string Message { get; set; }
    public string Name { get; set; }
    public string Age { get; set; }
    public void OnGet()
```

```
        {
            if (string.IsNullOrEmpty(HttpContext.Session.GetString(SessionKeyName)))
            {
                HttpContext.Session.SetString(SessionKeyName, "彭祖");
                HttpContext.Session.SetInt32(SessionKeyAge, 800);
            }
            var name = HttpContext.Session.GetString(SessionKeyName);
            var age = HttpContext.Session.GetInt32(SessionKeyAge);
            Name = name;
            Age = age.ToString();
        }
}
```

④ 在输入完以上代码后，会发现 GetString 方法下面有波浪线，如图 4.39 所示。

图 4.39 错误提示

⑤ 在 Visual Studio 2017 中使用 NuGet 安装 Microsoft.AspNetCore.Http.Extensions 包，该包中有现成的设置和读取整数和字符串值的方法。在 About.cshmtl.cs 的文件头部添加下面一行代码：

```
using Microsoft.AspNetCore.Http;
```

⑥ 在 Visual Studio 2017 中按 F5 键运行应用程序，在浏览器中浏览 About 页面，此时在 Visual Studio 2017 中会出现一个错误信息，如图 4.40 所示。

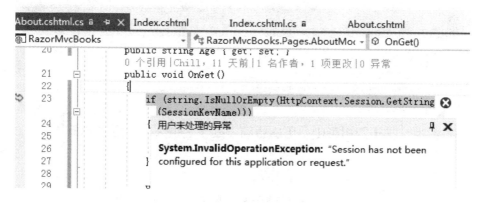

图 4.40 异常信息

⑦ 这是由于中间件的顺序不对,ASP.NET Core 的中间件的顺序很重要。需要修改一下 Startup.cs 文件中的代码,把"app.UseCookiePolicy();"和"app.UseSession();"移到"app.UseMvc"前面,代码如下:

```
public void Configure(IApplicationBuilder app, IHostingEnvironment env)
{
    if (env.IsDevelopment())
    {
        app.UseBrowserLink();
        app.UseDeveloperExceptionPage();
    }
    else
    {
        app.UseExceptionHandler("/Error");
    }
    app.UseStaticFiles();
    app.UseCookiePolicy();
    app.UseSession();
    app.UseMvc();
}
```

⑧ 在 About.cshtml 页面中添加以下语句:

`<h3>@Model.Name,生于四川彭山,历虞夏商,武丁之世灭之,活了 @Model.Age 岁。</h3>`

⑨ 在 Visual Studio 2017 中按 F5 键运行应用程序,在浏览器中浏览 About 页面,如图 4.41 所示。

图 4.41　浏览 About 页面

4.9 ASP.NET Core 中的缓存

4.9.1 缓存的基础知识

运用缓存可以显著提高应用的性能和可伸缩性，对于那种不经常更改的数据，使用缓存效果最佳。由缓存生成的数据副本的返回速度可以比从数据库中返回得更快。

ASP.NET Core 支持多种不同的缓存。开发人员可以选择将数据缓存在应用进程自身的内存中并基于 IMemoryCache，虽然这种方式的缓存具有最高的性能，但对于部署在集群式服务器中的应用会出现缓存数据不一致的情况。对于这种场景可以使用粘性会话来解决。

对于部署在集群式服务器上的场景，其他解决方案是将数据缓存于某一个独立的存储中心，以便让所有的 Web 服务器共享同一份缓存数据，因此将这种缓存形式称为"分布式缓存"。ASP.NET Core 为分布式缓存提供了两种原生的存储形式，一种是基于 NoSQL 的 Redis 数据库，另一种则是微软自家的关系型数据库 SQL Server。

ASP.NET Core 还借助一个中间件实现了"响应缓存"，该缓存会按照 HTTP 缓存规范对整个响应实施缓存，本书对响应缓存不做介绍。

内存中的缓存可以存储任何对象，分布式缓存仅存储字节数据。

4.9.2 将数据缓存在内存中

对于调用 I/O 操作来说，对内存访问的性能比对数据库和远程服务的访问性能要高不止一个数量级，所以将数据直接缓存在应用进程的内容中自然具有最佳的性能优势。

针对缓存的操作不外乎对缓存数据的存储与读取，这两个基本的操作都由 MemoryCache 对象来完成。如果在一个 ASP.NET Core 应用中，在启动 MemoryCache 服务时对其做了注册，那么就可以在任何地方获取该服务对象的设置和缓存数据，所以针对缓存的编程是非常简单的。

MemoryCache 对象位于 NuGet 包 Microsoft.Extensions.Caching.Memory 中，实现了 IMemoryCache 接口。由于是将缓存对象直接置于内存之中，中间并不涉及持久化存储的问题，自然也就无须考虑针对缓存对象的序列化问题，所以这种内存模式支持任意类型的缓存对象。

将数据缓存在内存中的步骤是：

① 在 Visual Studio 2017 中选择"工具"→"NuGet 包管理器"→"管理解决方案的 NuGet 程序包"菜单项，在 NuGet 包管理器界面搜索并安装 Microsoft.Extensions.Caching.Memory 包。

② 在 Startup.cs 文件的 ConfigureServices 方法中调用 ServiceCollection 的 AddMemoryCache 方法来完成针对 MemoryCache 服务的注册。在 Configure 方法中，通过调用 ApplicationBuilder 的 Run 方法注册了一个中间件，并对请求做简单的响应，代码如下：

```csharp
using System;
using System.Collections.Generic;
using System.Linq;
using System.Threading.Tasks;
using Microsoft.AspNetCore.Builder;
using Microsoft.AspNetCore.Hosting;
using Microsoft.Extensions.Configuration;
using Microsoft.Extensions.DependencyInjection;

namespace RazorMvcBooks
{
    public class Startup
    {
        public Startup(IConfiguration configuration)
        {
            Configuration = configuration;
        }
        public IConfiguration Configuration { get; }
        //This method gets called by the runtime. Use this method to add services to the
        //container.
        public void ConfigureServices(IServiceCollection services)
        {
            services.AddMemoryCache();
            services.AddMvc();
        }
        //This method gets called by the runtime. Use this method to configure the HTTP
        //request pipeline.
        public void Configure(IApplicationBuilder app, IHostingEnvironment env)
        {
            if (env.IsDevelopment())
            {
                app.UseBrowserLink();
                app.UseDeveloperExceptionPage();
            }
            else
            {
                app.UseExceptionHandler("/Error");
```

```
            }
            app.UseStaticFiles();
            app.UseMvc();
            app.UseMvcWithDefaultRoute();
        }
    }
}
```

③ 创建缓存帮助类,代码如下:

```
using Microsoft.Extensions.Caching.Memory;
using System;
using System.Collections;
using System.Collections.Generic;
using System.Linq;
using System.Reflection;
using System.Text.RegularExpressions;
using System.Threading.Tasks;

namespace RazorMvcBooks.Helpers
{
    public class MemoryCacheHelper
    {
        private static readonly MemoryCache Cache = new MemoryCache(new MemoryCacheOptions());
        /// <summary>
        /// 验证缓存项是否存在
        /// </summary>
        /// <param name="key">缓存 Key</param>
        /// <returns></returns>
        public bool IsExists(string key)
        {
            if (key == null)
                throw new ArgumentNullException(nameof(key));
            return Cache.TryGetValue(key, out _);
        }
        /// <summary>
        /// 添加缓存
        /// </summary>
        /// <param name="key">缓存 Key</param>
        /// <param name="value">缓存 Value</param>
        /// <param name="expiresSliding">滑动过期时长(如果在过期时间内有操作,则以当前时间点延长过期时间)</param>
        /// <param name="expiresAbsoulte">绝对过期时长</param>
        /// <returns></returns>
```

```csharp
public bool Add(string key, object value, TimeSpan expiresSliding, TimeSpan expiressAbsoulte)
{
    if (key == null)
        throw new ArgumentNullException(nameof(key));
    if (value == null)
        throw new ArgumentNullException(nameof(value));

    Cache.Set(key, value,new MemoryCacheEntryOptions().SetSlidingExpiration(expiresSliding)
        .SetAbsoluteExpiration(expiressAbsoulte));
    return IsExists(key);
}
/// <summary>
/// 添加缓存
/// </summary>
/// <param name = "key">缓存 Key</param>
/// <param name = "value">缓存 Value</param>
/// <param name = "expiresIn">缓存时长</param>
/// <param name = "isSliding">是否滑动过期(如果在过期时间内有操作,则以当前时间点延长过期时间)</param>
/// <returns></returns>
public bool Add(string key, object value, TimeSpan expiresIn, bool isSliding = false)
{
    if (key == null)
        throw new ArgumentNullException(nameof(key));
    if (value == null)
        throw new ArgumentNullException(nameof(value));

    Cache.Set(key, value, isSliding? new MemoryCacheEntryOptions().SetSlidingExpiration(expiresIn): new MemoryCacheEntryOptions().SetAbsoluteExpiration(expiresIn));

    return IsExists(key);
}
/// <summary>
/// 添加缓存,默认过期时间1小时
/// </summary>
/// <param name = "key">缓存 Key</param>
/// <param name = "value">缓存 Value</param>
/// <returns></returns>
public bool Add(string key, object value)
{
    return Add(key, value, TimeSpan.FromHours(1));
```

```csharp
}
#region 删除缓存
/// <summary>
/// 删除缓存
/// </summary>
/// <param name = "key">缓存 Key</param>
/// <returns></returns>
public void Remove(string key)
{
    if (key == null)
        throw new ArgumentNullException(nameof(key));

    Cache.Remove(key);
}
/// <summary>
/// 批量删除缓存
/// </summary>
/// <returns></returns>
public void RemoveAll(IEnumerable<string> keys)
{
    if (keys == null)
        throw new ArgumentNullException(nameof(keys));

    keys.ToList().ForEach(item => Cache.Remove(item));
}
#endregion

#region 获取缓存
/// <summary>
/// 获取缓存
/// </summary>
/// <param name = "key">缓存 Key</param>
/// <returns></returns>
public T Get<T>(string key) where T: class
{
    if (key == null)
        throw new ArgumentNullException(nameof(key));

    return Cache.Get(key) as T;
}
/// <summary>
/// 获取缓存
/// </summary>
/// <param name = "key">缓存 Key</param>
/// <returns></returns>
```

```csharp
public object Get(string key)
{
    if (key == null)
        throw new ArgumentNullException(nameof(key));

    return Cache.Get(key);
}
/// <summary>
/// 获取缓存集合
/// </summary>
/// <param name = "keys">缓存 Key 集合</param>
/// <returns></returns>
public IDictionary<string, object> GetAll(IEnumerable<string> keys)
{
    if (keys == null)
        throw new ArgumentNullException(nameof(keys));

    var dict = new Dictionary<string, object>();
    keys.ToList().ForEach(item => dict.Add(item, Cache.Get(item)));
    return dict;
}
#endregion

/// <summary>
/// 删除所有缓存
/// </summary>
public void RemoveCacheAll()
{
    var l = GetAllCacheKeys();
    foreach (var s in l)
    {
        Remove(s);
    }
}
/// <summary>
/// 删除匹配到的缓存
/// </summary>
/// <param name = "pattern"></param>
/// <returns></returns>
public void RemoveCacheRegex(string pattern)
{
    IList<string> l = SearchCacheRegex(pattern);
    foreach (var s in l)
    {
        Remove(s);
```

```csharp
        }
    }
    /// <summary>
    /// 搜索匹配到的缓存
    /// </summary>
    /// <param name="pattern"></param>
    /// <returns></returns>
    public IList<string> SearchCacheRegex(string pattern)
    {
        var cacheKeys = GetAllCacheKeys();
        var l = cacheKeys.Where(k => Regex.IsMatch(k, pattern)).ToList();
        return l.AsReadOnly();
    }
    /// <summary>
    /// 获取所有缓存键
    /// </summary>
    /// <returns></returns>
    public List<string> GetAllCacheKeys()
    {
        const BindingFlags flags = BindingFlags.Instance | BindingFlags.NonPublic;
        var entries = Cache.GetType().GetField("_entries", flags).GetValue(Cache);
        var cacheItems = entries as IDictionary;
        var keys = new List<string>();
        if (cacheItems == null) return keys;
        foreach (DictionaryEntry cacheItem in cacheItems)
        {
            keys.Add(cacheItem.Key.ToString());
        }
        return keys;
    }
}
```

④ 在 About.cshtml.cs 文件的 OnGet 方法中检测是否存在键值为"date1"的缓存。如果不存在,则创建并添加到缓存中;如果存在,则使用缓存中的日期和时间,代码如下:

```csharp
public void OnGet()
{
    string key = "date1";
    Account account = new Account
    {
        Name = "John Doe",
        Phone = "13300223344",
        CreateTime = new DateTime(2018, 7, 20, 0, 0, 0, DateTimeKind.Utc),
```

```
};
string json = JsonConvert.SerializeObject(account, Formatting.Indented);
Message = "Your application description page." + json;
if (!cacheHelper.IsExists(key))
{
    cacheHelper.Add("date1", DateTime.Now);
}
object dt = cacheHelper.Get("date1");
CacheDate = ((DateTime)dt).ToString("yyyy-MM-dd HH:mm:ss");
CurrDate = DateTime.Now.ToString("yyyy-MM-dd HH:mm:ss");
}
```

⑤ 在 About.cshtml 页面中显示当前的时间和缓存的时间，代码如下：

```
@page
@model AboutModel
@{
    ViewData["Title"] = "About";
}
<h2>@ViewData["Title"]</h2>
<h3>@Model.Message</h3>
<p>Use this area to provide additional information.</p>
<p>添加一行中文。</p>
<h3>当前时间:@Model.CurrDate;缓存时间:@Model.CacheDate</h3>
```

⑥ 当通过浏览器访问此页面时，会发现当前时间是动态变化的，而缓存时间则保持不变，如图 4.42 所示。

图 4.42　时间缓存

⑦ 在首页(Index.cshtml)中显示缓存时间。在 Index.cshtml 页面的最后添加如下代码：

```
<div class = "row">
<div class = "col-md-3">
<h3>缓存时间:@Model.CacheDate</h3>
</div>
</div>
```

⑧ 在 Index.cshtml.cs 页面的 OnGet 方法中添加如下代码：

```
public void OnGet()
{
    object dt = new Helpers.MemoryCacheHelper().Get("date1");
    if (dt != null)
    {
        CacheDate = ((DateTime)dt).ToString("yyyy-MM-dd HH:mm:ss");
    }
}
```

⑨ 在 Visual Studio 2017 中按 F5 键启动应用程序。首先在浏览器中显示首页，此时没有缓存时间，首页中的"缓存时间"显示为空，如图 4.43 中的 1 处所示。在浏览器中单击"About"，等待时间缓存，等待几秒之后，再次访问首页，此时首页中的"缓存时间"显示出刚才缓存的时间，如图 4.43 中的 2 处所示。

图 4.43 显示缓存时间

4.9.3 基于 SQL Server 的分布式缓存

ASP.NET Core 除了支持 Redis 这种主流的 NoSQL 数据库来支持分布式缓存外，也支持用 SQL Server 来实现分布式缓存。所谓的 SQL Server 的分布式缓存，实际上就是将标识缓存数据的字节数组存放在 SQL Server 数据库中某个具有固定结构的数据表中。具体实现方法是：

① 先创建一个缓存表，这可以通过一个名为 sql-cache 的工具来创建。在使用 sql-cache 工具创建缓存表之前，要通过 NuGet 安装 Microsoft.Extensions.Caching.SqlConfig.Tools 包。

② 在项目的根目录下找到项目文件 RazorMvcBooks.csproj，代码如下：

```
<Project Sdk = "Microsoft.NET.Sdk.Web">
<PropertyGroup>
<TargetFramework>netcoreapp2.0</TargetFramework>
</PropertyGroup>
<ItemGroup>
<PackageReference Include = "Microsoft.AspNetCore.All" Version = "2.0.0" />
<PackageReference Include = " Microsoft.Extensions.Caching.SqlConfig.Tools" Version = "2.0.0" />
</ItemGroup>
<ItemGroup>
<DotNetCliToolReference Include = "Microsoft.VisualStudio.Web.CodeGeneration.Tools" Version = "2.0.4" />
</ItemGroup>
</Project>
```

③ 在第二个<ItemGroup>中添加如下代码：

```
< DotNetCliToolReference Include = "Microsoft.Extensions.Caching.SqlConfig.Tools" Version = "2.0.0"/>
```

④ 在应用根目录中打开命令行窗口，在命令行窗口中通过执行"dotnet sql-cache create --help"命令来测试 SqlConfig.Tools 是否已经安装成功。从图 4.44 可以看出，该命名需要指定三个参数，分别表示缓存数据库的链接字符串、缓存表的 Schema 和名称。

⑤ 通过执行"dotnet sql-cache create "Data Source =(local)\sqlexpress;Initial Catalog=Test;Integrated Security=True;" dbo TestCache"命令来创建 SQL Server 缓存表。从图 4.45 中可以看出，在本机的一个名为 Test 的数据库中创建了一个名为 TestCache 的缓存表，该表采用 dbo 作为 Schema。

⑥ 创建的表格具有如图 4.46 所示的架构。

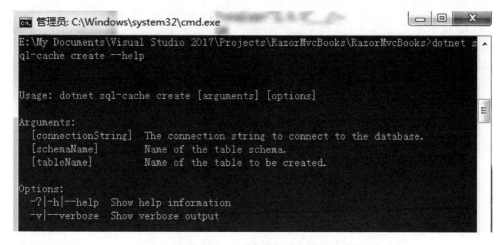

图 4.44 缓存表创建指令

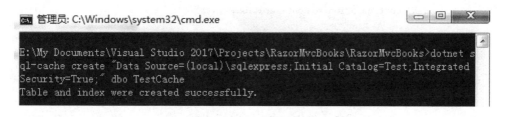

图 4.45 创建缓存表

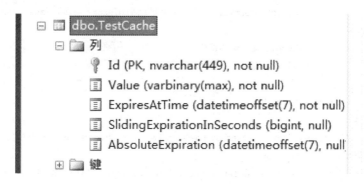

图 4.46 缓存表

⑦ 在 Visual Studio 2017 中通过 NuGet 安装 SQL Server 的分布式缓存实现包 Microsoft.Extensions.Caching.SqlServer。

⑧ 在 Startup.cs 文件的 ConfigureServices 方法中调用 IServiceCollection 的扩展方法 AddDistributedSqlServerCache 进行服务注册。在调用后一个方法时,通过设置 SqlServerCacheOptions 对象的三个属性的方式来指定缓存数据库的链接字符串、缓存表的 Schema 和名称,代码如下:

ASP.NET Core 应用开发入门教程

```csharp
using System;
using System.Collections.Generic;
using System.Linq;
using System.Threading.Tasks;
using Microsoft.AspNetCore.Builder;
using Microsoft.AspNetCore.Hosting;
using Microsoft.Extensions.Configuration;
using Microsoft.Extensions.DependencyInjection;

namespace RazorMvcBooks
{
    public class Startup
    {
        public Startup(IConfiguration configuration)
        {
            Configuration = configuration;
        }
        public IConfiguration Configuration { get; }
        // This method gets called by the runtime. Use this method to add services to the container.
        public void ConfigureServices(IServiceCollection services)
        {
            services.AddDistributedSqlServerCache(o =>
            {
                o.ConnectionString = "Server=.\\sqlexpress;Database=Test;Trusted_Connection=True;";
                o.SchemaName = "dbo";
                o.TableName = "TestCache";
            });
            services.AddMvc();
        }
        //This method gets called by the runtime. Use this method to configure the HTTP
        //request pipeline.
        public void Configure(IApplicationBuilder app, IHostingEnvironment env)
        {
            if (env.IsDevelopment())
            {
                app.UseBrowserLink();
                app.UseDeveloperExceptionPage();
            }
            else
            {
                app.UseExceptionHandler("/Error");
```

```
            }
            app.UseStaticFiles();
            app.UseMvc();
        }
    }
}
```

⑨ 创建 SQL Server 缓存帮助类，代码如下：

```
using Microsoft.Extensions.Caching.Distributed;
using System;
using System.Collections.Concurrent;
using System.Collections.Generic;
using System.Linq;
using System.Threading.Tasks;
using Microsoft.Extensions.Caching.SqlServer;
using System.Text.RegularExpressions;
using System.Reflection;
using System.Collections;

namespace RazorMvcBooks.Helpers
{
    public class SqlServerCacheHelper
    {
        private static SqlServerCacheOptions opt;
        private static readonly IDistributedCache Cache;
        static SqlServerCacheHelper()
        {
            opt = new SqlServerCacheOptions();
            opt.ConnectionString = "Server=.\\sqlexpress;Database=Test;Trusted_Connection=True;";
            opt.SchemaName = "dbo";
            opt.TableName = "TestCache";
            Cache = new SqlServerCache(opt);
        }
        /// <summary>
        /// 验证缓存项是否存在
        /// </summary>
        /// <param name="key">缓存 Key</param>
        /// <returns></returns>
        public bool IsExists(string key)
        {
            if (key == null)
```

```csharp
            throw new ArgumentNullException(nameof(key));
        byte[] bytes = Cache.Get(key);
        if (bytes == null)
        {
            return false;
        }
        if (bytes.Length <= 0)
        {
            return false;
        }
        return true;
    }
    /// <summary>
    /// 添加缓存
    /// </summary>
    /// <param name = "key">缓存 Key</param>
    /// <param name = "value">缓存 Value</param>
    /// <param name = "expiresSliding">滑动过期时长（如果在过期时间内有操作,则以当前时间点延长过期时间）</param>
    /// <param name = "expiressAbsoulte">绝对过期时长</param>
    /// <returns></returns>
    public bool Add(string key, byte[] value, TimeSpan expiresSliding, TimeSpan expiressAbsoulte)
    {
        if (key == null)
            throw new ArgumentNullException(nameof(key));
        if (value == null)
            throw new ArgumentNullException(nameof(value));
        Cache.Set(key, value, new DistributedCacheEntryOptions().SetSlidingExpiration(expiresSliding)
            .SetAbsoluteExpiration(expiressAbsoulte));
        return IsExists(key);
    }
    /// <summary>
    /// 添加缓存
    /// </summary>
    /// <param name = "key">缓存 Key</param>
    /// <param name = "value">缓存 Value</param>
    /// <param name = "expiresIn">缓存时长</param>
    /// <param name = "isSliding">是否滑动过期（如果在过期时间内有操作,则以当前时间点延长过期时间）</param>
    /// <returns></returns>
```

```csharp
public bool Add(string key, byte[] value, TimeSpan expiresIn, bool isSliding = false)
{
    if (key == null)
        throw new ArgumentNullException(nameof(key));
    if (value == null)
        throw new ArgumentNullException(nameof(value));
    Cache.Set(key, value, isSliding ? new DistributedCacheEntryOptions().SetSlidingExpiration(expiresIn) : new DistributedCacheEntryOptions().SetAbsoluteExpiration(expiresIn));
    return IsExists(key);
}
/// <summary>
/// 添加缓存,默认过期时间1小时
/// </summary>
/// <param name = "key">缓存 Key</param>
/// <param name = "value">缓存 Value</param>
/// <returns></returns>
public bool Add(string key, byte[] value)
{
    return Add(key, value, TimeSpan.FromHours(1));
}
#region 删除缓存
/// <summary>
/// 删除缓存
/// </summary>
/// <param name = "key">缓存 Key</param>
/// <returns></returns>
public void Remove(string key)
{
    if (key == null)
        throw new ArgumentNullException(nameof(key));
    Cache.Remove(key);
}
/// <summary>
/// 批量删除缓存
/// </summary>
/// <returns></returns>
public void RemoveAll(IEnumerable<string> keys)
{
    if (keys == null)
        throw new ArgumentNullException(nameof(keys));
```

```csharp
        keys.ToList().ForEach(item => Cache.Remove(item));
}
#endregion
#region 获取缓存
/// <summary>
/// 获取缓存
/// </summary>
/// <param name="key">缓存 Key</param>
/// <returns></returns>
public T Get<T>(string key) where T : class
{
    if (key == null)
        throw new ArgumentNullException(nameof(key));
    return Cache.Get(key) as T;
}
/// <summary>
/// 获取缓存
/// </summary>
/// <param name="key">缓存 Key</param>
/// <returns></returns>
public byte[] Get(string key)
{
    if (key == null)
        throw new ArgumentNullException(nameof(key));
    return Cache.Get(key);
}
/// <summary>
/// 获取缓存集合
/// </summary>
/// <param name="keys">缓存 Key 集合</param>
/// <returns></returns>
public IDictionary<string, object> GetAll(IEnumerable<string> keys)
{
    if (keys == null)
        throw new ArgumentNullException(nameof(keys));
    var dict = new Dictionary<string, object>();
    keys.ToList().ForEach(item => dict.Add(item, Cache.Get(item)));
    return dict;
}
#endregion
/// <summary>
/// 删除所有缓存
```

```csharp
/// </summary>
public void RemoveCacheAll()
{
    var l = GetAllCacheKeys();
    foreach (var s in l)
    {
        Remove(s);
    }
}
/// <summary>
/// 删除匹配到的缓存
/// </summary>
/// <param name = "pattern"></param>
/// <returns></returns>
public void RemoveCacheRegex(string pattern)
{
    IList<string> l = SearchCacheRegex(pattern);
    foreach (var s in l)
    {
        Remove(s);
    }
}
/// <summary>
/// 搜索匹配到的缓存
/// </summary>
/// <param name = "pattern"></param>
/// <returns></returns>
public IList<string> SearchCacheRegex(string pattern)
{
    var cacheKeys = GetAllCacheKeys();
    var l = cacheKeys.Where(k => Regex.IsMatch(k, pattern)).ToList();
    return l.AsReadOnly();
}
/// <summary>
/// 获取所有缓存键
/// </summary>
/// <returns></returns>
public List<string> GetAllCacheKeys()
{
    const BindingFlags flags = BindingFlags.Instance | BindingFlags.NonPublic;
    var entries = Cache.GetType().GetField("_entries", flags).GetValue(Cache);
```

```
            var cacheItems = entries as IDictionary;
            var keys = new List<string>();
            if (cacheItems == null) return keys;
            foreach (DictionaryEntry cacheItem in cacheItems)
            {
                keys.Add(cacheItem.Key.ToString());
            }
            return keys;
        }
    }
}
```

⑩ 在浏览器中浏览页面,操作与之前的类似。

⑪ 现在看一下 SQL Server 数据库中究竟包含了哪些缓存数据,此时只需直接在所在的数据库中查看对应的缓存表(TestCache)即可,如图 4.47 所示。

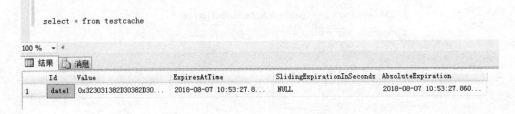

图 4.47 缓存表内容

第 5 章

Entity Framework Core

Entity Framework（EF）Core 是微软的一个 ORM 框架，是轻量化、可扩展和跨平台版的数据访问技术。EF Core 是一个对象关系映射（O/R Mapping）解决方案，以便于 .NET 开发人员使用 .NET 对象来处理数据库，减少数据访问代码的编写。EF Core 支持多个数据库引擎。

5.1 先决条件

为了开发 .NET Core 2.0 应用程序（包括面向 .NET Core 的 ASP.NET Core 2.0 应用程序），需要下载并安装适合所用平台的 .NET Core 2.0 SDK 工具包。

为了将 EF Core 2.0 或任何其他 .NET Standard 2.0 库用于非 .NET Core 2.0 版本的 .NET 平台（例如 .NET Framework 4.6.1 或更高版本），需要可识别 .NET Standard 2.0 及其兼容框架的 NuGet 程序包。通过下面几种方法可以获取此版本：

① 安装 Visual Studio 2017 的版本 15.7 或更高版本；
② 如果使用 Visual Studio 2015，则请下载 NuGet 客户端并升级至 3.6.0 版本。

5.2 Visual Studio 开发

使用 Visual Studio 可以开发许多不同类型的应用程序，这些应用程序面向 .NET Core、.NET Framework 或受 EF Core 支持的其他平台。

可通过以下两种方式并使用 Visual Studio 在应用程序中安装 EF Core 数据库支持程序。

5.2.1 使用 NuGet 的包管理器用户界面

使用 NuGet 的包管理器用户界面安装 EF Core 数据库支持程序的步骤是：
① 选择"项目"→"管理 NuGet 程序包"菜单项；
② 单击"浏览"或"更新"标签；
③ 选择 Microsoft.EntityFrameworkCore.SqlServer 包及所需版本，然后单击"安装"按钮。

5.2.2 使用 NuGet 的包管理器控制台

使用 NuGet 的包管理器控制台(PMC)安装 EF Core 数据库支持程序的步骤是：
① 选择"工具"→"NuGet 包管理器"→"程序包管理器控制台"菜单项；
② 在 PMC 中键入并运行以下命令安装 EF Core 2.0 包：

```
Install -Package Microsoft.EntityFrameworkCore.SqlServer
```

③ 也可以使用"Update -Package"命令将已安装的包更新至较新版本；
④ 若想指定安装特定版本，则可使用"- Version"修饰符，例如，若想安装 EF Core 2.0 包，则可将"- Version 2.0.0"追加到以上命令中。

5.3 创建数据库

在介绍项目示例之前，首先创建一个数据库。在创建数据库之前要有数据库软件，本书中使用的数据库软件是 SQL Server 2012 Express。当然也可以使用 SQL Server 2012 的其他版本，或者使用 SQL Server 2014 或 SQL Server 2016。创建数据库的步骤是：

① 打开 SQL Server Management Studio，在"对象资源管理器"中右击"数据库"，在弹出的快捷菜单中选择"新建数据库"。

② 在"新建数据库"对话框的"数据库名称"文本框中输入"EFCoreDemo"，然后单击"确定"按钮，如图 5.1 所示。

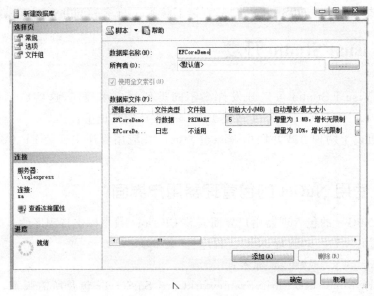

图 5.1 创建数据库

③ 在 SQL Server Management Studio 的工具栏上单击"新建查询"工具按钮，然后输入以下代码：

```
USE [EFCoreDemo]
GO
CREATE TABLE [dbo].[User](
    [Id] [int] IDENTITY(1,1) NOT NULL,
    [Account] [varchar](100) NOT NULL DEFAULT (' '),
    [Password] [varchar](255) NOT NULL DEFAULT (' '),
    [Name] [nvarchar](100) NOT NULL DEFAULT (' '),
    [Sex] [int] NOT NULL DEFAULT ((0)),
    [Status] [int] NOT NULL DEFAULT ((0)),
    [Type] [int] NOT NULL DEFAULT ((0)),
    [BizCode] [varchar](255) NOT NULL DEFAULT (' '),
    [CreateTime] [datetime] NOT NULL DEFAULT (getdate()),
    [CreateId] [int] NOT NULL DEFAULT ((0)),
    [Address] [nvarchar](220) NULL DEFAULT (' '),
    [Mobile] [nvarchar](50) NULL DEFAULT (' '),
    CONSTRAINT [PK_USER] PRIMARY KEY CLUSTERED (
        [Id] ASC
    )WITH (PAD_INDEX = OFF, STATISTICS_NORECOMPUTE = OFF, IGNORE_DUP_KEY = OFF, ALLOW_ROW_LOCKS = ON, ALLOW_PAGE_LOCKS = ON)ON [PRIMARY]
)ON [PRIMARY]
GO
```

④ 至此，已成功创建了名为 EFCoreDemo 的数据库和 User 表，如图 5.2 所示。

图 5.2 User 表

5.4 EF Core 的两种编程方式

像 SQL Server、MySql、Oracle 等这些传统的关系型数据库,都是体现实体与实体之间的联系。在以前开发时,可能先根据需求设计数据库,然后再写 Model 和业务逻辑,对于 Model 类,基本都是与表的字段对应着,而表中保存的每条记录又与类的实例对象对应着。有了这个对照关系,就使人思考能否只在一边设计,如在数据库中设计表或在 Visual Studio 2017 中设计 Model,然后直接生成另一边,从而可以节省很多时间成本,于是就有了对象关系映射,简称 ORM(Object Relation Mapping)。这样,就可以根据 Model 生成数据库,也可以根据数据库表生成 Model,而 Model 和数据库又是分离的,因此也可以根据 Model 生成不同类型(SQL Server\Oracle)的数据库,而不同类型的数据库(SQL Server\Oracle)也可以生成同样的 Model,它们共同的纽带就是 Mapping。

下面介绍两种 EF Core 的实体框架:Database First 和 Code First。

5.5 EF Core 2.0 Database First 的基本使用

"Database First"模式被称为"数据库优先",其前提是开发人员的应用已经有相应的数据库,同时可以使用 EF 设计工具根据数据库生成实体类,也可以使用 Visual Studio 模型设计器修改这些模型之间的对应关系。Database First 的使用方法是:

① 在 Visual Studio 2017 中打开 RazorMvcBooks 项目,然后在"解决方案资源管理器"中右击"RazorMvcBooks"项目,在弹出的快捷菜单中选择"添加"→"新建文件夹"菜单项,如图 5.3 所示。

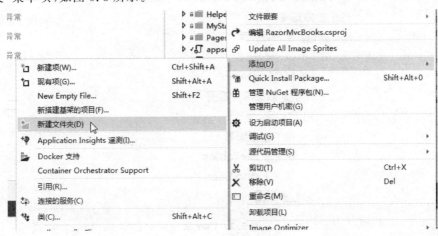

图 5.3 新建文件夹

② Visual Studio 2017 会在"解决方案资源管理器"中创建一个名为"新文件夹"的文件夹,把这个"新文件夹"改名为"Models",如图 5.4 所示。

图 5.4 Models 文件夹

③ 在 Visual Studio 2017 中选择"工具"→"NuGet 包管理器"→"管理解决方案的 NuGet 程序包"菜单项,在管理界面中安装 Microsoft.EntityFrameworkCore、Microsoft.EntityFrameworkCore.SqlServer 和 Microsoft.EntityFrameworkCore.Tools 三个包。

④ 在 Visual Studio 2017 中选择"工具"→"NuGet 包管理器"→"程序包管理器控制台"菜单项,如图 5.5 所示。

⑤ 在"程序包管理器控制台"窗口中输入以下命令:

Scaffold-DbContext "Data Source=.\sqlexpress;Initial Catalog=EFCoreDemo;User ID=sa;Password=密码" Microsoft.EntityFrameworkCore.SqlServer -OutputDir Models

此时会发现 Visual Studio 2017 自动生成了"User"实体类和一个"EFCoreDemoContext"数据库上下文操作类,如图 5.6 所示。

⑥ 下面来看一下在 Database First 模式下生成的 EFCoreDemoContext.cs 文件,以便与 Code First 模式中的相同文件进行比较。代码如下:

using System;
using Microsoft.EntityFrameworkCore;
using Microsoft.EntityFrameworkCore.Metadata;
namespace RazorMvcBooks.Models

图 5.5 程序包管理器控制台

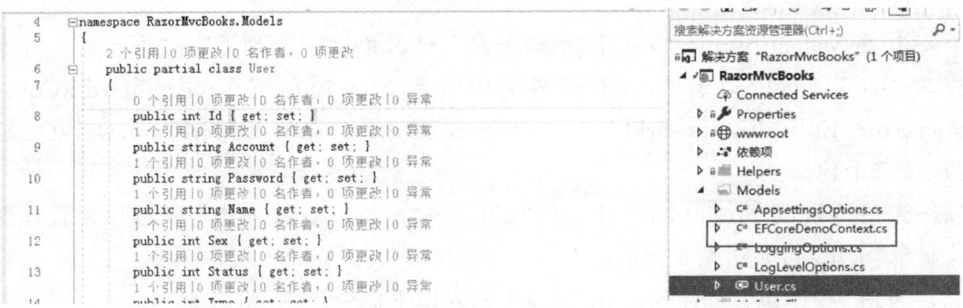

图 5.6 User 类

```
{
    public partial class EFCoreDemoContext : DbContext
    {
        public virtual DbSet<User> User { get; set; }
        protected override void OnConfiguring(DbContextOptionsBuilder optionsBuilder)
        {
        }
        protected override void OnModelCreating(ModelBuilder modelBuilder)
        {
            modelBuilder.Entity<User>(entity =>
            {
```

```csharp
entity.Property(e => e.Account)
    .IsRequired()
    .HasMaxLength(100)
    .IsUnicode(false)
    .HasDefaultValueSql("(' ')");
entity.Property(e => e.Address)
    .HasMaxLength(220)
    .HasDefaultValueSql("(' ')");
entity.Property(e => e.BizCode)
    .IsRequired()
    .HasMaxLength(255)
    .IsUnicode(false)
    .HasDefaultValueSql("(' ')");
entity.Property(e => e.CreateId).HasDefaultValueSql("((0))");
entity.Property(e => e.CreateTime)
    .HasColumnType("datetime")
    .HasDefaultValueSql("(getdate())");
entity.Property(e => e.Mobile)
    .HasMaxLength(50)
    .HasDefaultValueSql("(' ')");
entity.Property(e => e.Name)
    .IsRequired()
    .HasMaxLength(100)
    .HasDefaultValueSql("(' ')");
entity.Property(e => e.Password)
    .IsRequired()
    .HasMaxLength(255)
    .IsUnicode(false)
    .HasDefaultValueSql("(' ')");
entity.Property(e => e.Sex).HasDefaultValueSql("((0))");
entity.Property(e => e.Status).HasDefaultValueSql("((0))");
entity.Property(e => e.Type).HasDefaultValueSql("((0))");
});
    }
  }
}
```

下面简单介绍使用 EF 进行数据查询的步骤，通过下面的代码可以看到 EF 对数据的操作有多么优雅：

① 在 Visual Studio 2017 的"解决方案资源管理器"中打开 About.cshtml 文件，输入以下代码：

```
@page
@using Microsoft.Extensions.Configuration
@inject IConfiguration Configuration

@model AboutModel
@{
    ViewData["Title"] = "About";
}
<h2>@ViewData["Title"]</h2>
<h3>@Model.Message</h3>

<p>Use this area to provide additional information.</p>
<form method = "POST">
<div class = "form-group">
<input type = "submit" value = "新建默认用户" class = "btn btn-default" asp-page-handler = "Add" />
</div>
</form>
```

② 在 Visual Studio 2017 的"解决方案资源管理器"中打开 About.cshtml.cs 文件,输入以下代码:

```
using System;
using System.IO;
using System.Threading.Tasks;
using Microsoft.AspNetCore.Mvc;
using Microsoft.AspNetCore.Mvc.RazorPages;
using Microsoft.EntityFrameworkCore;
using Microsoft.Extensions.Configuration;
using Microsoft.Extensions.Logging;
using RazorMvcBooks.Models;

namespace RazorMvcBooks.Pages
{
    public class AboutModel:PageModel
    {
        private readonly ILogger m_logger;
        public AboutModel(ILogger<AboutModel> logger)
        {
            m_logger = logger;
        }
        public static AppsettingsOptions Appsettings { get; set; }
```

```csharp
public static IConfiguration Configuration { get; set; }
public string Message { get; set; }
public string Name { get; set; }
public string Code { get; set; }
public void OnGet()
{

}
public async Task<IActionResult> OnPostAddAsync()
{
    EFCoreDemoContext db = new EFCoreDemoContext();
    var model = new User
    {
        Account = "Test1",
        Name = "测试账号1",
        Password = "123456",
        Sex = 1,
        Status = 0,
        Type = 1,
        BizCode = "200000",
        CreateTime = DateTime.Now,
        CreateId = 0,
        Address = "上海",
        Mobile = "18911223344"
    };
    db.User.Add(model);
    await db.SaveChangesAsync();

    var users = await db.User.ToListAsync();
    if (users.Count > 0)
    {
        foreach (var item in users)
        {
            if (item.Account.Contains("Test1"))
            {
                Console.WriteLine("ID:{0}", item.Id);
                Console.WriteLine("Account:{0}", item.Account);
                Console.WriteLine("名称:{0}", item.Name);
                Console.WriteLine("地址:{0}", item.Address);
                Console.WriteLine("时间:{0}", item.CreateTime.ToString("yyyy-MM-dd HH:mm:ss"));
            }
        }
```

```
            }
            return RedirectToPage("./Index");
        }
    }
```

③ 在 Visual Studio 2017 中以项目形式启动应用程序,如图 5.7 所示。

图 5.7　以项目形式启动应用程序

④ 在浏览器中浏览 About 页面,单击"新建默认用户"按钮,如图 5.8 所示。

图 5.8　浏览 About 页面

⑤ 最终的运行结果如图 5.9 所示。

图 5.9　运行结果

注意:如果数据库表的结构发生改变,则只需在"程序包管理器控制台"窗口中输入以下命令:

Scaffold-DbContext "Data Source = .\sqlexpress;Initial Catalog = EFCoreDemo;User ID = sa;Password = 密码" Microsoft.EntityFrameworkCore.SqlServer -Force -OutputDir Models

此时会发现 Visual Studio 2017 自动重新生成了"User"实体类和一个"EFCore-DemoContext"数据库上下文操作类。如图 5.10 所示为 User 实体类在数据库变更之前和之后所生成的不同代码的比较。

```
数据库中User表变更前
public partial class User
{
    public int Id { get; set; }
    public string Account { get; set; }
    public string Password { get; set; }
    public string Name { get; set; }
    public int Sex { get; set; }
    public int Status { get; set; }
    public int Type { get; set; }
    public string BizCode { get; set; }
    public DateTime CreateTime { get; set; }
    public int CreateId { get; set; }
    public string Address { get; set; }
    public string Mobile { get; set; }
}
```

```
数据库中User表变更后
public partial class User
{
    public int Id { get; set; }
    public string Account { get; set; }
    public string Password { get; set; }
    public string Name { get; set; }
    public int Sex { get; set; }
    public int Status { get; set; }
    public int Type { get; set; }
    public DateTime CreateTime { get; set; }
    public int CreateId { get; set; }
    public string Address { get; set; }
    public string Mobile { get; set; }
}
```

图 5.10　User 实体类的变化

5.6　Entity Framework Core 的实体特性

在开始学习"Code First"模式如何使用之前，先来学习创建实体要用到的一些特性。

Entity Framework Core 使用一组基于实体类规则的约定来构建模型，可指定其他配置补充和/或替代由约定发现的内容。Entity Framework Core 提供了一系列特性（称为数据注释），可以将这些特性应用到实体类和属性中。特性重写了 EF 默认的约定。

Entity Framework Core 中有两个重要的特性类 System.ComponentModel.DataAnnotations（包含影响数据表列大小和可控性的特性，见表 5.1）和 System.ComponentModel.DataAnnotations.Schema（包含影响数据表的特性，见表 5.2）。

注意：特性仅仅提供了配置选项的子集，完整的配置可以在 EF Fluent API 中找到。

表 5.1　System.ComponentModel.DataAnnotations 特性

特性（Attribute）	说　明
Key	表示唯一标识实体的一个或多个属性，是数据表的主键
Timestamp	列的数据类型指定为行版本，不允许为空
ConcurrencyCheck	指定一个或多个属性参与乐观并发检查，当用户编辑或删除一个实体时，用来做并发检查
Required	指定数据字段值是必需的，不能为空
MinLength	指定属性中允许的数组或字符串数据的最小长度
MaxLength	指定属性中允许的数组或字符串数据的最大长度
StringLength	指定数据字段中允许的字符的最小长度和最大长度

表 5.2　System.ComponentModel.DataAnnotations.Schema 特性

特性（Attribute）	说　明
Table	指定类将映射成的数据表的名称
Column	指定属性将映射成的数据库的列名
ForeignKey	表示导航属性中用作外键的属性。可以将批注放在外键属性上，然后指定关联的导航属性名称；也可以将批注放在导航属性上，然后指定关联的外键名称
NotMapped	表示标注了该属性的字段将不会映射到数据表的列中
DatabaseGenerated	指定数据库生成属性值的方式。该属性是只读的，且可用来映射成自动增长列
InverseProperty	指定表示同一关系的另一端的导航属性的反向属性
ComplexType	表示该类是复杂类型。复杂类型是实体类型的非标量属性，实体类型允许在实体内组织标量属性。复杂类型没有键，实体框架不能脱离父对象来管理复杂类型

5.6.1　数据注释特性——Key

Key 特性可以被用到类的属性中，Entity Framework Core 的默认约定是：创建一个主键，主键名称是"Id"或类名＋"Id"。Key 特性重写了该默认约定，开发人员可以将 Key 特性应用到一个类的属性上，不管该属性的名字是什么，都可以创建一个主键，创建的过程是：

① 在 Visual Studio 2017 中创建一个名为 CodeFirstAttrDemo 的控制台项目。下面以该项目为例来了解其相关的特性。

② 在 Visual Studio 2017 中的"解决方案资源管理器"中右击"CodeFirstAttrDemo"项目，在弹出的快捷菜单中选择"添加"→"新建文件夹"菜单项，并把"新文件夹"重命名为"Models"。

③ 在 Visual Studio 2017 中选择"工具"→"NuGet 包管理器"→"管理解决方案

的 NuGet 程序包"菜单项,在管理界面中安装 Microsoft.EntityFrameworkCore、Microsoft.EntityFrameworkCore.SqlServer、Microsoft.EntityFrameworkCore.Tools 三个包。

④ 在 Visual Studio 2017 中的"解决方案资源管理器"中右击"Models"文件夹,在弹出的快捷菜单中选择"添加"→"类",把类名重命名为"Person",代码如下:

```
using System;
using System.Collections.Generic;
using System.ComponentModel.DataAnnotations;
using System.Text;
namespace CodeFirstAttrDemo.Models
{
    public class Person
    {
        public int Id { get; set; }
            [Key]
        public string UserId { get; set; }
        public string Name { get; set; }
        public string Mobile { get; set; }
        public int Age { get; set; }
        public string Address { get; set; }
        public DateTime CreateTime { get; set; }
    }
}
```

⑤ 上段代码把 Key 特性应用到了 Person 实体的 UserId 属性上,因此,在通过迁移(有关"迁移"的概念将在后面介绍)生成数据库之后会得到这个主键。

⑥ 在 Visual Studio 2017 中选择"菜单"→"NuGet 包管理器"→"程序包管理器控制台"菜单项,在打开的程序包管理器控制台依次执行以下命令:

```
Add-Migration InitialKey
Update-Database
```

⑦ 在 SQL Server Management Studio 中查看 Person 表,如图 5.11 所示。

⑧ 当然也可以创建复合主键,在两个属性上使用 Key 特性,使两个属性同时作为主键。下面先来看一个错误的例子,代码如下:

```
using System;
using System.Collections.Generic;
using System.ComponentModel.DataAnnotations;
using System.Text;
```

图 5.11 Person 表

```
namespace CodeFirstAttrDemo.Models
{
    public class Person
    {
        public int Id { get; set; }
            [Key]
        public string UserId { get; set; }
            [Key]
        public string Name { get; set; }
        public string Mobile { get; set; }
        public int Age { get; set; }
        public string Address { get; set; }
        public DateTime CreateTime { get; set; }
    }
}
```

⑨ 当然,在 Visual Studio 2017 中的程序包管理器控制台上执行以下命令后,程序包管理器控制台提示出错了,如图 5.12 所示。错误提示的大意是:不能创建复合主键,要使用复合主键请使用 Fluent API。

Add-Migration Key2

⑩ 在 Visual Studio 2017 中打开 EFCoreDemoContext.cs 文件,在 OnModel-Creating 方法中添加如下代码:

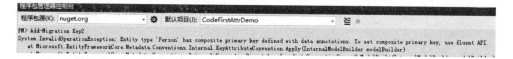

图 5.12　程序包管理器控制台错误提示

```
protected override void OnModelCreating(ModelBuilder modelBuilder)
{
    modelBuilder.Entity<Person>()
        .HasKey(p => new {p.UserId, p.Name});
}
```

⑪ 在 Visual Studio 2017 中的程序包管理器控制台上执行以下命令：

Add-Migration Key2
Update-Database

⑫ 在 SQL Server Management Studio 中查看 Person 表，如图 5.13 所示。

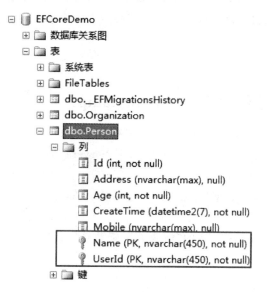

图 5.13　复合主键的 Person 表

注意：对于单个主键，如果属性是 int 型，则 Entity Framework Core 创建的主键是自动增长的；对于复合主键，得到的主键不是自动增长的。

Key 特性不仅仅可以应用到 int 类型的属性上，还可以应用到字符串、日期等类型上。

如果一开始使用 int 型创建了单主键，那么如果要修改成复合主键，则在执行迁移时会报错，如图 5.14 所示。如果要更改列的标识属性，则需要删除并重新创建列。

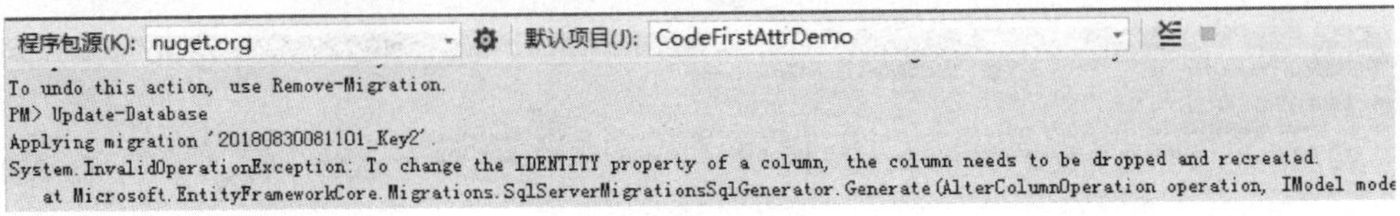

图 5.14 迁移错误信息

在介绍后续示例之前,先执行以下两步:

① 在 Visual Studio 2017 中打开 EFCoreDemoContext.cs 文件,把 EFCoreDemoContext 类中的 OnModelCreating 方法中的代码注释掉,代码如下:

```
protected override void OnModelCreating(ModelBuilder modelBuilder)
{
    //modelBuilder.Entity<Person>()
    //.HasKey(p => new {p.UserId, p.Name});
}
```

② 在 Visual Studio 2017 中选择"菜单"→"NuGet 包管理器"→"程序包管理器控制台"菜单项,在打开的程序包管理器控制台上依次执行以下命令:

Remove-Migration
Update-Database

5.6.2 数据注释特性——Timestamp

Timestamp 特性可以应用到实体类中和只有一个字节的数组的属性上,该特性给列设定的是 Tiemstamp 类型。对于 SQL Server,Timestamp 通常用于 byte[]属性上,此时其属性名(数据库列名)为 RowVersion。Timestamp 特性的作用是在每次插入或更新行时,由数据库生成一个新的值。在并发检查中,Entity Framework Core 会自动使用 Timestamp 类型的字段。Timestamp 特性的使用方法是:

① 在 Visual Studio 2017 的"解决方案资源管理器"中打开 Person.cs 文件,并添加一个字段 RowVersion,代码如下:

```
using System;
using System.Collections.Generic;
using System.ComponentModel.DataAnnotations;
using System.ComponentModel.DataAnnotations.Schema;
using System.Text;

namespace CodeFirstAttrDemo.Models
{
    public class Person
    {
```

```
        public int Id { get; set; }
            [Key]
        public string UserId { get; set; }
        public string Name { get; set; }
        public string Mobile { get; set; }
        public int Age { get; set; }
        public string Address { get; set; }
        public DateTime CreateTime { get; set; }
            [Timestamp]
        public byte[] RowVersion { get; set; }
    }
}
```

② 在 Visual Studio 2017 中选择"菜单"→"NuGet 包管理器"→"程序包管理器控制台"菜单项,在打开的程序包管理器控制台上依次执行以下命令:

```
Add-Migration RowVer
Update-Database
```

③ 在 SQL Server Management Studio 中查看 Person 表,如图 5.15 所示。

图 5.15　添加了 RowVersion 的 Person 表

5.6.3　数据注释特性——ConcurrencyCheck

ConcurrencyCheck 特性可以应用到实体类的属性上,用于实现乐观并发检查。当 Entity Framework Core 执行更新操作时,Entity Framework Core 将列的值放在

where 条件语句中,可以使用 CurrencyCheck 特性和已经存在的列做并发检查,而不是使用单独的 Timestamp 列来做并发检查。ConcurrencyCheck 特性的使用方法如下。

在 Visual Studio 2017 的"解决方案资源管理器"中打开 Person.cs 文件,在 Name 属性上添加 ConcurrencyCheck 特性。具体示例将在后文介绍。代码如下:

```
using System;
using System.Collections.Generic;
using System.ComponentModel.DataAnnotations;
using System.ComponentModel.DataAnnotations.Schema;
using System.Text;

namespace CodeFirstAttrDemo.Models
{
    public class Person
    {
        public int Id { get; set; }
            [Key]
        public string UserId { get; set; }
            [ConcurrencyCheck]
        public string Name { get; set; }
        public string Mobile { get; set; }
        public int Age { get; set; }
        public string Address { get; set; }
        public DateTime CreateTime { get; set; }
            [Timestamp]
        public byte[] RowVersion { get; set; }
    }
}
```

注意:Timestamp 特性只能被用到单字节属性的类中,但是 ConcurrencyCheck 特性可以在类的一个或多个任何类型的属性上应用。

5.6.4　数据注释特性——Required

Required 特性可以应用于实体类的属性上,也可以与 ASP.NET Core MVC 一起使用作为验证属性。Required 特性的使用方法是:

① 在 Visual Studio 2017 的"解决方案资源管理器"中打开 Person.cs 文件,在 Name 属性上添加 Required 特性,代码如下:

```
using System;
using System.Collections.Generic;
using System.ComponentModel.DataAnnotations;
using System.ComponentModel.DataAnnotations.Schema;
using System.Text;
```

```
namespace CodeFirstAttrDemo.Models
{
    public class Person
    {
        public int Id { get; set; }
            [Key]
        public string UserId { get; set; }
            [ConcurrencyCheck]
            [Required]
        public string Name { get; set; }
        public string Mobile { get; set; }
        public int Age { get; set; }
        public string Address { get; set; }
        public DateTime CreateTime { get; set; }
            [Timestamp]
        public byte[] RowVersion { get; set; }
    }
}
```

② 在 Visual Studio 2017 中选择"菜单"→"NuGet 包管理器"→"程序包管理器控制台"菜单项,在打开的程序包管理器控制台上依次执行以下命令:

```
Add – Migration Required
Update – Database
```

③ 在 SQL Server Management Studio 中查看 Person 表。从上段代码可以看出,已经将必填的特性应用到 Name 属性上了。添加 Required 特性前后的 Name 字段比较如图 5.16 所示。

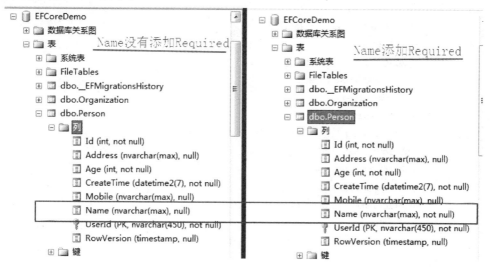

图 5.16　Name 字段必填

5.6.5 数据注释特性——MaxLength

MaxLength 特性可应用于实体类的字符串或数组类型属性,也可以与 ASP.NET Core MVC 一起使用作为验证属性。MaxLength 仅适用于数组数据类型,如 string 和 byte[]。

备注:默认情况下,应由数据库支持程序来选择适当的数据类型的属性。对于具有长度属性的数据类型,数据库支持程序通常选择所允许的最大长度数据的数据类型。例如,实体中的 string 默认使用 Microsoft SQL Servernvarchar(max)数据类型(或如果用作主键,则是 nvarchar(450))。

将数据传递到数据库支持程序之前,Entity Framework Core 不会执行任何最大长度校验。该校验是由数据库支持程序或数据库在适当的时候进行。例如,针对 SQL Server,超出最大长度会导致异常。

MaxLength 特性的使用方法是:

① 在 Visual Studio 2017 的"解决方案资源管理器"中打开 Person.cs 文件,在 Name 属性上添加 MaxLength(50)特性,代码如下:

```
using System;
using System.Collections.Generic;
using System.ComponentModel.DataAnnotations;
using System.ComponentModel.DataAnnotations.Schema;
using System.Text;
namespace CodeFirstAttrDemo.Models
{
    public class Person
    {
        public int Id { get; set; }
        [Key]
        public string UserId { get; set; }
        [ConcurrencyCheck]
        [Required]
        [MaxLength(50)]
        public string Name { get; set; }
        public string Mobile { get; set; }
        public int Age { get; set; }
        public string Address { get; set; }
        public DateTime CreateTime { get; set; }
        [Timestamp]
        public byte[] RowVersion { get; set; }
    }
}
```

② 在 Visual Studio 2017 中选择"菜单"→"NuGet 包管理器"→"程序包管理器控制台"菜单项,在打开的程序包管理器控制台上依次执行以下命令:

```
Add-Migration MaxLen
Update-Database
```

③ 在 SQL Server Management Studio 中查看 Person 表。从上段代码可以看出,已经将 MaxLength 特性应用到 Name 属性上,Name 的字段长度为 50。与图 5.16 中的 Name 字段比较,更新的字段长度如图 5.17 所示。

图 5.17 设置 Name 的字段长度

如果添加的数据长度大于设置值指定的大小,则实体框架将验证 MaxLength 特性的属性值。例如,如果添加了超过 50 个字符的姓名,那么 Entity Framework Core 就会给出 EntityValidationError 异常。

5.6.6 数据注释特性——MinLength

MinLength 特性可应用于实体类的字符串或数组类型属性,它是一个可验证的特性。该特性也可以与 ASP.NET Core MVC 一起使用,对数据库没有影响。MinLength 仅适用于数组数据类型,如 string 和 byte[]。如果将字符串或数组属性的值设置为小于 MinLength 属性中的指定长度,则 Entity Framework Core 将给出 EntityValidationError 异常。

MinLength 特性也可以与 MaxLength 特性一起使用。MinLength 的使用方法是:在 Visual Studio 2017 的"解决方案资源管理器"中打开 Person.cs 文件,在 Name 属性上添加 MinLength(2)特性,代码如下:

```csharp
using System;
using System.Collections.Generic;
using System.ComponentModel.DataAnnotations;
using System.ComponentModel.DataAnnotations.Schema;
using System.Text;

namespace CodeFirstAttrDemo.Models
{
    public class Person
    {
        public int Id { get; set; }
        [Key]
        public string UserId { get; set; }
        [ConcurrencyCheck]
        [Required]
        [MaxLength(50),MinLength(2)]
        public string Name { get; set; }
        public string Mobile { get; set; }
        public int Age { get; set; }
        public string Address { get; set; }
        public DateTime CreateTime { get; set; }
        [Timestamp]
        public byte[] RowVersion { get; set; }
    }
}
```

从上段代码可以看出,姓名不能少于2个字符,最多50个字符。

5.6.7 数据注释特性——Table

Table特性可以被用在实体类的类名上,Entity Framework Core默认的约定是使用类名作为数据库中创建的数据表的名称,Table特性可以重写该约定,只要指定了名字,Entity Framework Core就会根据Table特性里的名字创建数据表名称。Table特性的使用方法是:

① 在Visual Studio 2017的"解决方案资源管理器"中打开Person.cs文件,在类名Person上添加Table特性,代码如下:

```csharp
using System;
using System.Collections.Generic;
using System.ComponentModel.DataAnnotations;
using System.ComponentModel.DataAnnotations.Schema;
using System.Text;
```

```csharp
namespace CodeFirstAttrDemo.Models
{
    [Table("MyPerson")]
    public class Person
    {
        public int Id { get; set; }
            [Key]
        public string UserId { get; set; }
            [ConcurrencyCheck]
            [Required]
            [MaxLength(50),MinLength(2)]
        public string Name { get; set; }
        public string Mobile { get; set; }
        public int Age { get; set; }
        public string Address { get; set; }
        public DateTime CreateTime { get; set; }
            [Timestamp]
        public byte[] RowVersion { get; set; }
    }
}
```

② 在 Visual Studio 2017 中选择"菜单"→"NuGet 包管理器"→"程序包管理器控制台"菜单项,在打开的程序包管理器控制台上依次执行以下命令:

Add – Migration MyPerson

Update – Database

③ 在 SQL Server Management Studio 中查看 MyPerson 表,如图 5.18 所示。

图 5.18　表名变更

④ 同时还可以在代码中指定表的 schema,代码如下:

```csharp
using System;
using System.Collections.Generic;
using System.ComponentModel.DataAnnotations;
using System.ComponentModel.DataAnnotations.Schema;
using System.Text;

namespace CodeFirstAttrDemo.Models
{
    [Table("MyPerson",Schema = "EFCore")]
    public class Person
    {
        public int Id { get; set; }
            [Key]
        public string UserId { get; set; }
            [ConcurrencyCheck]
            [Required]
            [MaxLength(50),MinLength(2)]
        public string Name { get; set; }
        public string Mobile { get; set; }
        public int Age { get; set; }
        public string Address { get; set; }
        public DateTime CreateTime { get; set; }
            [Timestamp]
        public byte[] RowVersion { get; set; }
    }
}
```

⑤ 在 Visual Studio 2017 中选择"菜单"→"NuGet 包管理器"→"程序包管理器控制台"菜单项,在打开的程序包管理器控制台上依次执行以下命令:

Add－Migration MyPersonSchema
Update－Database

⑥ 在 SQL Server Management Studio 中查看 EFCore.MyPerson 表,如图 5.19 所示。

图 5.19 指定表的 schema

5.6.8 数据注释特性——Column

Column 特性可以用在实体类的属性上，Entity Framework Core 默认的约定是使用属性名称作为列名。不过，该特性也可以自定义数据库列名。Column 特性的使用方法是：

① 在 Visual Studio 2017 的"解决方案资源管理器"中打开 Person.cs 文件，在 Mobile 属性上添加 Column 特性，代码如下：

```
using System;
using System.Collections.Generic;
using System.ComponentModel.DataAnnotations;
using System.ComponentModel.DataAnnotations.Schema;
using System.Text;

namespace CodeFirstAttrDemo.Models
{
    [Table("MyPerson",Schema = "EFCore")]
    public class Person
    {
        public int Id { get; set; }
            [Key]
        public string UserId { get; set; }
            [ConcurrencyCheck]
            [Required]
            [MaxLength(50),MinLength(2)]
        public string Name { get; set; }
            [Column("Phone")]
        public string Mobile { get; set; }
        public int Age { get; set; }
        public string Address { get; set; }
        public DateTime CreateTime { get; set; }
            [Timestamp]
        public byte[] RowVersion { get; set; }
    }
}
```

② 在 Visual Studio 2017 中选择"菜单"→"NuGet 包管理器"→"程序包管理器控制台"菜单项，在打开的程序包管理器控制台上依次执行以下命令：

```
Add-Migration Col
Update-Database
```

③ 在 SQL Server Management Studio 中查看 EFCore.MyPerson 表,如图 5.20 所示。

图 5.20 变更列名

④ 当然也可以指定列的顺序(Order)和类型(Type),代码如下:

```
using System;
using System.Collections.Generic;
using System.ComponentModel.DataAnnotations;
using System.ComponentModel.DataAnnotations.Schema;
using System.Text;

namespace CodeFirstAttrDemo.Models
{
    [Table("MyPerson",Schema = "EFCore")]
    public class Person
    {
        public int Id { get; set; }
        [Key]
        public string UserId { get; set; }
        [ConcurrencyCheck]
        [Required]
        [MaxLength(50),MinLength(2)]
        public string Name { get; set; }
```

```
        [Column("Phone",TypeName = "char(20)",Order = 3)]
    public string Mobile { get; set; }
    public int Age { get; set; }
    public string Address { get; set; }
    public DateTime CreateTime { get; set; }
        [Timestamp]
    public byte[] RowVersion { get; set; }
    }
}
```

⑤ 在 Visual Studio 2017 中选择"菜单"→"NuGet 包管理器"→"程序包管理器控制台"菜单项,在打开的程序包管理器控制台上依次执行以下命令:

Add – Migration ColOrderType
Update – Database

⑥ 在 SQL Server Management Studio 中查看 EFCore.MyPerson 表,如图 5.21 所示。

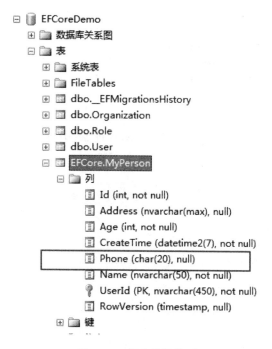

图 5.21　指定数据类型

5.6.9　数据注释特性——ForeignKey

ForeignKey 外键特性可以应用到类的属性上。Entity Framework Core 默认的约定是:对外键属性来说,假定外键属性的名称与主键属性是匹配的。ForeignKey

外键特性的使用方法是：

① 在 Visual Studio 2017 的"解决方案资源管理器"中右击"Models"文件夹，在弹出的快捷菜单中选择"添加"→"类"菜单项，把类名重命名为"Company"，代码如下：

```csharp
using System;
using System.Collections.Generic;
using System.ComponentModel.DataAnnotations;
using System.Text;

namespace CodeFirstAttrDemo.Models
{
    public class Company
    {
        [Key]
        public int Id { get; set; }
        [Required]
        [MaxLength(150), MinLength(2)]
        public string Name { get; set; }
        public string Mobile { get; set; }
        public string Address { get; set; }
        public DateTime CreateTime { get; set; }
    }
}
```

② 在 Visual Studio 2017"解决方案资源管理器"中打开 Person.cs 文件，新增两个属性 ForeignKeyId 和 CompanyInfo，并在 CompanyInfo 属性上添加 ForeignKey 特性，代码如下：

```csharp
using System;
using System.Collections.Generic;
using System.ComponentModel.DataAnnotations;
using System.ComponentModel.DataAnnotations.Schema;
using System.Text;

namespace CodeFirstAttrDemo.Models
{
    [Table("MyPerson", Schema = "EFCore")]
    public class Person
    {
        public int Id { get; set; }
        [Key]
        public string UserId { get; set; }
        [ConcurrencyCheck]
```

```
    [Required]
    [MaxLength(50),MinLength(2)]
public string Name { get; set; }
    [Column("Phone",TypeName = "char(20)",Order = 3)]
public string Mobile { get; set; }
public int Age { get; set; }
public string Address { get; set; }
public DateTime CreateTime { get; set; }
    [Timestamp]
public byte[] RowVersion { get; set; }
public int ForeignKeyId { get; set; }
    [ForeignKey("ForeignKeyId")]
public Company CompanyInfo { get; set; }
    }
}
```

③ 在 Visual Studio 2017 中选择"菜单"→"NuGet 包管理器"→"程序包管理器控制台"菜单项,在打开的程序包管理器控制台上依次执行以下命令:

Add – Migration ForeignKey

Update – Database

④ 在 SQL Server Management Studio 中查看 EFCore.MyPerson 表,如图 5.22 所示。

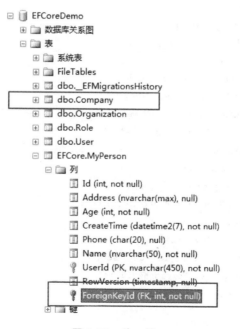

图 5.22 外　　键

5.6.10 数据注释特性——NotMapped

NotMapped 特性可以应用到实体类的属性上，Entity Framework Core 默认的约定是为所有带有 get 和 set 属性选择器的属性创建数据列。NotManpped 特性可以打破这个约定，如果将 NotMapped 特性应用到某个属性上，则 Entity Framework Core 就不会再为该属性在数据库的表中创建列了。

NotMapped 特性的使用方法是：

① 在 Visual Studio 2017 的"解决方案资源管理器"中打开 Person.cs 文件，在 Age 属性上添加 NotMapped 特性，代码如下：

```csharp
using System;
using System.Collections.Generic;
using System.ComponentModel.DataAnnotations;
using System.ComponentModel.DataAnnotations.Schema;
using System.Text;

namespace CodeFirstAttrDemo.Models
{
    [Table("MyPerson",Schema = "EFCore")]
    public class Person
    {
        public int Id { get; set; }
        [Key]
        public string UserId { get; set; }
        [ConcurrencyCheck]
        [Required]
        [MaxLength(50),MinLength(2)]
        public string Name { get; set; }
        [Column("Phone",TypeName = "char(20)",Order = 3)]
        public string Mobile { get; set; }
        [NotMapped]
        public int Age { get; set; }
        public string Address { get; set; }
        public DateTime CreateTime { get; set; }
        [Timestamp]
        public byte[] RowVersion { get; set; }
        public int ForeignKeyId { get; set; }
        [ForeignKey("ForeignKeyId")]
        public Company CompanyInfo { get; set; }
    }
}
```

② 在 Visual Studio 2017 中选择"菜单"→"NuGet 包管理器"→"程序包管理器控制台"菜单项，在打开的程序包管理器控制台上依次执行以下命令：

Add-Migration NotMapped
Update-Database

③ 在 SQL Server Management Studio 中查看 EFCore.MyPerson 表，如图 5.23 所示。

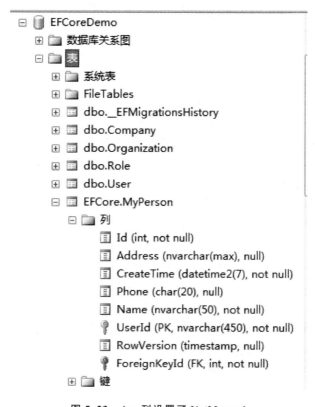

图 5.23　Age 列设置了 NotMapped

④ 注意到图 5.23 中，在 MyPerson 表中没有 Age 列。但是如果类的属性只有 get 属性选择器，或者只有 set 属性选择器，那么 Entity Framework Core 就不会为这样的类的属性创建数据列了。示例操作代码如下：

using System;
using System.Collections.Generic;
using System.ComponentModel.DataAnnotations;
using System.ComponentModel.DataAnnotations.Schema;
using System.Text;

```csharp
namespace CodeFirstAttrDemo.Models
{
    [Table("MyPerson",Schema = "EFCore")]
    public class Person
    {
        public int Id { get; set; }
            [Key]
        public string UserId { get; set; }
            [ConcurrencyCheck]
            [Required]
            [MaxLength(50),MinLength(2)]
        public string Name { get; set; }
            [Column("Phone",TypeName = "char(20)",Order = 3)]
        public string Mobile { get; set; }
            [NotMapped]
        public int Age { get; set; }
        public string Address { get; }
        private string m_idCard;
        public string IdCard { set {m_idCard = value; } }
        public DateTime CreateTime { get; set; }
            [Timestamp]
        public byte[] RowVersion { get; set; }
        public int ForeignKeyId { get; set; }
            [ForeignKey("ForeignKeyId")]
        public Company CompanyInfo { get; set; }
    }
}
```

⑤ 在 Visual Studio 2017 中选择"菜单"→"NuGet 包管理器"→"程序包管理器控制台"菜单项,在打开的程序包管理器控制台上依次执行以下命令:

```
Add - Migration NotGetOrSet
Update - Database
```

⑥ 在 SQL Server Management Studio 中查看 EFCore.MyPerson 表,会看到表中没有 Address 列,因为 Address 属性只有 get 属性选择器;表中也没有 IdCard 列,因为 IdCard 属性只有 set 属性选择器。从以上实践得知,Entity Framework Core 不会为只有 get 或 set 属性选择器的属性创建数据列,如图 5.24 所示。

图 5.24　EFCore.MyPerson 表

5.7　EF Core 2.0 Code First

"Code First"模式称为"代码优先"模式，是从 EF 4.1 开始新加入的功能。在使用"Code First"模式进行 EF 开发时，开发人员只需编写对应的实体类（其实就是领域模型的实现过程），然后自动生成数据库即可。这样设计的好处在于可以针对概念模型进行所有数据操作，而不必关心数据的存储关系，这样可以更加自然地采用面向对象的方式进行面向数据的应用程序开发。

"Code First"直接通过编码方式设计实体类（这也是为什么最开始"Code First"被叫作"Code Only"的原因）。另外需要注意的是，"Code First"并不代表一定必须通过数据类来定义模型，事实上也可以通过现有的数据库来生成实体类。

5.8　EF Core 2.0 Code First 创建数据库

5.8.1　创建实体

在开始学习"Code First"如何使用之前，先把 EFCoreDemo 数据库中的 User 表删除。创建实体的方法是：

① 在 Visual Studio 2017 中打开 RazorMvcBooks 项目。

② 在 Visual Studio 2017 中选择"工具"→"NuGet 包管理器"→"程序包管理器控制台"菜单项，如图 5.25 所示。

图 5.25　程序包管理器控制台

③ 在程序包管理器控制台中依次输入以下三条指令，安装 NuGet 包，如图 5.26 所示。

```
Install-Package Microsoft.EntityFrameworkCore-version 2.0.3
Install-Package Microsoft.EntityFrameworkCore.SqlServer-version 2.0.3
Install-Package Microsoft.EntityFrameworkCore.Tools-version 2.0.3
```

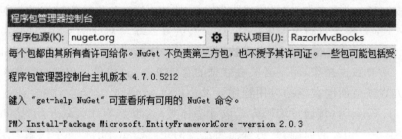

图 5.26　安装 NuGet 包

④ 添加两个类"Role"和"Organization"，可以看到这两个类只是简单的 C# 对象（POCO，Plain Old C# Object），这两个类基本与 EF 没有任何关系，代码如下：

```
using System;
using System.Collections.Generic;
using System.ComponentModel.DataAnnotations;
using System.ComponentModel.DataAnnotations.Schema;
```

```csharp
namespace RazorMvcBooks.Models
{
    public class Role
    {
        [Key]
        [DatabaseGeneratedAttribute(DatabaseGeneratedOption.Identity)]
        public int Id { get; set; }
        public string Name { get; set; }
        public int Status { get; set; }
        public int Type { get; set; }
        public DateTime CreateTime { get; set; }
        public string CreateId { get; set; }
        public int OrgId { get; set; }
        public string OrgCascadeId { get; set; }
        public string OrgName { get; set; }
    }
}


using System;
using System.Collections.Generic;
using System.ComponentModel.DataAnnotations;
using System.ComponentModel.DataAnnotations.Schema;

namespace RazorMvcBooks.Models
{
    public class Organization
    {
        [Key]
        [DatabaseGeneratedAttribute(DatabaseGeneratedOption.Identity)]
        public int Id { get; set; }
        public string CascadeId { get; set; }
        public string Name { get; set; }
        public string HotKey { get; set; }
        public int ParentId { get; set; }
        public int ParentName { get; set; }
        public int IsLeaf { get; set; }
        public int IsAutoExpand { get; set; }
        public string IconName { get; set; }
        public int Status { get; set; }
        public int Type { get; set; }
        public string BizCode { get; set; }
        public string CustomCode { get; set; }
```

```csharp
        public DateTime CreateTime { get; set; }
        public int CreateId { get; set; }
        public int SortNo { get; set; }
    }
}
```

⑤ 在 Visual Studio 2017 的"解决方案资源管理器"中的 Models 目录中添加 "EFCoreDemoContext"类，该类必须继承于 System.Data.Entity.DbContext 类，以赋予它数据操作能力，代码如下：

```csharp
using System;
using Microsoft.EntityFrameworkCore;
using Microsoft.EntityFrameworkCore.Metadata;

namespace RazorMvcBooks.Models
{
    public partial class EFCoreDemoContext : DbContext
    {
        public virtual DbSet<User> User { get; set; }
        public virtual DbSet<Role> Role { get; set; }
        public virtual DbSet<Organization> Organization { get; set; }

        public EFCoreDemoContext(DbContextOptions<EFCoreDemoContext> options) : base(options)
        {
        }
        protected override void OnConfiguring(DbContextOptionsBuilder optionsBuilder)
        {
        }
        protected override void OnModelCreating(ModelBuilder modelBuilder)
        {
            //省略
        }
    }
}
```

⑥ 在 Visual Studio 2017 的资源管理器中找到 appsettings.json 文件并双击打开，在文件中添加一个连接字符串，代码如下：

```json
{
    "Logging": {
        "IncludeScopes": false,
        "LogLevel": {
```

```
            "Default": "Warning",
            "Microsoft": "Warning"
        }
    },
    "ConnectionStrings": {
        "EFCoreDemoContext": "Server = .\\sqlexpress;Database = EFCoreDemo;Trusted_Connection = True;MultipleActiveResultSets = true"
    }
}
```

⑦ 在 Visual Studio 2017 的资源管理器中找到 startup.cs 文件并双击打开,在 startup.cs 文件的 ConfigureServices 方法中写入以下依赖注入容器注册数据库上下文的代码:

```
public void ConfigureServices(IServiceCollection services)
{
    services.AddDbContext<EFCoreDemoContext>(options => options.UseSqlServer(Configuration.GetConnectionString("EFCoreDemoContext")));
    services.AddMvc();
}
```

5.8.2 创建数据库

定义初始实体类后,就可以通过添加初始迁移来创建数据库了。在程序包管理器控制台中执行以下命令:

```
//其中 Initial 是版本名称
Add-Migration Initial
```

如图 5.27 所示,EF Core 将在项目中添加"Migrations"目录,并添加以下三个文件:

① 20180823101603_Initial.cs——主迁移文件,其中包含应用迁移所需的操作(在 Up()中)和还原迁移所需的操作(在 Down()中)。

② 20180823101603_Initial.Designer.cs——迁移元数据文件,其中包含 EF 所用的信息。

③ EFCoreDemoContextModelSnapshot.cs——当前模型的快照,用于确定添加下一个迁移时的更改内容。

文件名中的时间戳有助于保持文件按时间顺序排列,以便于查看更改进展。

提示:可以自由移动"迁移"文件并更改其命名空间。创建的新迁移与上一个迁移同级。

图 5.27 Migrations 目录

输入以下命令将迁移应用到数据库以创建数据库架构。

```
Update-Database
```

执行完以上操作后数据库就创建成功了,如图 5.28 所示。

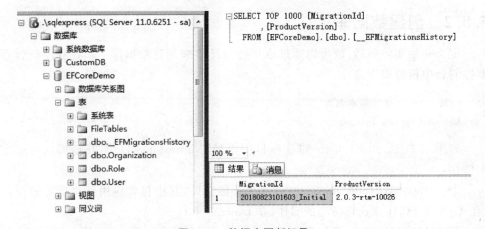

图 5.28 数据库更新记录

5.8.3 数据库修改

更改 EF Core 模型后,数据库架构将不再同步。通过迁移能够以递增的方式将架构的更改应用到数据库中,使其与 EF Core 模型保持同步,同时保留数据库中的现有数据。迁移名称的用途与版本控制系统中的提交消息类似。例如,如果对用户信息表的实体模型做了修改,则可以选择类似于 ModifyUser 的名称。数据库修改的方法是:

① 修改 User 类，代码如下：

```csharp
using System;
using System.Collections.Generic;
using System.ComponentModel.DataAnnotations;

namespace RazorMvcBooks.Models
{
    public partial class User
    {
        public int Id { get; set; }
            [MaxLength(150), Required]
        public string Account { get; set; }
            [MaxLength(250), Required]
        public string Password { get; set; }
            [MaxLength(100), Required]
        public string Name { get; set; }
        public int Sex { get; set; }
        public int Status { get; set; }
        public int Type { get; set; }
            [MaxLength(100)]
        public string BizCode { get; set; }
        public DateTime CreateTime { get; set; }
        public int CreateId { get; set; }
            [MaxLength(250)]
        public string Address { get; set; }
            [MaxLength(50)]
        public string Mobile { get; set; }
    }
}
```

② 在程序包管理器控制台上执行以下命令，会生成一个数据库修改代码文件，如图 5.29 所示。

Add-Migration ModifyUser

③ 在程序包管理器控制台上执行以下命令：

Update-Database

④ 从图 5.30 中可以看出，针对 User 类的修改已经更新到数据表中了。

图 5.29　数据库修改代码

图 5.30　User 表更新

5.8.4　还原迁移

如果已对数据库应用了一个迁移(或多个迁移)，但需要将其复原，则可使用与应用迁移相同的命令并指定要回退的迁移名称。还原迁移的方法是：

① 对于 User 类，将 Account 属性的 MaxLength(150)改为 MaxLength(100)，代码如下：

```
[MaxLength(100), Required]
```

```
public string Account { get; set; }
```

② 在程序包管理器控制台上依次执行以下命令：

```
Add-Migration ModifyRestoreUser
Update-Database
```

③ 在 SQL Server Management Studio 中查看 User 表，可见针对 User 类的修改已经更新到数据表中了，如图 5.31 所示。

④ 在程序包管理器控制台中输入以下命令执行指定迁移操作：

```
Update-Database ModifyUser
```

⑤ 在 SQL Server Management Studio 中查看 User 表，可见针对 User 类的修改已经更新到数据表中了，如图 5.32 所示。

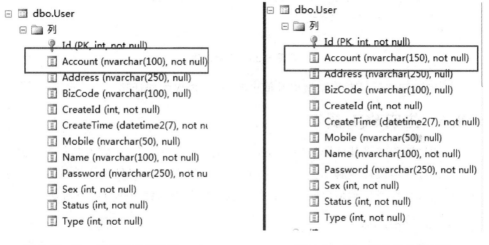

图 5.31　User 表字段长度还原　　　　图 5.32　User 表字段长度变化

5.8.5　删除迁移

也许在添加了迁移后又想对应用迁移前的 EF Core 模型做出其他更改，这时就要删除上一个迁移，具体方法是：

① 在程序包管理器控制台上依次执行以下命令：

```
Update-Database
Remove-Migration
```

② 先执行最后一次迁移的更新，然后删除最后一次迁移，此时会报出如下错误信息：

```
The migration '20180824054506_ModifyRestoreUser' has already been applied to the data-
```

base. Revert it and try again. If the migration has been applied to other databases, consider reverting its changes using a new migration.

从上面的错误信息来看，这是由于已经应用了最后一次迁移，所以无法删除迁移。

③ 在程序包管理器控制台上依次执行以下命令删除迁移，执行结果如图 5.33 所示。

```
Update-Database ModifyUser
Remove-Migration
```

```
PM> Update-Database ModifyUser
Done.
PM> Remove-Migration
Removing migration '20180824054506_ModifyRestoreUser'.
Reverting model snapshot.
Done.
```

图 5.33　删除迁移

④ 删除迁移后可对模型做出其他更改，然后再次添加迁移。

5.8.6　生成 SQL 脚本

当调试迁移或将其部署到生产数据库上时，生成一个 SQL 脚本是很有帮助的；之后可以进一步检查该脚本的准确性，并对其做出调整以满足生产数据库的需求。该脚本还可与部署技术结合使用。在程序包管理器控制台中执行以下命令：

```
Script-Migration
```

此命令有以下几个选项：

- from：迁移应是运行该脚本前应用到数据库的最后一个迁移。如果未应用任何迁移，则指定为 0（默认值）。
- to：迁移是运行该脚本后应用到数据库的最后一个迁移。它默认为项目中的最后一个迁移。

执行该命令后，会在项目的 obj\Debug\netcoreapp2.0 目录下生成一个脚本文件，如图 5.34 所示。

名称	修改日期	类型
3w4r1osn.sql	2018-08-24 14:10	SQL Script File
RazorMvcBooks.dll	2018-08-24 14:10	应用程序扩展
RazorMvcBooks.pdb	2018-08-24 14:10	程序调试数据库

图 5.34　生成脚本文件

5.8.7 创建存储过程

创建存储过程的代码如下：

```
CREATEPROCEDURE [dbo].[GetUserById]
@id int = 0
AS
BEGIN
if(@id = 0)
begin
select * from [User]
end
else
begin
select * from [User] where id = @id
end
END

GO
```

5.8.8 给数据库添加初始数据

给数据库添加初始数据的方法是：

① 在 Visual Studio 2017 的解决方案资源管理器中右击选中 Models 文件，然后在弹出的快捷菜单中选择创建一个新的类文件，并命名为 SeedData，如图 5.35 所示。

图 5.35 创建类

② 用下面的代码替换生成的代码：

```
using Microsoft.EntityFrameworkCore;
using Microsoft.Extensions.DependencyInjection;
using System;
using System.Collections.Generic;
```

```csharp
using System.Linq;
using System.Threading.Tasks;

namespace RazorMvcBooks.Models
{
    public class SeedData
    {
        public static void Initialize(IServiceProvider serviceProvider)
        {
            using(var context = new EFCoreDemoContext(serviceProvider.GetRequiredService<DbContextOptions<EFCoreDemoContext>>()))
            {
                // Look for any User.
                if (context.User.Any())
                {
                    return;   // DB has been seeded
                }
                context.User.AddRange(
                    new User
                    {
                        Account = "admin",
                        Password = "admin",
                        Name = "管理员",
                        Sex = 1,
                        Status = 1,
                        Type = 0,
                        BizCode = string.Empty,
                        CreateId = 0,
                        CreateTime = DateTime.Now,
                        Address = string.Empty,
                        Mobile = string.Empty
                    },
                    new User
                    {
                        Account = "test",
                        Password = "test",
                        Name = "Test",
                        Sex = 0,
                        Status = 1,
                        Type = 0,
                        BizCode = string.Empty,
                        CreateId = 0,
```

```csharp
        CreateTime = DateTime.Now,
        Address = "上海黄浦",
        Mobile = "58805505"
},
new User
{
        Account = "wang",
        Password = "wang",
        Name = "王五",
        Sex = 1,
        Status = 1,
        Type = 0,
        BizCode = string.Empty,
        CreateId = 0,
        CreateTime = DateTime.Now,
        Address = "上海松江",
        Mobile = "13358805505"
},
new User
{
        Account = "shSale",
        Password = "shsale",
        Name = "张三",
        Sex = 0,
        Status = 1,
        Type = 0,
        BizCode = string.Empty,
        CreateId = 0,
        CreateTime = DateTime.Now,
        Address = "上海奉贤",
        Mobile = "13900805505"
},
new User
{
        Account = "bjSale",
        Password = "bjsale",
        Name = "西门庆",
        Sex = 1,
        Status = 1,
        Type = 0,
        BizCode = string.Empty,
        CreateId = 0,
```

```
                    CreateTime = DateTime.Now,
                    Address = "北京朝阳",
                    Mobile = "18900804444"
                }
            );
            context.SaveChanges();
        }
    }
}
```

③ 在 Visual Studio 2017 的"解决方案资源管理器"中打开 Program.cs 文件,然后找到 Main 方法,在该方法程序的最后面添加 SeedData.Initialize()方法,代码如下:

```
using System;
using System.Collections.Generic;
using System.IO;
using System.Linq;
using System.Threading.Tasks;
using Microsoft.AspNetCore;
using Microsoft.AspNetCore.Hosting;
using Microsoft.EntityFrameworkCore;
using Microsoft.Extensions.Configuration;
using Microsoft.Extensions.DependencyInjection;
using Microsoft.Extensions.Logging;
using RazorMvcBooks.Models;

namespace RazorMvcBooks
{
    public class Program
    {
        public static void Main(string[] args)
        {
            // BuildWebHost(args).Run();
            var host = BuildWebHost(args);
            using(var scope = host.Services.CreateScope())
            {
                var services = scope.ServiceProvider;
                try
                {
                    var context = services.GetRequiredService<EFCoreDemoContext>();
                    // requires using Microsoft.EntityFrameworkCore;
```

```csharp
                context.Database.Migrate();
                // Requires using RazorPagesMovie.Models;
                SeedData.Initialize(services);
            }
            catch (Exception ex)
            {
                var logger = services.GetRequiredService<ILogger<Program>>();
                logger.LogError(ex, "数据库数据初始化错误.");
            }
        }
        host.Run();
    }

    public static IWebHost BuildWebHost(string[] args) =>
        WebHost.CreateDefaultBuilder(args)
        .ConfigureAppConfiguration((hostingContext, config) =>
        {
            var env = hostingContext.HostingEnvironment;
            config.AddJsonFile("appsettings.json", optional: true, reloadOnChange: true);
        })
        .ConfigureLogging((hostingContext, logging) =>
        {
            logging.AddConfiguration(hostingContext.Configuration.GetSection("Logging"));
            logging.AddConsole(options => options.IncludeScopes = true);
            logging.AddDebug();
        })
        .UseStartup<Startup>()
        .Build();
    }
}
```

④ 在 SQL Server Management Studio 中查看 User 表，没有数据，如图 5.36 所示。

⑤ 强制应用程序初始化（调用 Startup 类中的方法），这样 SeedData 方法能够正常运行。如果要强制初始化，则必须先停止 IIS，然后再重新启动。可以使用以下方法初始化应用程序：

ⓐ 在通知区域中右击 IIS Express 系统托盘图标，在弹出的快捷菜单中选择"退出"或"停止站点"，如图 5.37 所示。

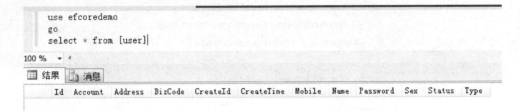

图 5.36 查询 User 表

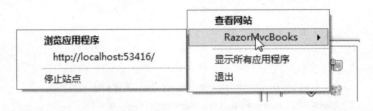

图 5.37 停止网站

ⓑ 如果是在非调试模式下运行 Visual Studio 2017，则按 F5 键以在调试模式下运行。

ⓒ 如果是在调试模式下运行 Visual Studio 2017，则先停止调试程序，再按 F5 键。

⑥ 再次在 SQL Server Management Studio 中查看 User 表，由图 5.38 可知，在代码中写的初始数据已经写入数据库。

	Id	Account	Address	BizCode	CreateId	CreateTime	Mobile	Name	Password	Sex	Status	Type
1	1	admin			0	2018-08-24 17:15:41.1865887		管理员	admin	1	1	0
2	2	test	上海黄浦		0	2018-08-24 17:15:41.2025896	58805505	Test	test	0	1	0
3	3	wang	上海松江		0	2018-08-24 17:15:41.2025896	13358805505	王五	wang	1	1	0
4	4	shSale	上海奉贤		0	2018-08-24 17:15:41.2025896	13900805505	张三	shsale	0	1	0
5	5	bjSale	北京朝阳		0	2018-08-24 17:15:41.2025896	18900804444	西门庆	bjsale	1	1	0

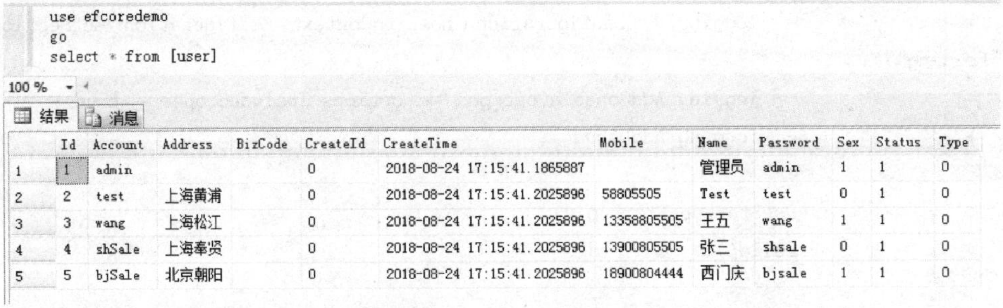

图 5.38 User 表中的数据

经过上面的准备工作后，就可以对数据进行增加、删除、修改、查询操作了。

5.9 用 EF Core 2.0 Code First 查询数据

Entity Framework Core 使用语言集成查询（LINQ）来查询数据库中的数据。通过 LINQ 可使用 C#（或其他的 .NET 语言）基于派生的上下文和实体类来编写强类型查询。查询时,会将 LINQ 查询的一种表示形式传递给数据库支持程序,进而转换为特定于数据库的查询语言(例如,适用于关系数据库的 SQL)。

5.9.1 查询的工作原理

下面介绍几种查询过程:

① LINQ 查询由 Entity Framework Core 处理,用于生成已准备好由数据库支持程序处理的表示形式。查询结果将被缓存,以便每次执行查询时无须再次查询。

② LINQ 查询的 Entity Framework Core 处理结果会被传递到数据库支持程序中,然后执行以下操作:

ⓐ 数据库支持程序评估确定哪些部分可以在数据库中查询;

ⓑ 查询的这些部分将被转换为数据库特定的查询语言(例如,SQLServer);

ⓒ 一个或多个查询会被发送到数据库并返回结果集(结果是数据库中的值,而不是实体实例)中。

③ 如果是跟踪查询,则对于返回的结果集,EF 会先检查结果数据在上下文实例更改跟踪器中是否已存在实体:

ⓐ 如果存在实体,则会返回现有实体;

ⓑ 如果不存在实体,则会创建新实体、设置更改跟踪并返回该新实体。

④ 如果是非跟踪查询,则 EF 会先检查结果数据在此查询的结果集中是否已存在实体:

ⓐ 如果存在实体,则会返回现有实体;

ⓑ 如果不存在实体,则会创建新实体并返回该新实体。

说明:非跟踪查询使用弱引用跟踪已返回的实体。如果具有相同标识的上一个结果超出范围,并运行了垃圾回收,则可能获得新的实体实例。

5.9.2 执行查询

当调用 LINQ 运算符时,只会在内存中生成"查询"的表示形式。只有在使用结果时,"查询"才会被发送到数据库中。触发将"查询"发送到数据库的常见操作如下:

① 在 for 循环中循环访问结果;

② 使用 ToList、ToArray、Single、Count 等运算符;

③ 将查询结果数据绑定到 UI。

警告:始终验证用户输入:当 EF 抵御 SQL 注入攻击时,不会执行输入的任何常规验证。因此,如果传递到 API、用于 LINQ 查询、分配给实体属性等的值来自不受信任的源,则应按照每个应用程序的要求执行相应的验证,这包括用于动态构造查询的所有用户输入。即使在使用 LINQ 时,如果接受用于生成表达式的用户输入,则也需要确保只能构造预期表达式。

5.9.3 基本查询

基本查询的步骤是：

① 在 Visual Studio 2017 的"解决方案资源管理器"中右击 Pages 文件夹，在弹出的快捷菜单中选择"添加"→"新建文件夹"菜单项，创建一个新文件夹，并把文件夹命名为 Users。

② 在 Visual Studio 2017 的"解决方案资源管理器"中右击 Users 文件夹，在弹出的快捷菜单中选择"添加"→"Razor 页面"菜单项，在弹出的对话框中选择"使用实体框架生成 Razor 页面（CRUD）"，如图 5.39 所示。

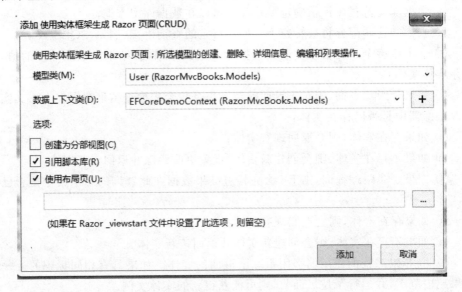

图 5.39　使用实体框架生成 Razor 页面（CRUD）

③ Visual Studio 2017 会自动生成创建、删除、详细信息、编辑和列表页面，如图 5.40 所示。

④ 在 Visual Studio 2017 的"解决方案资源管理器"中打开 _Layout.cshtml 文件，找到菜单部分的代码，并添加如下菜单代码：

```
<div class = "navbar-collapse collapse">
<ul class = "nav navbar-nav">
<li><a asp-page = "/Index">Home</a></li>
<li><a asp-page = "/About">About</a></li>
<li><a asp-page = "/Contact">Contact</a></li>
<li><a asp-page = "/Users/Index">用户信息</a></li>
</ul>
</div>
```

图 5.40 框架生成的页面

5.9.4 异步查询

当在数据库中查询时,异步查询可避免线程阻塞。异步查询可以避免 Windows 应用程序 UI 假死,还可以增加 Web 应用程序中的吞吐量,它可以在等待数据库操作完成时释放当前线程到线程池,以使该线程去处理其他请求。

警告:

① EF Core 不支持在同一上下文实例上运行多个并行操作,应始终等待操作完成,然后再开始下一个操作,这通常是通过在每个异步操作上使用 await 关键字完成的。

② Entity Framework Core 提供了一组异步扩展方法,可以用作执行查询并返回结果的 LINQ 方法的替代方法,包括 ToListAsync()、ToArrayAsync()、SingleAsync()等。对于部分 LINQ 运算符(如 Where(...)、OrderBy(...)等)没有对应的异步版本,因为这些 LINQ 运算符仅用于生成 LINQ 表达式树,而未将"查询"发送到数据库中执行。

在 Visual Studio 2017 的"解决方案资源管理器"中打开 Index.cshtml.cs 文件,添加创建异步查询的代码如下:

```
public async Task OnGetAsync()
{
    User = await _context.User.ToListAsync();
}
```

5.9.5 加载所有数据

加载所有数据的步骤是：

① 在 Visual Studio 2017 的“解决方案资源管理器”中打开 Index. cshtml. cs 文件，把上面的异步查询方法改成同步查询方法，代码如下：

```
public void  OnGet()
{
    User =   _context.User.ToList();
}
```

② 在 Visual Studio 2017 中按 F5 键运行应用程序，在浏览器中单击“用户信息”。当 Index 页面发出请求时，OnGetAsync 或 OnGet 方法向 Razor 页面返回用户信息列表。在本例中，OnGetAsync 或 OnGet 方法将返回数据库的 User 表中所有的用户信息，并以列表的形式显示出来，如图 5.41 所示。

<table>
<tr><th>Account</th><th>Password</th><th>Name</th><th>Sex</th><th>Status</th><th>Type</th><th>BizCode</th><th>CreateTime</th><th>CreateId</th><th>Address</th><th>Mobile</th><th></th></tr>
<tr><td>admin</td><td>admin</td><td>管理员</td><td>1</td><td>1</td><td>0</td><td></td><td>2018-08-24 17:15:41</td><td>0</td><td></td><td></td><td>Edit | Details | Delete</td></tr>
<tr><td>test</td><td>test</td><td>Test</td><td>0</td><td>1</td><td>0</td><td></td><td>2018-08-24 17:15:41</td><td>0</td><td>上海黄浦</td><td>58805505</td><td>Edit | Details | Delete</td></tr>
<tr><td>wang</td><td>wang</td><td>王五</td><td>1</td><td>1</td><td>0</td><td></td><td>2018-08-24 17:15:41</td><td>0</td><td>上海松江</td><td>13358805505</td><td>Edit | Details | Delete</td></tr>
<tr><td>shSale</td><td>shsale</td><td>张三</td><td>0</td><td>1</td><td>0</td><td></td><td>2018-08-24 17:15:41</td><td>0</td><td>上海奉贤</td><td>13900805505</td><td>Edit | Details | Delete</td></tr>
<tr><td>bjSale</td><td>bjsale</td><td>西门庆</td><td>1</td><td>1</td><td>0</td><td></td><td>2018-08-24 17:15:41</td><td>0</td><td>北京朝阳</td><td>18900804444</td><td>Edit | Details | Delete</td></tr>
</table>

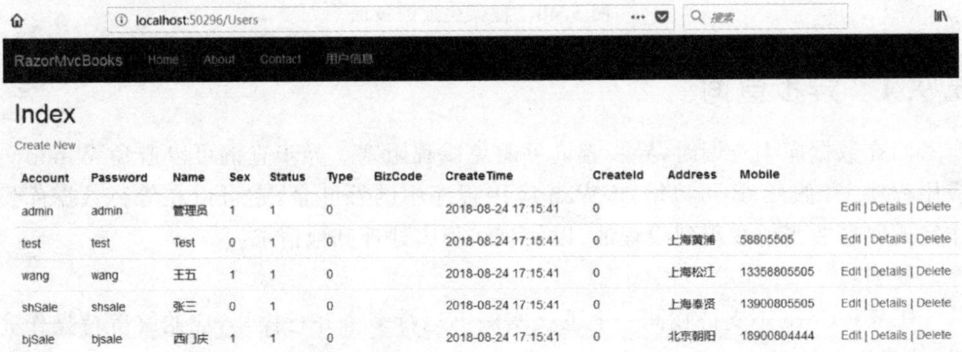

图 5.41 User 列表

5.9.6 加载单个实体

加载单个实体的步骤是：

① 在 Visual Studio 2017 的“解决方案资源管理器”中打开 Details. cshtml. cs 文件，在 OnGetAsync 方法中添加实现加载单个实体的操作，代码如下：

```
public async Task<IActionResult> OnGetAsync(int? id)
{
    if (id == null)
    {
        return NotFound();
    }
    User = await  _context.User.SingleOrDefaultAsync(m => m.Id == id);
```

```
    if (User == null)
    {
        return NotFound();
    }
    return Page();
}
```

② 在 Visual Studio 2017 中按 F5 键运行应用程序，在浏览器中单击"用户信息"，然后单击列表页面中用户"test"最右侧的"Details"，浏览器会跳转到用户详细信息页面，如图 5.42 所示。

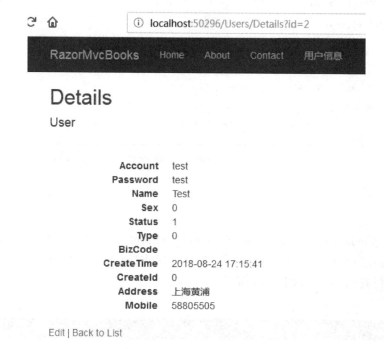

图 5.42　User 详细信息

5.9.7　条件查询

条件查询的步骤是：

① 在 Visual Studio 2017 的"解决方案资源管理器"中打开 Index.cshtml 文件，添加查询条件，代码如下：

```
@page
@model RazorMvcBooks.Pages.Users.IndexModel

@{
```

```
        ViewData["Title"] = "Index";
}
<h2>Index</h2>
<form>
<p>
    用户名称：<input type = "text" name = "SearchString">
<input type = "submit" value = "查询" />
</p>
</form>
```

② 在 Visual Studio 2017 的"解决方案资源管理器"中打开 Index.cshtml.cs 文件，修改 OnGetAsync 方法，添加查询条件，代码如下：

```
public async Task OnGetAsync(string SearchString)
{
    var users = from m in _context.User
    select m;
    if (!String.IsNullOrEmpty(SearchString))
    {
        users = users.Where(s => s.Name.Contains(SearchString));
    }
    User = await users.ToListAsync();
}
```

③ 在 Visual Studio 2017 中按 F5 键运行应用程序，在浏览器中浏览"用户信息"页面，然后在"用户名称"文本框中输入"Test"，单击"查询"按钮，应用程序会根据输入条件进行查询，结果如图 5.43 所示。

图 5.43 用户查询

5.9.8 使用 SQL 语句查询

通过 Entity Framework Core 可以在使用关系型数据库的 SQL 查询语句的数据库中进行查询，该方法在无法使用 LINQ 进行查询时，或者在使用 LINQ 查询导致数据库查询效率很低时非常好用。使用 SQL 语句查询字符串（如 FromSql 和 ExecuteSqlCommand）的 API 允许将值作为参数传递，除了验证用户输入外，始终为

SQL 语句查询/存储过程中所要传递的条件进行参数化传递。如果使用字符串拼接生成查询字符串的任何部分,则必须验证所有输入可以抵御 SQL 注入攻击。

使用原生 SQL 查询时需注意以下几个限制:
① SQL 查询只能用于返回属于用户模型的实体类型或临时类型。
② SQL 查询必须返回实体或查询类型的所有属性的数据。
③ 在结果集中,列名必须与属性所映射到的列名的名称匹配。
④ SQL 查询不能包含相关数据。但是,在许多情况下可以使用 Include 运算符在查询顶部编写代码以返回相关数据。
⑤ 传递到此方法的 SELECT 语句通常可以与 LINQ 运算符进行组合,可以在原始 SQL 查询后面紧跟着 LINQ 运算符进行过滤和排序操作。
⑥ 除了 SELECT 以外,其他 SQL 语句自动识别为不可编写代码。因此,存储过程的完整结果将始终返回到客户端,且在内存中可以使用任何 LINQ 运算符进行计算。

5.9.9 基本 SQL 查询

Entity Framework Core 允许通过 FromSql 扩展方法直接使用 SQL 语句或存储过程进行 LINQ 查询。查询方法是:在 Visual Studio 2017 的"解决方案资源管理器"中打开 Index.cshtml.cs 文件,修改 OnGetAsync 方法,使用 SQL 语句进行查询,代码如下:

```
public async Task OnGetAsync(string searchString)
{
    User = await _context.User.FromSql("Select * from [User]").ToListAsync();
}
//执行存储过程
public async Task OnGetAsync(string searchString)
{
    User = await _context.User.FromSql("EXECUTE dbo.GetUserById").ToListAsync();
}
```

5.9.10 传递参数

能够使用 SQL 语句进行查询,就能接收查询条件。因为在使用这些查询条件时必须考虑到 SQL 注入攻击的可能性,所以这些查询条件需要参数化。可以将参数占位符包含在 SQL 查询字符串中,然后提供参数值作为其他参数。提供的任何参数值都将自动转换为 DbParameter。

通过下面的一个将参数传递给存储过程的示例,来了解如何将查询条件参数化,代码如下:

```
public async Task OnGetAsync(string searchString)
{
    int id = 1;
    User = await _context.User.FromSql("EXECUTE dbo.GetUserById {0}", id).ToListAsync();
}
```

EF Core 2.0 及更高版本支持在字符串中直接插入查询条件,代码如下:

```
public async Task OnGetAsync(string searchString)
{
    int id = 1;
    User = await _context.User.FromSql($"EXECUTE dbo.GetUserById {id}").ToListAsync();
}
```

还可以构造 DbParameter 并将其作为参数值提供。这样,可以在 SQL 查询字符串中使用命名参数,代码如下:

```
public async Task OnGetAsync(string searchString)
{
    var id = new SqlParameter("Id", 1);
    User = await _context.User.FromSql("EXECUTE dbo.GetUserById @id", id).ToListAsync();
}
```

5.9.11　使用 SQL 查询,用 LINQ 编写条件排序

在使用 SQL 查询的基础上,使用 LINQ 编写筛选条件和排序,代码如下:

```
public async Task OnGetAsync(string searchString)
{
    User = await _context.User.FromSql("select * from [User]")
        .Where(u => u.Id < 4)
        .OrderByDescending(u => u.Name)
        .ToListAsync();
}
```

在 Visual Studio 2017 中按 F5 键运行应用程序,在浏览器中浏览"用户信息"页面,结果如图 5.44 所示。

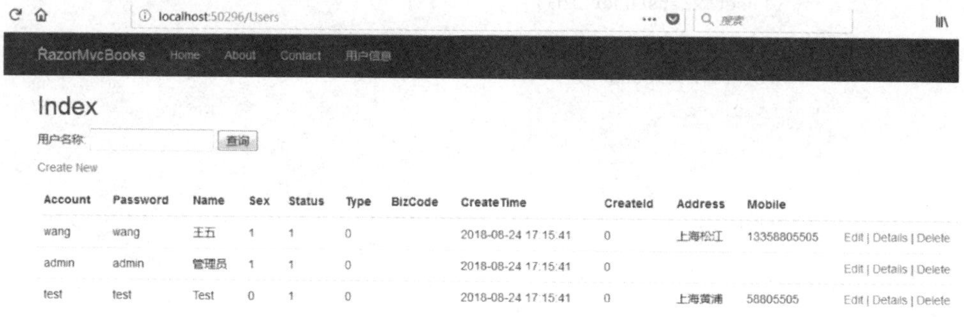

图 5.44　用户列表排序

5.9.12　跟踪与非跟踪查询

跟踪行为用来在上下文跟踪对象是否有更新或改变,要想了解对象是否有过更改,需要检索每个对象在上下文中的状态。

1. 跟踪查询

当一个数据库上下文检索数据表行并建立一个实体对象用来表示其本身时,默认的跟踪是内存里的实体与数据库里的实体保持同步,内存里的数据就像放在缓存中一样,用来更新实体数据。默认情况下,Entity Framework Core 查询是跟踪查询,返回的实体类型在数据库上下文中保留实体的相应信息。对这些实体实例进行的修改,可通过 SaveChanges()方法把这些数据保存到数据库中。Entity Framework Core 还会修复从跟踪查询中获取的实体和先前已加载到数据库上下实例中的实体之间的导航属性。

在以下示例中,对用户编辑页面中的用户信息所做的任何修改,都会通过 SaveChangesAsync()方法永久保存到数据库中,代码如下:

```
public async Task<IActionResult> OnPostAsync()
{
    if (!ModelState.IsValid)
    {
        return Page();
    }
    _context.Attach(User).State = EntityState.Modified;
    try
    {
        await _context.SaveChangesAsync();
    }
    catch (DbUpdateConcurrencyException)
    {
```

```
        if (!UserExists(User.Id))
        {
            return NotFound();
        }
        else
        {
            throw;
        }
    }
    return RedirectToPage("./Index");
}
```

2. 非跟踪查询

在可能只需要读取数据结果进行数据集合显示,而不需要对集合进行修改并更新到数据库时,非跟踪查询的性能可能会更好,因为它可以更快速地执行查询,并且无须更改跟踪信息。

可以交换的单个非跟踪查询的代码如下:

```
public void OnGet()
{
    //非跟踪查询
    User = _context.User.AsNoTracking().ToList();
}
```

还可以在上下文实例级别更改默认跟踪行为,代码如下:

```
public void OnGet()
{
    //非跟踪查询
    _context.ChangeTracker.QueryTrackingBehavior = QueryTrackingBehavior.NoTracking;
    User = _context.User.ToList();
}
```

备注:非跟踪查询仍在执行查询中执行标识解析。如果结果集多次包含相同的实体,则每次会在结果集中返回实体类的相同实例,并使用弱引用跟踪已返回的实体。如果具有相同标识的上一个结果超出范围,并遇到垃圾回收机制执行,则可能会获得新的实体实例。

5.10　EF Core 2.0 Code First 保存数据

每个上下文实例都有一个 ChangeTracker,它负责跟踪需要写入数据库的更改。在更改实体类的实例时,这些更改会记录在 ChangeTracker 中,然后在调用

SaveChanges 时被写入数据库。数据支持程序负责将更改转换为特定于数据库的操作（例如，关系型数据库的 INSERT、UPDATE 和 DELETE 命令）。

通过下面的示例来了解如何使用上下文和实体类去添加、修改和删除数据。

5.10.1 添加数据

添加数据的操作是：

① 在 Visual Studio 2017 的"解决方案资源管理器"中双击打开 Pages/Users/Create.cshtml 文件，文件的内容如下：

```
@page
@model RazorMvcBooks.Pages.Users.CreateModel
@{
    ViewData["Title"] = "Create";
}
<h2>Create</h2>
<h4>User</h4>
<hr/>
<div class="row">
    <div class="col-md-4">
        <form method="post">
            <div asp-validation-summary="ModelOnly" class="text-danger"></div>
            <div class="form-group">
                <label asp-for="User.Account" class="control-label"></label>
                <input asp-for="User.Account" class="form-control" />
                <span asp-validation-for="User.Account" class="text-danger"></span>
            </div>
            <div class="form-group">
                <label asp-for="User.Password" class="control-label"></label>
                <input asp-for="User.Password" class="form-control" />
                <span asp-validation-for="User.Password" class="text-danger"></span>
            </div>
            <div class="form-group">
                <label asp-for="User.Name" class="control-label"></label>
                <input asp-for="User.Name" class="form-control" />
                <span asp-validation-for="User.Name" class="text-danger"></span>
            </div>
            <div class="form-group">
                <label asp-for="User.Sex" class="control-label"></label>
                <input asp-for="User.Sex" class="form-control" />
                <span asp-validation-for="User.Sex" class="text-danger"></span>
            </div>
            <div class="form-group">
                <label asp-for="User.Status" class="control-label"></label>
                <input asp-for="User.Status" class="form-control" />
                <span asp-validation-for="User.Status" class="text-danger"></span>
```

```html
        </div>
        <div class="form-group">
            <label asp-for="User.Type" class="control-label"></label>
            <input asp-for="User.Type" class="form-control" />
            <span asp-validation-for="User.Type" class="text-danger"></span>
        </div>
        <div class="form-group">
            <label asp-for="User.BizCode" class="control-label"></label>
            <input asp-for="User.BizCode" class="form-control" />
            <span asp-validation-for="User.BizCode" class="text-danger"></span>
        </div>
        <div class="form-group">
            <label asp-for="User.CreateTime" class="control-label"></label>
            <input asp-for="User.CreateTime" class="form-control" />
            <span asp-validation-for="User.CreateTime" class="text-danger"></span>
        </div>
        <div class="form-group">
            <label asp-for="User.CreateId" class="control-label"></label>
            <input asp-for="User.CreateId" class="form-control" />
            <span asp-validation-for="User.CreateId" class="text-danger"></span>
        </div>
        <div class="form-group">
            <label asp-for="User.Address" class="control-label"></label>
            <input asp-for="User.Address" class="form-control" />
            <span asp-validation-for="User.Address" class="text-danger"></span>
        </div>
        <div class="form-group">
            <label asp-for="User.Mobile" class="control-label"></label>
            <input asp-for="User.Mobile" class="form-control" />
            <span asp-validation-for="User.Mobile" class="text-danger"></span>
        </div>
        <div class="form-group">
            <input type="submit" value="Create" class="btn btn-default" />
        </div>
    </form>
    </div>
</div>
<div>
    <a asp-page="Index">Back to List</a>
</div>
@section Scripts {
    @{await Html.RenderPartialAsync("_ValidationScriptsPartial");}
}
```

② 在 Visual Studio 2017 的"解决方案资源管理器"中打开 Pages/Users/Create.cshtml.cs 文件,该文件中的代码是由 Visual Studio 2017 自动生成的,主要包含 OnPostAsync 方法,该方法中使用 DbSet.Add 方法添加实体类的对象,调用 SaveChanges 方法将数据插入到数据库中,代码如下:

```
public async Task<IActionResult> OnPostAsync()
{
    if (!ModelState.IsValid)
    {
        return Page();
    }
    _context.User.Add(User);
    await _context.SaveChangesAsync();
    return RedirectToPage("./Index");
}
```

③ 在 Visual Studio 2017 中按 F5 键运行应用程序,在浏览器中浏览"用户信息"页面,单击列表页面上的"Create New"链接,进入到"Create"页面,输入相应的用户信息,然后单击"Create"按钮,如图 5.45 所示。

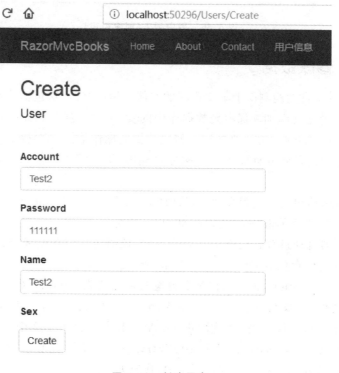

图 5.45　创建用户

④ 浏览器自动跳转到用户信息列表页面，此时可以看到刚才创建的"Test2"用户已经成功插入到数据库中，如图 5.46 所示。

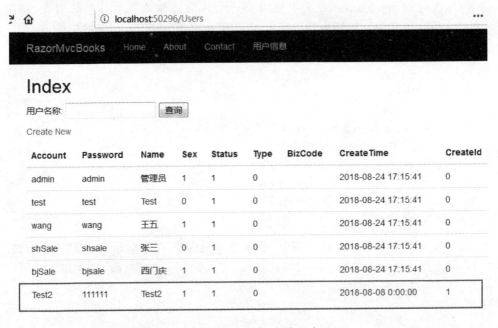

图 5.46　用户列表（创建用户）

5.10.2　修改数据

EF Core 将自动检测由上下文跟踪的实体所做的修改，这包括从数据库加载/查询的实体，以及之前添加并保存到数据库中的实体，这可通过只修改相应属性的值，然后调用 SaveChanges 方法来实现。修改数据的操作是：

① 在用户信息列表页面中选择刚才创建的"Test2"用户信息，然后单击"Edit"链接。

② 在编辑用户信息列表页面中，修改"Test2"用户的"Name"值，将"Test2"修改为"测试用户"，然后单击"Save"按钮，如图 5.47 所示。

③ 浏览器自动跳转到用户信息列表页面，此时可以看到刚才修改的内容"测试用户"已经成功保存到数据库中，如图 5.48 所示。

④ 在 Visual Studio 2017 的"解决方案资源管理器"中打开 Pages/Users/Edit.cshtml.cs 文件，该文件中的代码是由 Visual Studio 2017 自动生成的，主要包含 OnPostAsync 方法，该方法中使用 DbSet.Attach 方法将上下文中跟踪的实体对象的状态置为 UnChanged，使用 EntityEntry.State 属性设置实体的状态，调用 SaveChanges 方法将数据保存到数据库中，代码如下：

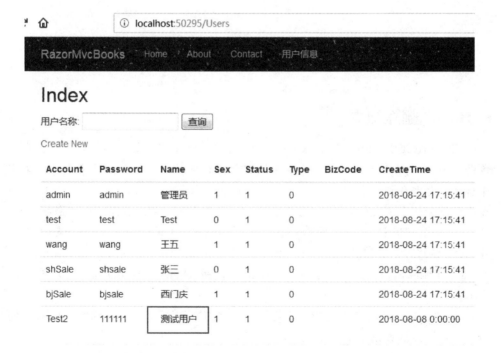

图 5.47 用户更新

图 5.48 用户列表(用户更新)

```csharp
public async Task<IActionResult> OnPostAsync()
{
    if (!ModelState.IsValid)
    {
        return Page();
    }
    _context.Attach(User).State = EntityState.Modified;
    try
    {
        await _context.SaveChangesAsync();
    }
    catch (DbUpdateConcurrencyException)
    {
        if (!UserExists(User.Id))
        {
            return NotFound();
        }
        else
        {
            throw;
        }
    }
    return RedirectToPage("./Index");
}
```

5.10.3 删除数据

EF Core 使用 DbSet.Remove 方法删除实体类的对象。如果实体对象已存在于数据库中，则将在执行 SaveChanges 方法时删除该实体对象对应的原数据。如果实体对象没有保存到数据库（即跟踪为"Added"）中，则在调用 SaveChanges 方法时，该实体对象会从上下文中删除，并且不会插入到数据库中。删除数据的操作是：

① 在用户信息列表页面中，选择刚才创建的"Test2"用户信息，然后单击"Delete"链接。

② 在删除用户信息页面中，单击"Delete"按钮，删除 Test2 用户信息，如图 5.49 所示。

③ 浏览器自动跳转到用户信息列表页面，此时可以看到 Name 为"测试用户"的用户信息已经从数据库中删除，如图 5.50 所示。

④ 在 Visual Studio 2017 的"解决方案资源管理器"中打开 Pages/Users/Delete.cshtml.cs 文件，该文件中的代码是由 Visual Studio 2017 自动生成的，主要包含 OnPostAsync 方法，该方法中使用 DbSet.Remove 方法将上下文中跟踪的实体

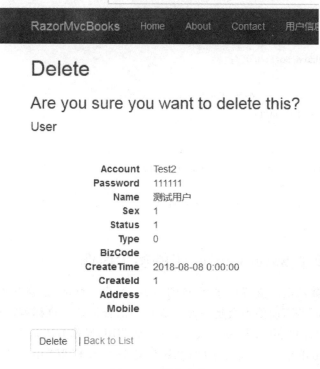

图 5.49　删除用户

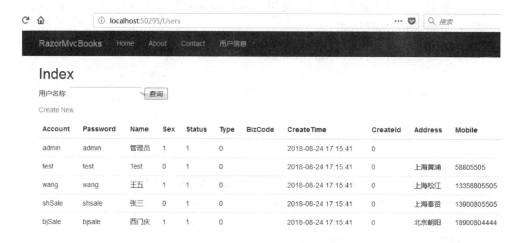

图 5.50　用户列表(删除用户)

对象的状态置为 Deleted,并调用 SaveChanges 方法将数据从数据库中删除,代码如下:

```
public async Task<IActionResult> OnPostAsync(int? id)
{
    if (id == null)
    {
        return NotFound();
    }
    User = await _context.User.FindAsync(id);
    if (User != null)
    {
        _context.User.Remove(User);
        await _context.SaveChangesAsync();
    }
    return RedirectToPage("./Index");
}
```

5.10.4 单个 SaveChanges 中的多个操作

对于大多数数据库支持程序，SaveChanges 是事务性的，这意味着所有操作要么都成功，要么都失败，而绝不会是一部分成功一部分失败。这样就可以把多个新增/更新/删除操作合并，由一个 SaveChanges 方法调用，操作方法是：

① 在 Visual Studio 2017 的"解决方案资源管理器"中打开 Pages/Users/Delete.cshtml.cs 文件，并新增一个方法 OnPostMultOperAsync，在该方法中将新增、更新、删除三种操作合并在一起，使用 SaveChanges 方法实现数据库中数据的变化（新增、更新、删除），代码如下：

```
public async Task<IActionResult> OnPostMultOperAsync(int? id)
{
    if (id == null)
    {
        return NotFound();
    }
    //新增
    _context.User.Add(new User{Account = "Test2",Password = "222222",Name = "Test2",Sex = 0,Status = 0,Type = 0,CreateId = 0,CreateTime = DateTime.Now });
    //更新
    var updUser = _context.User.Find(4);
    updUser.Name = "张不三";
    _context.Attach(updUser).State = EntityState.Modified;
    User = await _context.User.FindAsync(id);
    if (User != null)
    {
```

```
        //删除
        _context.User.Remove(User);
        await _context.SaveChangesAsync();
    }
    return RedirectToPage("./Index");
}
```

② 在 Visual Studio 2017 的"解决方案资源管理器"中打开 Pages/Users/Delete.cshtml 文件,新增一个按钮来调用 OnPostMultOperAsync 方法,代码如下:

```
<form method="post">
<input type="hidden" asp-for="User.Id" />
<input type="submit" value="Delete" class="btn btn-default" /> |
<a asp-page="./Index">Back to List</a> |
<input type="submit" value="多个操作合一" asp-route-id="@Model.User.Id" class="btn btn-default" asp-page-handler="MultOper" />
</form>
```

③ 在 Visual Studio 2017 中按 F5 键运行应用程序,在浏览器中单击"用户信息"显示用户信息页面,然后选择一条用户信息,单击"Delete"链接,如图 5.51 所示。

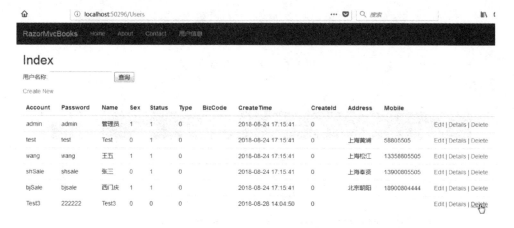

图 5.51　删除操作

④ 浏览器跳转到删除页面,单击"多个操作合一"按钮,如图 5.52 所示。

⑤ 浏览器跳转到用户信息列表页面,这时会发现新增了一个 Name 为"Test2"的用户信息,同时将 Name 为"张三"的内容修改为了"张不三",将 Name 为"Test3"的用户信息删除了。可见,三个不同的操作由一个 SaveChangesAsync 方法调用实现,如图 5.53 所示。

ASP.NET Core 应用开发入门教程

图 5.52　多个操作合一

图 5.53　用户列表(多个操作合一)

5.10.5 异步保存

异步保存与异步查询类似。当将数据写入数据库时,异步保存可避免线程阻塞。异步保存可以避免 Windows 应用程序 UI 假死,还可以增加 Web 应用程序中的吞吐量,它可以在等待数据库操作完成时释放当前线程到线程池,以使该线程去处理其他请求。

EF Core 不支持在同一上下文实例上运行多个并行操作,而是始终等待操作完成再开始下一个操作,这通常是通过在每个异步操作上使用 await 关键字来完成的。

Entity Framework Core 提供了 DbContext.SaveChangesAsync() 作为 DbContext.SaveChanges() 的异步替代方法。

5.10.6 使用事务

事务是指作为单个逻辑工作单元执行的一系列操作,要么完全执行,要么完全不执行。事务处理可以确保只有在事务性单元内的所有操作都成功完成时才会把数据更新到数据库中。通过将一组相关操作组合为一个要么全部成功要么全部失败的单元,可以简化错误恢复并使应用程序更加可靠。一个逻辑工作单元要想成为事务,必须满足所谓的 ACID(原子性、一致性、隔离性和持久性)属性。

5.10.7 默认事务

默认情况下,如果数据库支持程序支持事务的话,则对 SaveChanges() 方法应用事务处理,即 SaveChanges() 包含的所有操作如果都成功,则把数据变化更新到数据库中;如果有一部分更新失败,则回滚事务且所有更新都不会更新到数据库中。

对于大多数应用程序,SaveChanges() 方法的默认事务已经足够。如果应用程序有特殊要求,则应该手动控制事务。不过 SaveChanges() 方法的默认事务不适合大批量数据的更新和插入操作,因为大批量数据使用该方法更新数据库的效率低下。

5.10.8 显式事务

可以使用 DbContext.Database API 来开始、提交和回滚事务。

Database.BeginTransaction() 是在一个已存在的 DbContext 上下文中启动和完成 Transactions 的一种简单方式,它允许多个操作组合存在于相同的 Transaction 中,所以要么提交要么全部作为一体回滚,同时也允许更加容易地去显式指定 Transaction 的隔离级别。

Database.BeginTransaction 有两种重载:一种是显式指定隔离级别,另一种是无参数,而是使用来自于底层数据库提供的默认隔离级别,两种都返回一个 DbContextTransaction 对象,该对象提供了事务提交(Commint)和回滚(RollBack)的方法,直接表现在底层数据库上的事务提交和事务回滚上。

DbContextTransaction 一旦被提交或回滚就会被回收,所以使用它的简单方式就是使用 using(){}语法,当 using 构造块完成时会自动调用 Dispose()方法。

以下示例显示了两个 SaveChanges()操作和一个查询操作在一个事务中的执行。

注:并非所有数据库支持程序都支持事务。在调用事务 API 时,某些支持程序可能会引发异常或不执行任何操作。

执行显式事务的操作是:

① 在 Visual Studio 2017 的"解决方案资源管理器"中双击打开 Pages/Users/Delete.cshtml.cs 文件,添加一个 OnPostTransAsync 方法进行显式事务调用,把新增、更新、查询、删除四种操作放在一个显式事务里,代码如下:

```csharp
public async Task<IActionResult> OnPostTransAsync(int? id)
{
    if (id == null)
    {
        return NotFound();
    }
    //使用事务
    using (var transaction = _context.Database.BeginTransaction())
    {
        //新增
        _context.User.Add(new User { Account = "Test3", Password = "111111", Name = "Test3", Sex = 0, Status = 0, Type = 0, CreateId = 0, CreateTime = DateTime.Now });
        _context.SaveChanges();
        //修改
        var updUser = _context.User.Find(2);
        updUser.Name = "李四";
        _context.Attach(updUser).State = EntityState.Modified;
        _context.SaveChanges();
        var item = _context.User.Where(u => u.Id == 1).First();
        //查询
        if (item != null)
        {
            Console.WriteLine("ID:{0}", item.Id);
            Console.WriteLine("Account:{0}", item.Account);
            Console.WriteLine("名称:{0}", item.Name);
            Console.WriteLine("地址:{0}", item.Address);
            Console.WriteLine("时间:{0}", item.CreateTime.ToString("yyyy-MM-dd HH:mm:ss"));
        }
        User = await _context.User.FindAsync(id);
        if (User != null)
        {
            //删除
```

```
            _context.User.Remove(User);
            await _context.SaveChangesAsync();
        }
        //提交事务
        transaction.Commit();
    }
    return RedirectToPage("./Index");
}
```

② 在 Visual Studio 2017 的"解决方案资源管理器"中打开 Pages/Users/Delete.cshtml 文件,新增一个按钮来调用 OnPostTransAsync 方法,代码如下:

```
<form method="post">
<input type="hidden" asp-for="User.Id" />
<input type="submit" value="Delete" class="btn btn-default" /> |
<a asp-page="./Index">Back to List</a> |
<input type="submit" value="多个操作合一" asp-route-id="@Model.User.Id" class="btn btn-default" asp-page-handler="MultOper" /> |
<input type="submit" value="显式事务" asp-route-id="@Model.User.Id" class="btn btn-default" asp-page-handler="Trans" />
</form>
```

③ 在 Visual Studio 2017 中按 F5 键运行应用程序,在浏览器中单击"用户信息"显示用户信息页面,然后选择"Test2"用户信息并单击 Delete 链接,如图 5.54 所示。

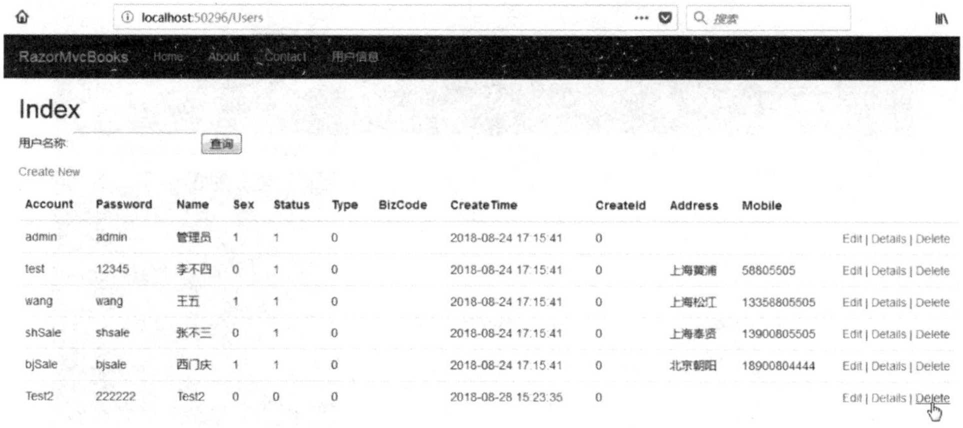

图 5.54 删除操作

④ 浏览器跳转到删除页面,然后单击"显式事务"按钮,如图 5.55 所示。

⑤ 浏览器跳转到用户信息列表页面,可以看到新增了一个 Name 为"Test3"的用户信息,同时将 Name 的值"李不四"修改为了"李四",将 Name 为"Test2"的用户信息删除了。四个不同的操作由一个事务中的 SaveChangesAsync 方法调用实现,

如图 5.56 所示。

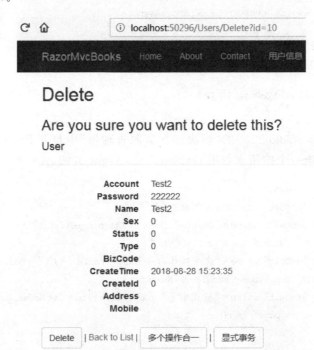

图 5.55 显示事务

图 5.56 三合一操作

5.11　EF Core 2.0 Code First 处理并发冲突

数据库并发指的是多个进程或用户同时访问或更改数据库中的相同数据。并发控制指的是在发生并发更改时确保数据一致性的特定机制。

EF Core 实现乐观并发控制,意味着它将允许多个进程或用户独立进行更改而不产生同步或锁定的开销。在理想情况下,这些更改将不会相互影响,因此都能成功。在最坏的情况下,两个或更多进程将尝试进行冲突更改,其中只有一个进程应该成功。

5.11.1　并发冲突

在以下情况下,会发生并发冲突:
① 用户导航到实体的编辑页面。
② 第一个用户的更改还未写入数据库之前,另一个用户更新同一实体。
如果未启用并发检测,则当发生并发更新时会出现以下情况:
① 最后一个更新优先,也就是最后一个更新的值保存至数据库中。
② 第一个并发更新将会丢失。

5.11.2　乐观并发

乐观并发指允许发生并发冲突,并在并发冲突发生时做出正确反应。例如,管理员访问用户信息编辑页面,将"Test"用户的 Password 修改为"123456",如图 5.57 所示。

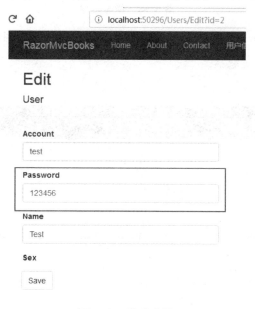

图 5.57　修改密码

在管理员单击 Save 按钮之前，Test 用户访问了相同页面，并将 Name 修改为"李不四"，如图 5.58 所示。

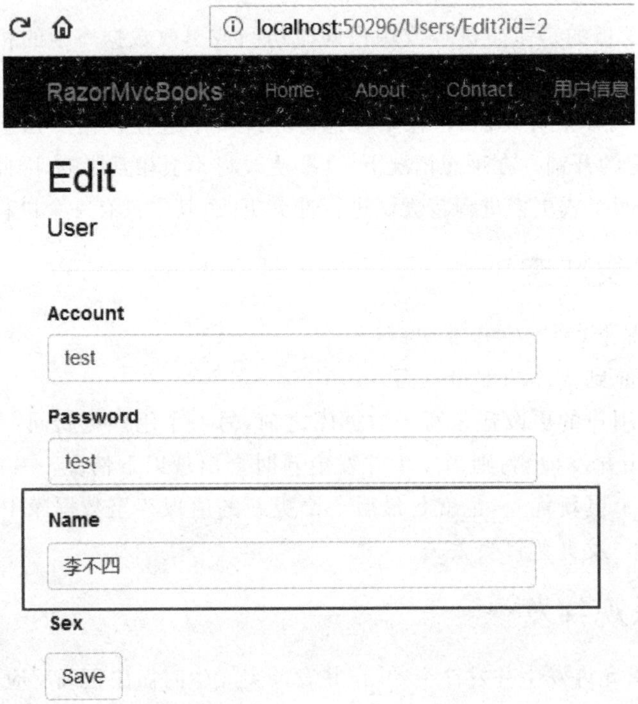

图 5.58　修改名称

Test 用户先单击 Save 按钮，并在浏览器显示索引页时看到他的修改，如图 5.59 所示。

图 5.59　用户列表 1

管理员单击编辑页面上的 Save 按钮，但页面上的 Test 用户的 Name 值仍显示为"Test"，如图 5.60 所示。

图 5.60　用户列表 2

乐观并发包括以下选项：

① 可以跟踪用户已修改的属性，并仅更新数据库中相应的列。

在这种情况下，数据不会丢失。两个用户更新了不同的属性。下次有人浏览用户信息时，将看到管理员和 Test 两个人的更改。这种更新方法可以减少导致数据丢失的冲突数。这种方法需要维持大量的状态，以便跟踪所有提取值和当前值；它还增加了应用复杂性，并可能影响应用的性能。因此一般不适用于 Web 应用。

② 可让管理员的更改覆盖 Test 的更改。

就如上面示例所示。管理员在完成保存之后，当浏览用户信息时，看到"123456"和"Test"。这种方法称为"客户端优先"或"最后一个优先"方案(客户端的所有值优先于存储的数据值)。如果不对并发处理进行任何编码，则自动执行"客户端优先"。

③ 可以阻止在数据库中更新对 Test 用户的更改。

这种方法需要显示错误信息，显示当前数据和数据库中的数据，允许用户重新修改并保存。这种方法称为"存储优先"方案(数据存储值优先于客户端提交的值)。

下面来看看"存储优先"方案，此方案可确保用户在未收到警报时不去覆盖任何更改。

5.11.3　检测并发冲突

可以使用 5.6.3 小节中介绍的 ConcurrencyCheck 特性，把该特性应用到实体类中需要配置为并发令牌的属性上，就可以实现乐观并发控制，即每当在 SaveChanges 方法执行更新或删除操作时，都会将实体类中已经应用 ConcurrencyCheck 特性的属

性对应的数据,与通过 EF Core 读取数据库表中对应属性的原始值进行比较,比较的结果是:

① 如果这些值匹配,则可以完成该操作。

② 如果这些值不匹配,则 EF Core 会假设另一个用户已执行了冲突操作,因此中止当前事务。另一个用户已执行与当前操作冲突的操作称为并发冲突。

在关系型数据库中,EF Core 包括对任何 UPDATE 或 DELETE 语句的 WHERE 子句中的并发令牌值的检查。执行这些语句后,EF Core 会读取受影响的行数。如果未影响任何行,则将检测到并发冲突,并且 EF Core 会引发 DbUpdateConcurrencyException。例如,ConcurrencyCheck 特性在属性级别上检测并发冲突,该特性可应用于模型上的多个属性。在 User 实体类的 Name 上使用 ConcurrencyCheck 特性将 Name 属性配置为并发令牌,代码如下:

```csharp
using System;
using System.Collections.Generic;
using System.ComponentModel.DataAnnotations;
namespace RazorMvcBooks.Models
{
    public partial class User
    {
        public int Id { get; set; }
        [MaxLength(100), Required]
        public string Account { get; set; }
        [MaxLength(250), Required]
        public string Password { get; set; }
        [ConcurrencyCheck]
        [MaxLength(100), Required]
        public string Name { get; set; }
        public int Sex { get; set; }
        public int Status { get; set; }
        public int Type { get; set; }
        [MaxLength(100)]
        public string BizCode { get; set; }
        public DateTime CreateTime { get; set; }
        public int CreateId { get; set; }
        [MaxLength(250)]
        public string Address { get; set; }
        [MaxLength(50)]
        public string Mobile { get; set; }
    }
}
```

在使用 ConcurrencyCheck 特性之后,再进行修改操作时,EF Core 生成的 SQL 语句将在 WHERE 子句中进行并发检查,代码如下:

```
UPDATE [User] SET [Name] = @p1
WHERE [Id] = @p0 AND [Name] = @p2;
```

5.11.4 解决并发冲突

再次进行上面示例的操作,如果尝试保存对用户信息的 Name 值所做的修改,则将引发异常。

此时,应用程序可能只需通知用户由于发生冲突更改而导致更新未成功,然后再继续操作;也可能需要提示用户保证此记录是以用户当前修改的内容为准,并把当前用户修改的内容保存到数据库中;也可能提示以数据库中的数据为准,而当前用户的修改无效。

有三组值可用于帮助解决并发冲突:
- "当前值"是应用程序尝试写入数据库的值。
- "原始值"是在进行任何修改之前最初从数据库中检索的值。
- "数据库值"是当前存储在数据库中的值。

处理并发冲突的常规方法是:
① 在 SaveChanges 方法执行期间捕获 DbUpdateConcurrencyException。
② 使用 DbUpdateConcurrencyException.Entries 为受影响的实体准备一组新更改。
③ 刷新并发令牌的原始值以反映数据库中的当前值。
④ 重试该过程,直到不发生任何冲突。

在下面的示例中,将 Name 设置为并发令牌。如果发生冲突,则抛出异常,具体操作如下:

① 在 Visual Studio 2017 的"解决方案资源管理器"中双击打开 Pages/Users/Edit.cshtml.cs 文件,对 OnPostAsync 方法进行修改,代码如下:

```
public async Task<IActionResult> OnPostAsync()
{
    if (!ModelState.IsValid)
    {
        return Page();
    }
    _context.Attach(User).State = EntityState.Modified;
    try
    {
        await _context.SaveChangesAsync();
    }
```

```csharp
            catch (DbUpdateConcurrencyException ex)
            {
                if (!UserExists(User.Id))
                {
                    return NotFound();
                }
                else
                {
                    foreach (var entry in ex.Entries)
                    {
                        if (entry.Entity is User)
                        {
                            //当前值
                            var proposedValues = entry.CurrentValues;
                            //数据库值
                            var databaseValues = entry.GetDatabaseValues();
                            foreach (var property in proposedValues.Properties)
                            {
                                var proposedValue = proposedValues[property];
                                var databaseValue = databaseValues[property];
                                //更新当前值
                                // proposedValues[property] = databaseValue;
                            }
                            //更新原始值来保证修改成功
                            entry.OriginalValues.SetValues(databaseValues);
                        }
                        else
                        {
                            throw new NotSupportedException("Don't know how to handle concurrency conflicts for" + entry.Metadata.Name);
                        }
                    }
                    throw new NotSupportedException("当前数据与数据库中的值不相符,不允许修改!");
                }
            }
            return RedirectToPage("./Index");
        }
```

② 在OnPostAsync方法中的"if (entry.Entity is User)"行上设置断点。

③ 在Visual Studio 2017中按F5键运行应用程序。在浏览器中单击"用户信息"显示用户信息页面,然后选择一条用户信息并单击"Edit"链接。在编辑页面中把Name值由"李不四"修改为"test",然后单击"Save"按钮,如图5.61所示。

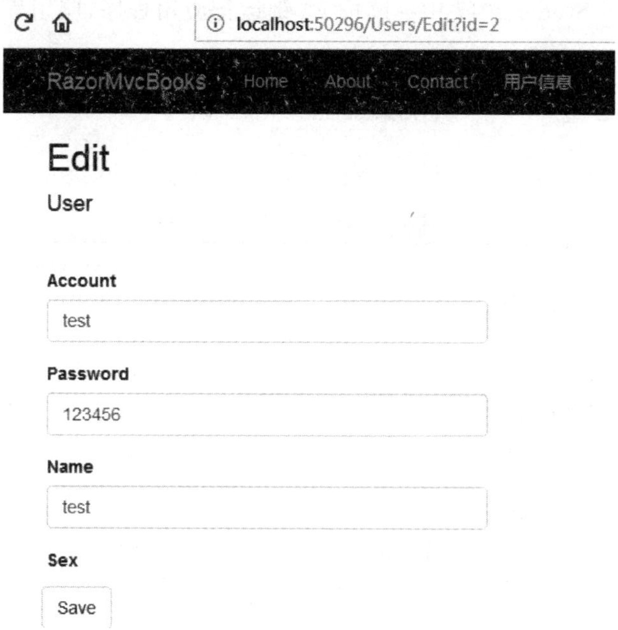
localhost:50296/Users/Edit?id=2
RazorMvcBooks
Home
About
Contact
用户信息
Edit
User
Account
test
Password
123456
Name
test
Sex
Save

图 5.61 修改用户信息

④ 由于在 Name 属性上设置了 ConcurrencyCheck，因此应用程序会抛出 DbUpdateConcurrencyException 异常，Visual Studio 2017 进入断点，如图 5.62 所示。

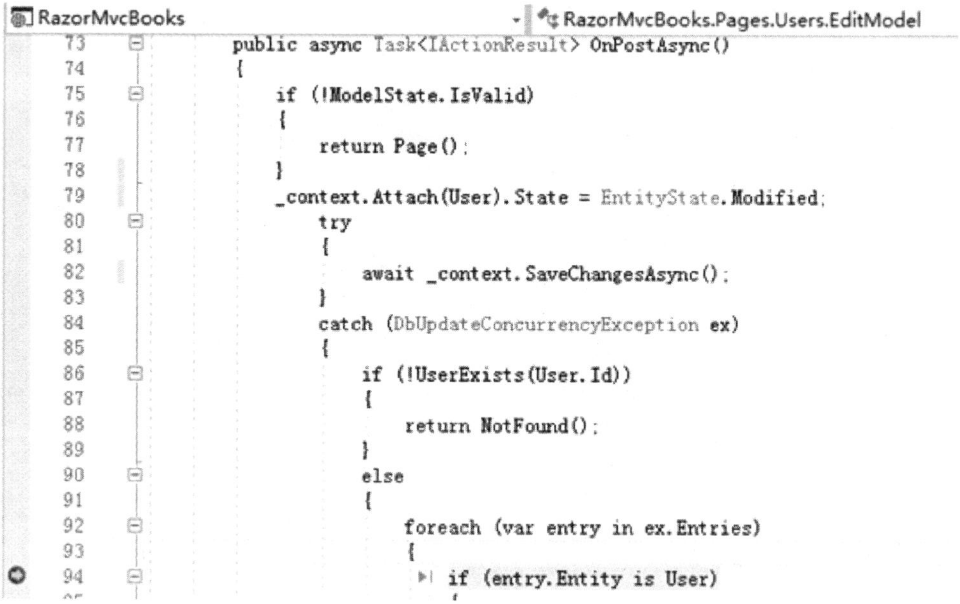
RazorMvcBooks
RazorMvcBooks.Pages.Users.EditModel
public async Task<IActionResult> OnPostAsync()
{
if (!ModelState.IsValid)
{
return Page();
}
_context.Attach(User).State = EntityState.Modified;
try
{
await _context.SaveChangesAsync();
}
catch (DbUpdateConcurrencyException ex)
{
if (!UserExists(User.Id))
{
return NotFound();
}
else
{
foreach (var entry in ex.Entries)
{
if (entry.Entity is User)

图 5.62 中 断

⑤ 在 Visual Studio 2017 中继续按 F5 键运行应用程序，应用程序会抛出异常，如图 5.63 所示。

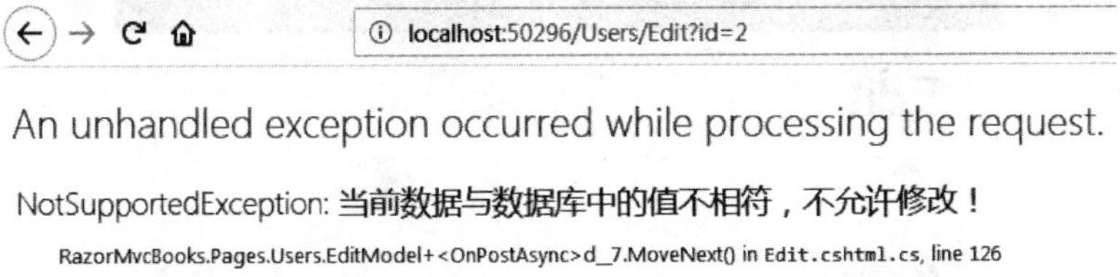
localhost:50296/Users/Edit?id=2
An unhandled exception occurred while processing the request.
NotSupportedException: 当前数据与数据库中的值不相符，不允许修改！
RazorMvcBooks.Pages.Users.EditModel+<OnPostAsync>d_7.MoveNext() in Edit.cshtml.cs, line 126

图 5.63 抛出异常

5.11.5 使用时间戳和行级版本号

使用时间戳和行级版本号的操作是：

① 在 Visual Studio 2017 的"解决方案资源管理器"中双击打开 Models/User.cs 文件，对 User 实体添加跟踪属性 RowVersion，并在其上添加 Timestamp 特性，删除 Name 属性上的 ConcurrencyCheck 特性，代码如下：

```
using System;
using System.Collections.Generic;
using System.ComponentModel.DataAnnotations;

namespace RazorMvcBooks.Models
{
    public partial class User
    {
        public int Id { get; set; }
            [MaxLength(100), Required]
        public string Account { get; set; }
            [MaxLength(250), Required]
        public string Password { get; set; }
            [MaxLength(100), Required]
        public string Name { get; set; }
        public int Sex { get; set; }
        public int Status { get; set; }
        public int Type { get; set; }
            [MaxLength(100)]
        public string BizCode { get; set; }
        public DateTime CreateTime { get; set; }
        public int CreateId { get; set; }
            [MaxLength(250)]
        public string Address { get; set; }
```

```
        [MaxLength(50)]
        public string Mobile { get; set; }
        [Timestamp]
        public byte[] RowVersion { get; set; }
    }
}
```

② 在 Visual Studio 2017 中选择"菜单"→"NuGet 包管理器"→"程序包管理器控制台"菜单项,在打开的程序包管理器控制台上依次执行以下命令:

Add – Migration RowVer
Update – Database

③ 在 SQL Server Management Studio 中查看 User 表,如图 5.64 所示。

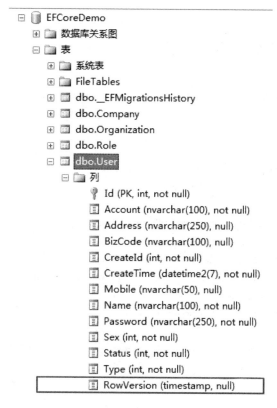

图 5.64　添加 RowVersion 字段

④ 在 Visual Studio 2017 的"解决方案资源管理器"中双击打开 Pages/Users/Edit.cshtml.cs 文件,对 OnPostAsync 方法进行修改。Entity Framework Core 使用包含原始 RowVersion 值的 WHERE 子句生成 SQL UPDATE 命令。如果没有行受到 UPDATE 命令影响(即没有行具有原始的 RowVersion 值),则将引发

DbUpdateConcurrencyException 异常，代码如下：

```
public async Task<IActionResult> OnPostAsync()
{
    if (!ModelState.IsValid)
    {
        return Page();
    }
    var updUser = _context.User.AsNoTracking().Where(u => u.Id == User.Id).First();
    // 如果为 null，则当前用户信息已经被删除
    if (updUser == null)
    {
        return HandleDeletedUser();
    }
    _context.Attach(User).State = EntityState.Modified;
    if (await TryUpdateModelAsync<User>(
        User,
        "User",
        s => s.Name, s => s.Password, s => s.Account, s => s.Sex))
    {
        try
        {
            await _context.SaveChangesAsync();
            return RedirectToPage("./Index");
        }
        catch (DbUpdateConcurrencyException ex)
        {
            var exceptionEntry = ex.Entries.Single();
            var clientValues = (User)exceptionEntry.Entity;
            var databaseEntry = exceptionEntry.GetDatabaseValues();
            if (databaseEntry == null)
            {
                ModelState.AddModelError(string.Empty, "保存失败！.当前用户信息已经被删除");
                return Page();
            }
            var dbValues = (User)databaseEntry.ToObject();
            setDbErrorMessage(dbValues, clientValues, _context);
            //用数据库中的 RowVersion 值设置为当前实体对象客户端界面中的 RowVersion
            //值。用户下次单击"保存"按钮时，将仅捕获最后一次显示编辑页后发生的并
            //发错误
            User.RowVersion = (byte[])dbValues.RowVersion;
```

```csharp
        //ModelState 具有旧的 RowVersion 值,因此需使用 ModelState.Remove 语句。
        //在 Razor 页面中,当两者都存在时,字段的 ModelState 值优于模型属性值
        ModelState.Remove("User.RowVersion");
        }
    }
    return Page();
}
private PageResult HandleDeletedUser()
{
    User deletedDepartment = new User();
    ModelState.AddModelError(string.Empty,"保存失败!.当前用户信息已经被删除!");
    return Page();
}
```

⑤ 在 Edit.cshtml.cs 文件中添加 setDbErrorMessage 方法,为每列添加自定义错误消息,当这些列中的数据库值与客户端界面上的值不同时,给出相应的错误信息,代码如下:

```csharp
private void setDbErrorMessage(User dbValues, User clientValues, EFCoreDemoContext context)
{
    if (dbValues.Name != clientValues.Name)
    {
        ModelState.AddModelError("User.Name", $"数据库值:{dbValues.Name}");
    }
    if (dbValues.Account != clientValues.Account)
    {
        ModelState.AddModelError("User.Account", $"数据库值:{dbValues.Account}");
    }
    if (dbValues.Password != clientValues.Password)
    {
        ModelState.AddModelError("User.Password", $"数据库值:{dbValues.Password}");
    }
    if (dbValues.Sex != clientValues.Sex)
    {
        ModelState.AddModelError("User.Sex", $"数据库值:{dbValues.Sex}");
    }
    ModelState.AddModelError(string.Empty,
        "您尝试编辑的记录" +
        "被另一个用户修改了" +
        "编辑操作被取消,数据库中的当前值" +
```

"已经显示。如果仍想编辑此记录,请单击" +
""保存"按钮。");
}

⑥ 在 Visual Studio 2017 的"解决方案资源管理器"中双击打开 Pages/Users/Edit.cshtml 文件,在〈form method="post"〉标签下面添加隐藏的行版本,同时也必须添加 RowVersion,以便回发绑定值,代码如下:

〈input type="hidden" asp-for="User.RowVersion" /〉

⑦ 在 Visual Studio 2017 中按 F5 键运行应用程序。使用两个浏览器打开编辑用户信息的编辑界面,此时两个浏览器显示的用户信息是一样的。在浏览器 1 中的用户信息界面上,将 Password 的数据由"111111"修改为"333333",然后单击"保存"按钮,如图 5.65 所示。

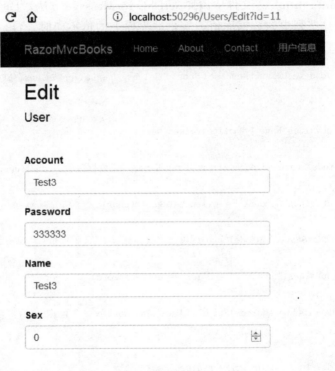

图 5.65 修改密码

⑧ 在浏览器 2 中,将 Sex 的值由 0 修改为 1,如图 5.66 所示。

⑨ 单击"保存"按钮,此时由于客户端界面上的信息与数据库中的值不一样,所以会出现错误提示信息,如图 5.67 所示。

⑩ 再次单击"保存"按钮,保存浏览器 2 中输入的值,在用户列表页面中可以看到保存的值,如图 5.68 所示。

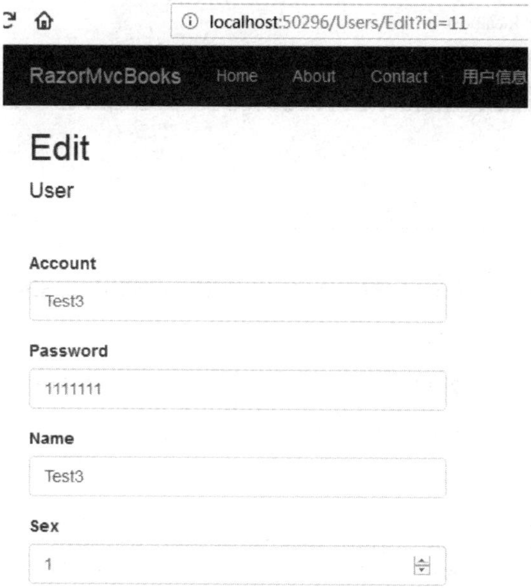

图 5.66　修改性别

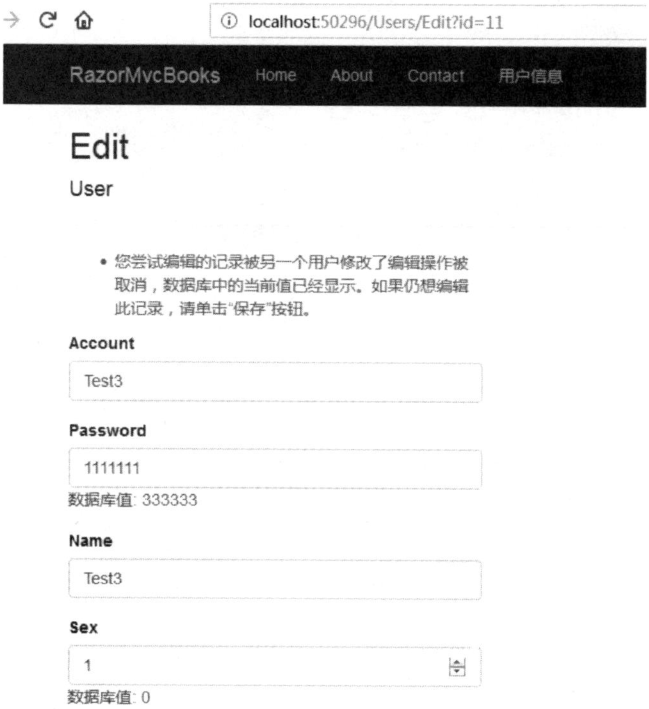

图 5.67　错误提示

图 5.68 用户列表

第 6 章

ASP.NET Core MVC

6.1 ASP.NET Core MVC 概述

ASP.NET Core MVC 使用"模型-视图-控制器"设计模式构建 Web 应用和 API 的丰富框架。

6.1.1 什么是 MVC 模式

模型-视图-控制器（MVC，Model-View-Controller）体系结构模式将应用程序分成 3 个主要组件：模型、视图和控制器。此模式体现了关注点分离这一基本方针。使用此模式，用户的请求被路由到控制器，后者负责使用模型来执行用户操作和/或检索查询结果。控制器负责选择要显示给用户的视图，并为其提供所需的任何模型数据。图 6.1 显示了 3 个主要组件及其相互引用关系。

这种责任划分有助于根据复杂性来缩放应用程序，因为这更易于编码、调试和测试有单一作业的某些内容（模型、视图或控制器）。但这会加大更新、测试和调试代码的难度，该代码在这 3 个领域的 2 个或多个领域间存在依赖关系。例如，用户界面逻辑的变更频率往往高于业务逻辑。如果将表示代码和业务逻辑组合在单个对象中，则每次更改用户界面时都必须修改包含业务逻辑的对象。这常常会引发错误，并且需要在每次进行细微的用户界面更改后重新测试业务逻辑。

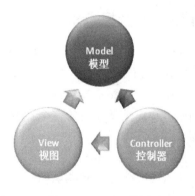

图 6.1 MVC 三要素

视图和控制器均依赖于模型。但是，模型既不依赖于视图，也不依赖于控制器。这是分离的一个关键优势。这种分离允许模型独立于可视化展示进行构建和测试。

1. 模型责任

MVC 应用程序中的模型（M）表示应用程序状态和业务功能的封装。模型也可以是使用了某个工具的一种数据访问层，该工具可以是 EntityFrameworkCore。强类型视图通常使用 ViewModel 类型,其包含了要在该视图上显示的数据。控制器根据一个或多个模型创建 ViewModel 实例,并进行数据填充。

2. 视图责任

视图(V)负责通过用户界面展示内容,并捕获最终用户的交互操作(如鼠标或键盘操作)。通过使用 Razor 视图引擎在 HTML 标记中嵌入 .NET 代码。视图仅包含最少的逻辑,并且这些逻辑都必须与展示内容相关。如果在视图文件中需要执行大量的逻辑才能从模型显示数据,则请考虑使用视图组件、视图模型或视图模板来简化视图。

3. 控制器责任

控制器(C)是处理用户交互、使用模型、处理应用程序逻辑并最终选择要呈现的视图的组件。在 MVC 应用程序中,视图显示信息,并将捕获到的用户交互操作直接转发给控制器,控制器处理用户的输入和交互并做出响应。在 MVC 模式中,控制器是初始入口点,根据用户的输入负责选择要使用的模型类型和要呈现的视图,它控制应用程序对用户交互操作做出响应。

控制器不应有太多的责任而变得过于复杂。为避免控制器逻辑变得过于复杂,请使用单一责任原则将业务逻辑从控制器移到领域模型。

如果发现控制器操作方法中经常执行一些相同或类似的代码,则可以依照"不要自我重复"原则,将这些代码移到过滤器中。

6.1.2 什么是 ASP.NET Core MVC

ASP.NET Core MVC 框架是轻量级、开源、高度可测试的框架,并针对 ASP.NET Core 进行了优化。

ASP.NET Core MVC 提供一种基于模式,使用干净的关注分离的方式构建动态网站。它提供对标记的完全控制,支持友好的测试驱动开发并使用最新的 Web 标准。

ASP.NET Core MVC 提供以下生成 Web API 和 Web 应用所需的功能:

① MVC 模式使 Web API 和 Web 应用可测试。

② ASP.NET Core 2.0 中新增的 Razor 页面是基于页面的编程模型,可简化 Web UI 的生成并提高工作效率。

③ Razor 标记提供了适用于 Razor 页面和 MVC 视图的高效语法。

④ 标记帮助程序使服务器端代码可以在 Razor 文件中参与创建和呈现 HTML 元素。

⑤ 内置的多数据格式和内容协商支持使 Web API 可访问多种客户端,包括浏览器和移动设备。

⑥ 模型绑定自动将 HTTP 请求中的数据映射到操作方法参数中。

⑦ 模型验证自动执行客户端和服务器端验证。

在继续后面的学习之前，先做一些准备工作，创建一个新项目，本章中的示例都将在该项目中实现。具体步骤如下：

① 启动 Visual Studio 2017，在顶部菜单栏选择"文件"→"新建"→"项目"菜单项。

② 在"新建项目"对话框中会显示几个项目模板。模板包含给定项目类型所需的基本文件和设置。在"新建项目"对话框左侧的窗格中，展开"Visual C♯"，然后选择".NET Core"。在中间窗格中，选择"ASP.NET Core Web 应用程序"，在名称文本框中输入"RazorMvcDemo"，然后单击"确定"按钮，显示如图6.2所示对话框。

图 6.2　创建 ASP.NET Core Web 应用程序

③ 在 Visual Studio 2017 的"解决方案资源管理器"中查看"RazorMvcDemo"项目的结构，如图6.3所示。

④ 在程序包管理器控制台中依次输入以下三条指令，安装 NuGet 包。

```
Install-Package Microsoft.EntityFrameworkCore -version 2.0.3
Install-Package Microsoft.EntityFrameworkCore.SqlServer -version 2.0.3
Install-Package Microsoft.EntityFrameworkCore.Tools -version 2.0.3
```

⑤ 在 Visual Studio 2017 的"解决方案资源管理器"中的"Models"目录下创建"Cargo"和"EFCoreDemoContext"两个类文件。"EFCoreDemoContext"类必须继承于"System.Data.Entity.DbContext"类以赋予其数据操作能力。程序代码如下：

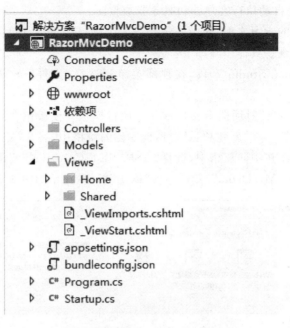

图 6.3 RazorMvcDemo 项目结构

```
using System;
using System.Collections.Generic;
using System.ComponentModel.DataAnnotations;
using System.ComponentModel.DataAnnotations.Schema;
using System.Linq;
using System.Threading.Tasks;

namespace RazorMvcDemo.Models
{
    public class Cargo
    {
        [Key]
        [DatabaseGeneratedAttribute(DatabaseGeneratedOption.Identity)]
        public int Id { get; set; }
        public string Name { get; set; }
        public string Code { get; set; }
        public string Country { get; set; }
        public int MinStockQty { get; set; }
        public decimal GrossWt { get; set; }
        public string Unit { get; set; }
    }
}
```

```csharp
using Microsoft.EntityFrameworkCore;
using System;
using System.Collections.Generic;
using System.Linq;
using System.Threading.Tasks;

namespace RazorMvcDemo.Models
{
    public partial class EFCoreDemoContext : DbContext
    {
        public virtual DbSet<Cargo> Cargo { get; set; }
        public EFCoreDemoContext(DbContextOptions<EFCoreDemoContext> options): base(options)
        {
        }
    }
}
```

⑥ 在 Visual Studio 2017 的资源管理器中找到 appsettings.json 文件并双击打开,在文件中添加一个连接字符串,代码如下:

```json
{
    "Logging": {
        "IncludeScopes": false,
        "LogLevel": {
            "Default": "Warning",
            "Microsoft": "Warning"
        }
    },
    "ConnectionStrings": {
        "EFCoreDemoContext": "Server=.\\sqlexpress;Database=EFCoreDemo;Trusted_Connection=True;MultipleActiveResultSets=true"
    }
}
```

⑦ 在 Visual Studio 2017 的资源管理器中找到 startup.cs 文件并双击打开,在 ConfigureServices 方法中写入依赖注入容器注册数据库上下文的代码,具体代码如下:

```csharp
public void ConfigureServices(IServiceCollection services)
{
    services.AddDbContext<EFCoreDemoContext>(options => options.UseSqlServer(Configuration.GetConnectionString("EFCoreDemoContext")));
    services.AddMvc();
}
```

⑧ 在 Visual Studio 2017 的程序包管理器控制台上依次执行以下命令：

Add-Migration RazorDemoInit
Update-Database

⑨ 在 Visual Studio 2017 的"解决方案资源管理器"中右击选中 Views 文件夹，在弹出的快捷菜单中选择"添加"→"新建文件夹"菜单项，创建一个新文件夹，并把文件夹名称命名为 Cargos。

⑩ 在 Visual Studio 2017 的"解决方案资源管理器"中右击选中 Controllers 文件夹，在弹出的快捷菜单中选择"添加"→"控制器"菜单项，在弹出的对话框中选择"视图使用 Entity Framework 的 MVC 控制器"，同时会在 Views 目录中创建 Cargoes 目录，如图 6.4 所示。

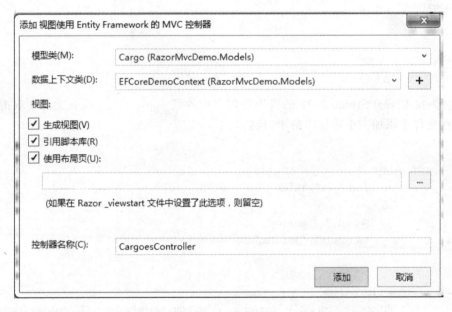

图 6.4　创建控制器

⑪ 在 SQL Server Management Studio 中执行以下 SQL 语句，插入几条测试数据：

```
insert cargo(name,code,country,grosswt,minstockqty,unit)
select '小方巾,宜品牌','T0001',142,0.1,100,'011'
union
select '垫子','T0002',142,0.3,80,'007'
union
select '杯子,宜品牌','T0003',142,0.2,120,'011'
```

6.2 ASP.NET Core 中的路由

6.2.1 路 由

ASP.NET Core MVC 建立在 ASP.NET Core 的路由之上。ASP.NET Core 路由是一个功能强大的 URL 映射组件,用于生成易于理解和可搜索的 URL,使用它可根据搜索引擎优化(SEO)规则对 URL 进行有效设计,使 URL 易于被主流的搜索引擎收录,而不用考虑如何组织 Web 服务器上的文件。可以使用支持路由值约束、默认值和可选值的路由模板语法来定义路由。

路由功能负责将传入请求映射到路由处理程序。路由在 ASP.NET 应用中定义,并在应用启动时进行配置。路由通过对请求的拦截和对请求 URL 的解析,得到以 Controller 和 Action 名称为核心的值,然后这些值便可用于处理请求。使用 ASP.NET 应用中的路由信息,路由功能还能生成要映射到路由处理程序的 URL。因此,路由可以根据 URL 查找路由处理程序,或者根据路由处理程序信息查找给定路由处理程序对应的 URL。

路由中保留了如表 6.1 所列的关键字,它们不能用作路由名称或参数。

表 6.1 保留关键字

序 号	关键字	序 号	关键字
1	action	4	handler
2	area	5	page
3	controller		

6.2.2 路由基础知识

路由主要实现以下功能:
① 将传入请求映射到路由处理程序中;
② 生成 URL 用来响应控制器操作。

通常情况下,每个应用都含有一个路由集合。当请求到达时,将按顺序处理路由集合。传入请求通过对路由集合中的每个可用路由调用 RouteAsync 方法来查找与请求 URL 匹配的路由。与此相反,响应可根据路由信息使用路由来生成 URL(例如,用于重定向或链接),因此,要避免在 URL 中进行硬编码,以提高可维护性。

路由通过路由中间件连接到中间件管道。ASP.NET Core MVC 向中间件管道添加路由,作为其配置的一部分。

1. 将 URL 映射到路由处理程序

传入的请求将进入路由集合,路由中间件将对路由集合中的每个路由调用异步方法 RouteAsync。IRouter 实例将通过将 RouteContext.Handler 设置为非空的 RequestDelegate 来选择是否处理请求。如果路由为请求设置处理程序，则路由处理将停止，处理程序将被调用以处理该请求。如果所有的路由都执行了，且请求未找到任何处理程序，那么中间件将调用 next 方法，使请求管道中的下一个中间件被调用。

RouteAsync 的主要输入是与当前请求关联的 RouteContext.HttpContext；在一个路由成功匹配之后，输出 RouteContext.Handler 和 RouteContext.RouteData。

在 RouteAsync 方法执行期间，在路由匹配成功的情况下，会得到一个用于封装路由信息的 RouteData 对象，同时把该对象设置到 RouteContext.RouteData 属性上。RouteContext.RouteData 包含了以下 3 个重要的匹配结果信息，如图 6.5 所示：

① RouteData.Values 是路由对象通过对请求（URL）解析得到的。

② RouteData.DataTokens 是一个路由与 URL 匹配之后直接附加到路由对象上的自定义变量。

③ RouteData.Routers 是一个成功匹配请求的路由列表。

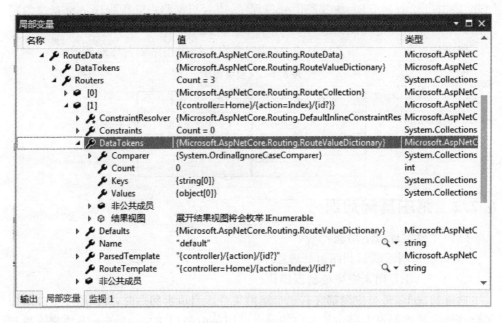

图 6.5　RouteData

2. 创建路由

通过基于约定的路由，可以对应用程序可接受的 URL 格式以及每个格式映射到指定控制器上的特定的操作方法进行全局定义。当接收传入请求时，路由引擎分析 URL 并将其匹配到所定义的路由上，然后调用关联控制器的操作方法。

路由提供 Route 类作为 IRouter 的标准实现。当 GetVirtualPath 方法被执行时,如果定义在 URL 模板中的变量与指定变量列表相匹配,则用指定的路由变量值替换 URL 模板中的变量占位符以生成一个 URL。

默认情况下,针对请求 URL 的路由,可通过调用 MapRoute 方法或定义在 IRouteBuilder 接口上的一个类似的扩展方法来完成。所有的这些方法都将创建一个 Route 的实例,并将其添加到路由集合中。

请求的地址与对应的 URL 模板的匹配规则将通过下面一个简单的 MapRoute 调用示例来说明:

```
app.UseMvc(routes =>
{
    routes.MapRoute(
        name: "default",
        template: "{controller=Home}/{action=Index}/{id?}");
});
```

此模板将匹配类似"/Cargoes/Edit/2"的 URL 路径,如图 6.6 所示,并提取路由值{controller = Cargoes, action = Edit, id = 2},将其替换为相应的路由参数。对于上述这个 URL 模板来说,通过分隔符"/"对其进行拆分得到 3 个基本的字符串,并将它们称为"段"。对于组成某个段的内容,又可以分为"路由参数"和"文本",前者采用大括号"{}"对变量名进行封装表示,后者则代表单纯的文字。

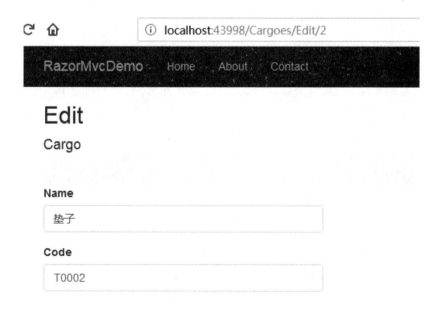

图 6.6　路由地址

对于一个试图进行匹配的 URL 地址来说,匹配成功需要具备两个基本条件,即

URL 地址中包含的段的数量与 URL 模板中的数量相同，以及对应的文本段内容也一样。如果 URL 模板中有默认值，则当进行 URL 地址匹配时，若 URL 地址只能匹配 URL 模板的前面部分，而后面部分均为路由参数段且是具体对应的默认值或可选路由参数（例如上例 id 中的路由参数），则这种情况下依然被认为匹配成功。例如：/Cargoes/Index。

下面把上例中的路由代码修改如下：

```
routes.MapRoute(
    name: "default",
    template: "{controller=Home}/{action=Index}/{id:int}");
```

此模板将匹配类似"/Cargoes/Edit/2"而不是"/Cargoes/Edit/Cup"的 URL 路径。路由参数定义"{id:int}"是为 id 路由参数定义的路由约束。在此示例中，路由值 id 必须可转换为整数。

MapRoute 的其他重载方法可以接受 constraints、dataTokens 和 defaults 的值，这些参数值被定义为 object 类型。这些参数的典型用法是传递匿名类型化对象，其中匿名类型的属性名匹配路由参数名称。下面是一个添加路由约束和数据令牌的示例：

```
routes.MapRoute(
    name: "us_english_cargoes",
    template: "en-US/Cargoes/{id}",
    defaults: new { controller = "Cargoes", action = "Edit" },
    constraints: new { id = new IntRouteConstraint() },
    dataTokens: new { locale = "en-US" });
```

此模板与"/en-US/Cargoes/1"等 URL 路径相匹配，如图 6.7 所示，并提取路由值{controller＝Cargoes，action＝Edit，id＝1}和数据令牌{locale＝en-US}。

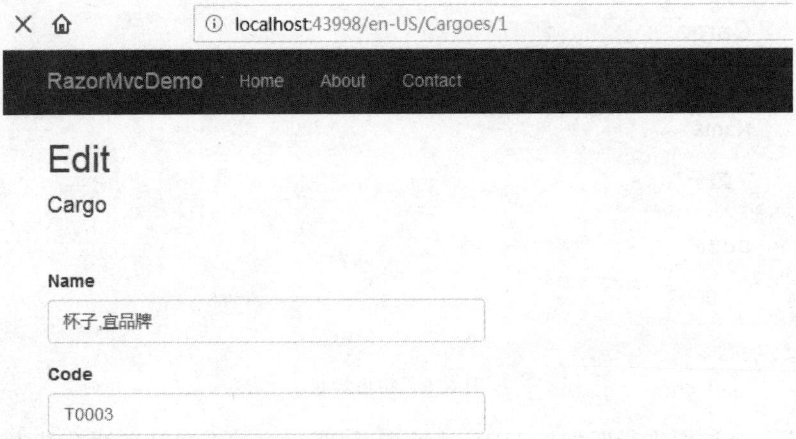

图 6.7　URL 路径

下面对特性路由进行简单介绍。特性路由是一种指定路由的方法,可将路由信息通过注解添加到控制类或操作方法上。

6.2.3 路由模板

下面针对上例中的路由模板来学习一下路由模板的相关知识。先来看下面这个路由模板:

{controller=Home}/{action=Index}/{id?}

路由模板中的大括号({})定义了路由参数的边界。可以在一个路由模板中定义多个路由参数,但它们必须用"/"分开,例如"{controller=Home}{action=Index}"不是一个有效路由,因为在"{controller}"与"{action}"之间没有"/"。这些路由参数必须有一个名称,并可以有附加指定的属性。

路由参数(例如,{action})可以有默认值,定义的方式是在参数名称后定义默认值,并用等号"="分开。例如,{controller=Home}将 Home 作为 controller 的默认值。如果在 URL 中路由参数没有值,则将使用默认值。除默认值外,路由参数可定义为可选的(通常是在参数名称后加一个"?",比如"id?")。"可选"与"默认值"之间的区别是,使用默认值的路由参数始终会生成一个值;而可选的参数只有在 URL 提供了值时才会有值。如果要在路由参数中使用"{或}",则可通过重复该字符({{或}})对其进行转义。常用的路由模板及其示例和说明如表 6.2 所列。

表 6.2 常用的路由模板、示例、说明

路由模板	匹配 URL 示例	说　明
Index	/Index	仅匹配单个路径 /Index
{Page=Home}	/	匹配并将 Page 设置为 Home
{Page=Home}	/Contact	匹配并将 Page 设置为 Contact
{controller}/{action}/{id?}	/Products/List	映射到 Products 控制器和 List 操作
{controller}/{action}/{id?}	/Products/Details/123	映射到 Products 控制器和 Details 操作;将 id 设置为 123
{controller=Home}/{action=Index}/{id?}	/	映射到 Home 控制器和 Index 方法;忽略 id

一般来说,使用路由最简单的方法是使用路由模板。除此之外,还可在路由模板外指定约束和默认值。

6.2.4 路由约束

路由参数也可以有约束,为了确保用户在请求中填写的值是程序所需要的类型,可以通过在路由参数名后增加一个冒号":"和约束名来定义一个内联约束(例如{id:int})。如果请求地址中的内容不符合相关变量段的约束条件,则意味着对应的路由对象与之不匹配。

在实际应用中,应该避免使用约束来做验证,这样做意味着非法输入找不到一个具体的目标文件而返回 404 错误结果,而不是一个含有相应错误消息的 400 状态码的错误。路由约束用来消除类似路由间的歧义,而不是用来验证特定路由的输入。表 6.3 演示了某些路由约束及其预期的行为。

表 6.3　演示某些路由约束及其预期行为

约束	示例	匹配项示例	说明
int	{id:int}	123 456 789, -123 456 789	匹配任何整数
bool	{active:bool}	true, false	匹配 true 或 false(区分大小写)
datetime	{dob:datetime}	2018-12-31, 2018-12-31 7:32pm	匹配有效的 datetime 值
decimal	{price:decimal}	49.99, -1 000.01	匹配有效的 decimal 值
double	{weight:double}	1.234, -1 001.01e8	匹配有效的 double 值
float	{weight:float}	1.234, -1 001.01e8	匹配有效的 float 值
guid	{id:guid}	CD2C1638-1638-72D5-1638-DEADBEEF1638, {CD2C1638-1638-72D5-1638-DEADBEEF1638}	匹配有效的 guid 值
long	{ticks:long}	123 456 789, -123 456 789	匹配有效的 long 值
minlength(value)	{username:minlength(4)}	Rick	字符串必须至少为 4 个字符
maxlength(value)	{filename:maxlength(8)}	Richard	字符串不得超过 8 个字符
length(length)	{filename:length(12)}	somefile.txt	字符串必须正好为 12 个字符
length(min,max)	{filename:length(8,16)}	somefile.txt	字符串必须至少为 8 个字符,且不得超过 16 个字符
min(value)	{age:min(18)}	19	整数值必须至少为 18
max(value)	{age:max(120)}	91	整数值不得超过 120
range(min,max)	{age:range(18,120)}	91	整数值必须至少为 18,且不得超过 120

续表 6.3

约　束	示　例	匹配项示例	说　明
alpha	{name:alpha}	Rick	字符串必须由一个或多个字母字符(a—z,区分大小写)组成
regex(expression)	{ssn:regex(^\\d{{3}}-\\d{{2}}-\\d{{4}}$)}	123-45-6789	字符串必须匹配正则表达式(参见6.2.5 小节有关定义正则表达式的提示)
required	{name:required}	Rick	用于强制在 URL 生成过程中存在非参数值

6.2.5　正则表达式

ASP.NET Core 框架将向正则表达式构造函数添加 RegexOptions.IgnoreCase｜RegexOptions.Compiled｜RegexOptions.CultureInvariant。

路由中使用的正则表达式通常以"^"字符(匹配字符串的起始位置)开头,以"$"字符(匹配字符串的结束位置)结尾。"^"和"$"字符可确保正则表达式匹配整个路由参数值。如果没有"^"和"$"字符,那么正则表达式将匹配字符串内的所有子字符串,这些字符串通常不是你想要的。表 6.4 显示了部分示例,并说明了它们为何匹配或未能匹配。

表 6.4　正则表达式

表达式	字符串	匹配	注　释
[a—z]{2}	hello	是	子字符串匹配
[a—z]{2}	123abc456	是	子字符串匹配
[a—z]{2}	mz	是	匹配表达式
[a—z]{2}	MZ	是	不区分大小写
^[a—z]{2}$	hello	否	参阅上述"^"和"$"的说明
^[a—z]{2}$	123abc456	否	参阅上述"^"和"$"的说明

若想将参数限制为一组已知的可能值,则可使用正则表达式。例如,{action:regex(^(Index|edit|create)$)}仅将 action 路由值匹配到 Index、edit 或 create,结果如图 6.8 所示。具体路由代码如下:

```
routes.MapRoute(
    name: "default",
    template: "{controller = Home}/{action:regex(^(Index|edit|create)$)}/{id?}"
);
```

图 6.8　正则表达式匹配路由

如图 6.9 所示,当将路由值匹配到 delete 时,没有任何显示。

图 6.9　路由值无法匹配

6.3　ASP.NET Core 中的模型绑定

6.3.1　模型绑定简介

　　ASP.NET Core MVC 中的模型绑定可将客户端请求数据(窗体值、路由数据、查询字符串参数、HTTP 头)转换到控制器可以处理的对象中。因此,控制器逻辑不必找出传入的请求数据,而只需对其操作方法参数中的数据进行处理即可。这些参数可能是简单类型的参数,如字符串、整数或浮点数,也可能是复杂类型的参数。这是 MVC 的一项强大功能,不管数据的大小和复杂性,将传入数据映射到对应位置都是经常重复的情况。MVC 通过抽象绑定解决了这一问题,使开发者不必在每个应用中反复编写类似的代码。

6.3.2　模型绑定的工作原理

　　当 MVC 收到 HTTP 请求时,会将此请求路由到控制器的某个特定操作方法上。MVC 首先基于路由数据中的内容来决定要运行的操作方法,然后将 HTTP 请求中的值绑定到该操作方法的参数上。现以下列 URL 为例进行解释:

http://localhost:43998/Cargoes/edit/2

由于路由模板为 {controller = Home}/{action = Index}/{id?},因此,当请求 Cargoes/edit/2 传入时,这个路由模板将请求匹配到 Cargoes 控制器及其 Edit 操作方法上。此外,MVC 用参数(id)的名称在请求(Cargoes/edit/2)中找 id 值,并转换成 Eidt 方法的 id 参数。操作方法的代码如下所示:

```
public async Task<IActionResult> Edit(int? id)
{
    if (id == null)
    {
        return NotFound();
    }
    var cargo = await _context.Cargo.SingleOrDefaultAsync(m => m.Id == id);
    if (cargo == null)
    {
        return NotFound();
    }
    return View(cargo);
}
```

由于模型绑定请求名为 id 的键,而窗体值中没有任何名为 id 的键,因此,它将转到路由值以查找该键。在本示例中,该键是一个匹配项,当发生绑定时,该键值被转换为整数 2。

注意:URL 路由中的字符串不区分大小写。

MVC 尝试按名称将请求数据绑定到操作参数上,并使用参数名称和其公共可设置属性的名称查找每个参数的值。在以上示例中,唯一的操作参数名为 id,MVC 会将此参数绑定到路由值中具有相同名称的值上。除路由值外,MVC 还会绑定来自请求各个部分的数据,并按一定顺序执行此操作。模型绑定数据源如表 6.5 所列。

表 6.5 模型绑定数据源

客户端请求数据	说 明
Form values	这些是使用 POST 方法进入 HTTP 请求的窗体值(包括 jQuery POST 请求)
Route values	路由提供的路由值集
Query strings	URL 的查询字符串部分

注意:窗体值、路由数据和查询字符串均存储为键/值对。

此示例中 Edit 方法使用的是简单类型。所谓简单类型指任何 .NET 基元类型或包含字符串类型转换器的类型。如果操作方法(如 Add(Cargo c))的参数是复杂类型的类,那么 MVC 的模型绑定也能进行处理,MVC 模型绑定使用反射和递归来遍历复杂类型以寻找匹配的属性。下面先来看一段 MVC 生成的页面代码:

```
<form action = "/cargoes/Edit" method = "post">
    <div class = "form-group">
        <label class = "control-label" for = "Name">Name</label>
        <input class = "form-control" type = "text" id = "Name" name = "Name" value = "" />
    </div>
    <div class = "form-group">
        <label class = "control-label" for = "Code">Code</label>
        <input class = "form-control" type = "text" id = "Code" name = "Code" value = "" />
    </div>
</form>
```

通过上面的代码可以看到,每一个表单的输入项都匹配 Cargo 的属性,Code 值的 input 元素名称就是 Code,Name 值的 input 元素名称就是 Name,MVC 会根据命名约定进行模型绑定。为了进行绑定,类必须具有公开的默认构造函数,并且要绑定的成员必须是公共可写属性。MVC 运行环境中的模型绑定器会首先检查 Cargo 类,然后查找能用于绑定的所有 Cargo 属性。按照 MVC 中的约定,模型绑定器自动将请求的值转换并写入一个 Cargo 对象中。通俗一点来说,当模型绑定器看到Cargo 的 Code 属性时,它就在请求中查找名为"Code"的参数。请注意,这里是说"在请求中"而不是"在表单集合中"。最终结果是模型绑定 Cargo 对象,如图 6.10 所示。

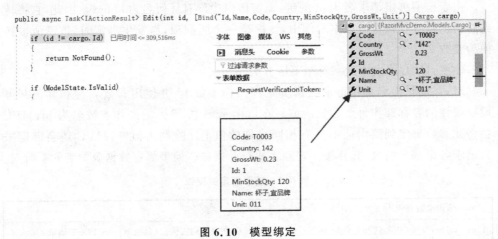

图 6.10　模型绑定

当一个参数被绑定时,模型绑定将停止查找具有该名称的值并开始绑定下一个参数。

如果绑定失败,MVC 不会引发错误。在接受用户输入的每个操作时均应检查 ModelState.IsValid 属性。

注意:控制器的 ModelState 属性中的每个 Entry 都是一个包含了 Errors 属性的 ModelStateEntry。

此外,MVC 在执行模型绑定时必须考虑以下一些特殊数据类型:

- IFormFile、IEnumerable〈IFormFile〉：一个或多个通过 HTTP 请求上传的文件。
- CancellationToken：用于在异步控制器中取消活动。

这些类型可以绑定到操作参数或一个类的属性上。

模型绑定完成后将进行验证。默认模型的绑定适合大多数开发场景；它还可以扩展，若有特殊需求，则可自定义内置行为。

6.3.3 数组绑定

上面学习了模型绑定的工作原理，下面通过一个数组绑定的示例来加深理解。如果绑定的目标对象是一个简单类型的数组，则匹配的同名数据项将会被提取出来并转换成目标对象的元素，作为输入数据源。下面修改 Edit.cshtml 页面中的代码如下：

```
@{
    ViewData["Title"] = "数组模型绑定";
}
<h2>数组模型绑定</h2>
<h3>@ViewBag.Tays</h3>
<form asp-action="Edit">
    <div class="form-group">
        <input name="Tay" type="text" />
    </div>
    <div class="form-group">
        <input name="Tay" type="text" />
    </div>
    <div class="form-group">
        <input name="Tay" type="text" />
    </div>
    <div class="form-group">
        <input type="submit" value="Save" class="btn btn-default" />
    </div>
</form>
```

现在，在 HomeController 中定义一个如下的 Action 的方法 Edit，它具有一个数组类型的参数 Tay，元素类型为字符串。如果针对该 Action 的请求中包含了如上〈input〉元素的表单，那么上面一组〈input〉元素的值会被绑定到这个匹配的参数上。Edit 方法的代码如下：

```
[HttpPost]
public IActionResult Edit(string[] Tay)
{
```

```
if (Tay.Length>0)
{
    foreach(var item in Tay)
    {
        ViewBag.Tays += string.Format("Tay:{0};",item);
    }
}
else
{
    ViewBag.Tays = "没有数组";
}
return View();
}
```

6.3.4 返回带格式的数据

ASP.NET Core MVC 可以返回带格式的数据，包括 JSON、XML 和许多其他格式。ASP.NET Core MVC 运行时向格式化程序委托读取流的责任，MVC 会使用一组已配置的格式化程序，并基于请求数据的内容类型对请求数据进行处理。默认情况下，MVC 包括用于处理 JSON 数据的 JsonInputFormatter 类（默认基于 JSON.NET 进行格式化，如图 6.11 所示），但可以添加用于处理 XML 和其他自定义格式的其他格式化程序。

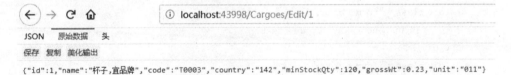

图 6.11　JSON 格式数据

如果想要使用 XML 或其他格式，则必须在 Startup.cs 文件中配置该格式，方法是：

① 使用 NuGet 获取对 Microsoft.AspNetCore.Mvc.Formatters.Xml 的引用。

② 修改 Startup.cs 的 ConfigureService 方法中的代码，此处将添加 XML 格式化程序作为 MVC 为此应用提供的服务。通过传递给 AddMvc 方法的 options 参数，可在应用启动后添加和管理筛选器、格式化程序和其他 MVC 系统选项。然后，将 Consumes 特性应用于控制器类或操作方法，以使用所需的格式，代码如下：

```
public void ConfigureServices(IServiceCollection services)
{
    services.AddMvc()
```

```
        .AddXmlSerializerFormatters();
}
```

③ 修改控制器中的 Edit 方法,代码如下:

```
public IActionResult Edit(int? id)
{
    if (id == null)
    {
        return NotFound();
    }
    var cargo = _context.Cargo.SingleOrDefaultAsync(m => m.Id == id);
    if (cargo == null)
    {
        return NotFound();
    }
    return Ok(cargo.Result);
}
```

④ 在浏览器中进行浏览,结果如图 6.12 所示。

图 6.12 XML 格式数据

6.4 ASP.NET Core MVC 中的模型验证

6.4.1 模型验证简介

在应用将数据存储到数据库之前必须先验证数据,检查数据是否存在潜在的安全威胁,以确保数据已设置适当的类型和大小格式,且必须符合相关规则。进行验证的过程可能有些单调乏味,却必不可少。ASP.NET Core MVC通过使用数据注释验证特性修饰模型对象的属性来支持验证。验证特性是在数据值发布到服务器前在客户端上进行检查,并在调用控制器操作前在服务器上进行检查。

框架处理客户端和服务器上的验证请求数据。在模型类型上指定的验证逻辑作为非介入式注释添加到所呈现的视图上,并使用jQuery验证在浏览器中强制执行。

在调用每个控制器操作之前都会执行模型验证,由操作方法负责检查ModelState.IsValid并做出正确反应。在许多情况下,正确的反应是返回错误响应,理想状况下会详细说明模型验证失败的原因。

模型状态表示已提交的HTML表单值中的验证错误,错误数上限默认为200个,可以通过向Startup.cs文件中的ConfigureServices方法插入以下代码来配置该数字:

```
services.AddMvc(options => options.MaxModelValidationErrors = 50);
```

6.4.2 验证特性

验证特性用于配置实体验证,在System.ComponentModel.DataAnnotations命名空间下定义了一系列具体的验证特性,它们可以直接应用在实体类的某个属性上对目标成员进行验证,包括诸如数据类型或必填字段之类的约束。其他类型的验证包括诸如对模型数据强制实施业务规则,比如对信用卡、电话号码或电子邮件地址的验证。通常可以在模型对象的属性上应用相应的验证特性来定义相应的验证规则。

下面是一个应用中已经添加验证特性的Cargo类,该模型类用于存储货物信息,其中有的属性是必需属性,有的属性对于字符串长度有要求。此外,还有一个针对GrossWt属性设置的对0~9 999.99数值范围的限制,以及一个自定义验证特性。

```
using System;
using System.Collections.Generic;
using System.ComponentModel.DataAnnotations;
using System.ComponentModel.DataAnnotations.Schema;
using System.Linq;
using System.Threading.Tasks;

namespace RazorMvcDemo.Models
{
    public class Cargo
    {
        [Key]
        [DatabaseGeneratedAttribute(DatabaseGeneratedOption.Identity)]
        public int Id { get; set; }
        [Required]
```

```
        [StringLength(100)]
        public string Name { get; set; }
        [Required]
        public string Code { get; set; }
        [StringLength(10)]
        public string Country { get; set; }
        [CargoMinQty(10)]
        public int MinStockQty { get; set; }
        [Range(0, 9999.99)]
        public decimal GrossWt { get; set; }
        [MinLength(3)]
        public string Unit { get; set; }
    }
}
```

通过读取整个模型即可显示有关此应用的数据规则,从而使代码维护变得更轻松。下面是几个常用的内置验证特性:

- [CreditCard]:验证属性是否具有信用卡格式。
- [Compare]:验证某个模型中的两个属性是否匹配。
- [EmailAddress]:验证属性是否具有电子邮件格式。
- [Phone]:验证属性是否具有电话格式。
- [Range]:验证属性值是否落在给定范围内。
- [RegularExpression]:验证数据是否与指定的正则表达式匹配。
- [Required]:将属性设置为必需属性。
- [StringLength]:验证字符串属性是否最多具有给定的最大长度。
- [Url]:验证属性是否具有 URL 格式。

6.4.3 自定义验证

在某些情况下,内置验证特性可能无法满足业务所需的功能,自己的业务验证规则特定于自己的业务,如确保字段是必填字段且符合一系列条件值。这时,就可以通过继承 ValidationAttribute 来创建自定义验证特性。MVC 支持由 ValidationAttribute 派生的所有用于验证的特性。在继承 ValidationAttribute 之后需要重写 IsValid 方法以便定义验证逻辑。IsValid 方法使用两个参数,第一个是名为 value 的对象,第二个是名为 validationContext 的 ValidationContext 对象。Value 引用自定义验证程序要验证的字段中的实际值。

在下面的示例中,一项业务规则规定,用户不能将安全库存设置为低于 10。[CargoMinQty]属性会检查安全库存,如果安全库存低于 10,则验证失败。此属性采用一个表示安全库存的整数参数用于验证数据。可以在该属性的构造函数中捕获

该参数的值,代码如下:

```csharp
using Microsoft.AspNetCore.Mvc.ModelBinding.Validation;
using System;
using System.Collections.Generic;
using System.ComponentModel.DataAnnotations;
using System.Linq;
using System.Threading.Tasks;

namespace RazorMvcDemo.Models
{
    public class CargoMinQtyAttribute:ValidationAttribute, IClientModelValidator
    {
        private int _qty;
        public CargoMinQtyAttribute(int qty)
        {
            _qty = qty;
        }
        protected override ValidationResult IsValid(object value, ValidationContext validationContext)
        {
            Cargo cargo = (Cargo)validationContext.ObjectInstance;
            if (cargo.MinStockQty>_qty)
            {
                return new ValidationResult(GetErrorMessage());
            }
            return ValidationResult.Success;
        }
        private string GetErrorMessage()
        {
            return $"最低安全库存必须大于{_qty}.";
        }
    }
}
```

上面代码中的 cargo 变量表示一个 Cargo 对象,其中包含要验证的表单提交中的数据。在此例中,验证代码会根据规则检查 CargoMinQtyAttribute 类的 IsValid 方法中的安全库存数量。验证成功时,IsValid 返回 ValidationResult.Success 代码。验证失败时,IsValid 返回 ValidationResult 和错误消息,代码如下:

```csharp
private string GetErrorMessage()
{
    return $"最低安全库存必须大于{_qty}.";
}
```

当用户修改 MinStockQty 字段并提交表单时，CargoMinQtyAttribute 的 IsValid 方法将验证该安全库存是否小于 10。与所有内置特性一样，将 CargoMinQtyAttribute 应用于属性（如 MinStockQty）以确保执行验证，如图 6.13 所示。

图 6.13　自定义验证

6.4.4　客户端验证

客户端验证使用户不必将时间浪费于等待服务器往返。从商业角度而言，即使每次只有几分之一秒，但如果每天有几百次，也会耗费大量的时间和成本，带来很多不必要的烦恼。客户端验证能够节省时间，提高用户的工作效率和投入产出比。

使用客户端验证功能时，必须在视图中引用以下 JavaScript 脚本，具体代码如下：

〈script src = "https://ajax.aspnetcdn.com/ajax/jQuery/jquery - 2.2.0.min.js"〉〈/script〉

〈script src = "https://ajax.aspnetcdn.com/ajax/jquery.validate/1.16.0/jquery.validate.min.js"〉〈/script〉

〈script src = "https://ajax.aspnetcdn.com/ajax/jquery.validation.unobtrusive/3.2.6/jquery.validate.unobtrusive.min.js"〉〈/script〉

jQuery 非介入式验证脚本是一个基于 jQuery Validate 插件的自定义 Microsoft 前端库。如果没有 jQuery 非介入式验证，则必须在两个位置编写相同的验证逻辑代码：一个位置是在实体属性上的服务器端验证特性中，另一个位置是在客户端脚本中。

MVC 的标记助手和 HTML 帮助程序将实体属性中的验证特性和类型元数据以表单元素属性（HTML 5 "data -"）的方式输出到最终生成的 HTML 中；然后，jQuery 非介入式验证分析这些"data -"属性并将逻辑传递给 jQuery Validate，从而保证服务器端与客户端采用相同的验证规则。可以使用相关标记助手在客户端上显

示验证错误,示例代码如下:

```
<div class = "form-group">
    <label asp-for = "Code" class = "control-label"></label>
    <input asp-for = "Code" class = "form-control"/>
    <span asp-validation-for = "Code" class = "text-danger"></span>
</div>
```

标记助手将上面的代码生成以下 HTML 程序。请注意,HTML 输出中的"data-"特性与 Code 属性的验证特性相对应。从下面的 HTML 中可以看到,Code 属性对应<input>元素生成了一个"data-val"属性和以"data-val-"为前缀的属性,前者表示是否需要对用户输入的值进行验证,后者代表相应的验证规则。下面的"data-val-required"属性包含在用户未填写 Code 字段时将显示的错误消息中。jQuery 非介入式验证将此值传递给 jQuery Validate required()方法,该方法随后在随附的<span>元素中显示验证失败后的错误消息。代码如下:

```
<form action = "/cargoes/Edit/1" method = "post">
    <divc lass = "form-group">
    <label class = "control-label" for = "Code">Code</label>
    <input class = "form-control" type = "text" data-val = "true" data-val-required = "The Code field is required." id = "Code" name = "Code" value = "T0003" />
    <span class = "text-danger field-validation-valid" data-valmsg-for = "Code" data-valmsg-replace = "true"></span>
    </div>
</form>
```

如果没有通过客户端验证,则客户端将无法提交表单,直到表单中的内容通过了客户端的验证为止。"提交"按钮运行的 JavaScript 脚本的功能是:要么提交表单,要么显示错误消息。客户端验证界面如图 6.14 所示。

图 6.14 客户端验证

MVC 用基于属性的.NET 数据类型来确定类型特性值(可使用[DataType]特性进行重写)。[DataType]基本特性不执行真正的服务器端验证。浏览器选择自己的错误消息,并根据需要显示这些错误,但 jQuery 非介入式验证包可以重写消息,并使它们与其他消息的显示保持一致。当用户应用[DataType]子类时,最常发生这种情况。

6.4.5 远程验证

远程验证是一项非常好的功能,当需要在客户端上使用服务器上的数据进行验证时,客户端通过 Ajax 请求向服务器端验证数据。例如,应用可能需要验证某个电子邮件或用户名是否已被使用,并且必须为此查询大量数据。为验证一个或几个字段而下载大量数据会占用过多资源,并且还有可能暴露敏感信息。一种替代方法是发出往返请求来验证字段。

可以分两步实现远程验证。第一步,必须将[Remote]特性添加到相应模型类的属性上。[Remote]特性采用多个重载,可用于将客户端的 JavaScript 脚本定向到要调用的相应代码上。下面的示例代码指向 Cargoes 控制器的 VerifyCode 操作方法。

```
[Remote(action: "VerifyCode", controller: "Cargoes")]
publicstring Code { get; set; }
```

第二步,按照[Remote]特性中的定义,将验证代码放入相应的操作方法中。jQuery Validate remote()方法的文档有以下论述:服务器端响应必须是使用默认错误消息的 JSON 字符串,如果验证通过,则返回值必须为 true;如果验证不通过,则返回值可以为 false、undefined 或 null。如果服务器端响应是一个字符串,例如,"货物代码 {code} 已经存在",则此字符串将显示为自定义错误消息,以取代默认错误消息。

VerifyCode()方法的定义遵循以下规则:如果货物代码已被占用,则它会返回验证错误消息;如果货物代码可用,则它返回 true,并将结果包装在 JsonResult 对象中。然后,客户端可以使用返回的值,继续进行下一步操作或根据需要显示错误。远程验证的示例代码如下:

```
[AcceptVerbs("Get", "Post")]
public IActionResult VerifyCode(string code)
{
    var cargos = _context.Cargo.Where(m => m.Code == code);
    if (cargos.Count() >= 1)
    {
        return Json($"货物代码 {code} 已经存在.");
    }
    return Json(true);
}
```

现在，当用户输入货物代码时，视图中的 JavaScript 脚本会发出远程调用，以了解该货物代码是否已被占用，如果是，则显示错误消息；如果不是，用户就可以像往常一样提交表单。远程验证界面如图 6.15 所示。

图 6.15　远程验证

以上验证方法存在一个缺点，即如果对货物信息进行修改，则也会报出图 6.15 中的错误。为了避免此错误，就要用到[Remote]特性的 AdditionalFields 属性，该属性可用于根据服务器上的数据来验证字段组合。例如，如果上面的 Cargo 类中使用两个附加属性进行验证，一个为 Code，另一个为 Id，那么货物代码只校验除当前已经在修改的货物信息之外的记录。可按照以下代码定义新特性：

```
[Remote(action: "VerifyCode", controller: "Cargoes", AdditionalFields = "Id")]
public string Code { get; set; }
```

AdditionalFields 可能已显式设置为字符串"Id"。然后，用于执行验证的操作方法必须使用两个参数，一个用于 Code 的值，一个用于 Id 的值。改进的代码如下：

```
[AcceptVerbs("Get", "Post")]
public IActionResult VerifyCode(string code,int id)
{
    var cargos = _context.Cargo.Where(m => m.Code == code);
    if (cargos.Count()>1)
    {
        return Json( $ "货物代码 {code} 已经存在.");
    }
    if (cargos.Count() == 1)
    {
        Cargo c = cargos.First();
        if (c.Id != id)
        {
```

```
                return Json($"货物代码{code}已经存在.!.");
            }
        }
        return Json(true);
    }
```

现在，当用户输入货物代码时，JavaScript 脚本会发出远程调用，以了解该货物代码是否已被占用：

- 如果被占用，则显示一条错误消息，如图 6.16 所示；
- 如果未被占用，则用户可以提交表单。

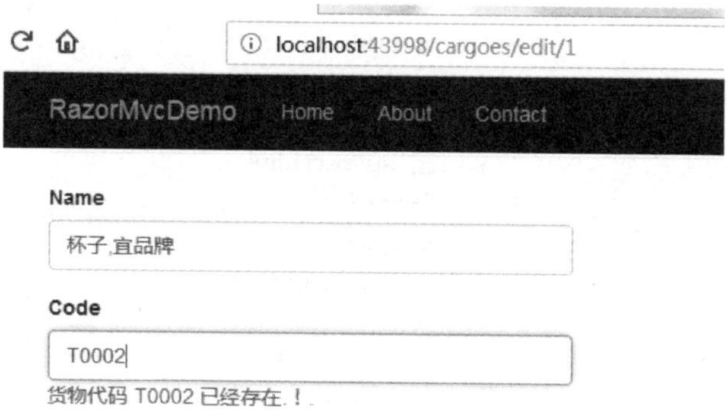

图 6.16　改进的远程验证

6.5　ASP.NET Core MVC 中的视图

6.5.1　Razor 视图引擎

ASP.NET Core MVC 视图使用 Razor 视图引擎呈现视图。在"模型-视图-控制器(MVC)"模式中，视图用来处理应用的数据表示和用户交互。Razor 是一种精简、富有表现力且流畅的模板标记语言，最大限度地减少了语法和额外的字符。Razor 用于在服务器上动态生成 Web 内容，且可以完全将服务器代码与客户端内容和代码混合。使用 Razor 视图引擎可以定义布局、分部视图和视图组件。

在 ASP.NET Core MVC 中，视图在 Razor 标记中使用 C#语言的.cshtml 文件。通常，视图文件会被分组到以每个应用的控制器命名的文件夹中，此文件夹存储在应用根目录的"Views"文件夹中，如图 6.17 所示。

主页控制器由"Views"文件夹内的"Home"文件夹表示。"Home"文件夹包含"关于"、"联系人"和"索引"（主页）网页的视图。当用户请求这三个网页中的一个时，主页控制器中的控制器操作决定使用三个视图中的哪一个来生成网页并将其返回给用户。

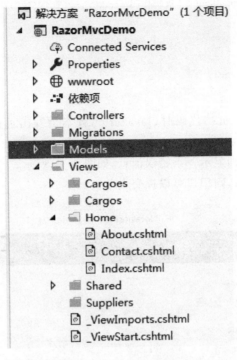

图 6.17　Views 文件夹

　　使用布局提供一致的网页外观并减少代码重复。布局通常包含页眉、导航和菜单元素以及页脚。页眉和页脚通常包含许多元数据元素的标记以及脚本和样式表的链接。

　　分部视图通过管理视图的可重用部分来减少代码重复。例如，分部视图可用于在多个视图中出现的博客网站上的作者简介。作者简介是普通的视图内容，不需要执行代码，可以仅通过模型绑定就能在网页中生成作者简介的内容，因此，这种内容类型使用分部视图是一个理想的选择。

　　视图组件与分部视图的相似之处在于它们可以减少重复性代码，但视图组件还适用于需要在服务器上运行代码才能呈现网页的视图内容。当呈现的内容需要与数据库交互时（例如网站购物车），视图组件就非常有用了。为了生成网页输出，视图组件不局限于模型绑定。

6.5.2　使用视图的好处

　　控制器与模型是开发一个可维护 Web 应用程序的基础，但是用户在使用浏览器访问 Web 应用程序时，首先看到的是视图，用户与 Web 应用程序的交互过程都是从视图开始。ASP.NET MVC 视图采用 Razor 语法在 HTML 标记和服务端逻辑之间进行轻松切换。可以通过布局和共享指令或局部视图对应用的用户界面中重复的外观轻松地进行复用。

视图通常按应用功能进行分组,因此更易于维护;应用程序的应用功能松散耦合,因此可以生成和更新独立于业务逻辑和数据访问组件的应用视图;也可以只修改应用的视图,而不必更新应用的其他部分;通过视图更容易测试应用的用户界面部分。

6.5.3　创建视图

属于某个控制器的视图创建在 Views/[ControllerName]文件夹中,在控制器之间共用的视图则放在 /Views/Shared 文件夹中。将视图文件命名为与其关联的控制器操作一样的名字,并添加 .cshtml 文件扩展名。例如,为 Home 控制器的 About 操作创建一个视图,则应该在 /Views/Home 文件夹下创建一个 About.cshtml 文件,代码如下:

```
@{
    ViewData["Title"] = "About";
}
<h2>@ViewData["Title"].</h2>
<h3>@ViewData["Message"]</h3>

<p>Use this area to provide additional information.</p>
```

6.5.4　控制器如何指定视图

视图通常以 ViewResult 的形式从操作中返回,这是一种 ActionResult。操作方法可以直接创建并返回一个 ViewResult,但一般不会这样做。

当操作方法返回一个视图时,会先去查找视图文件,然后根据默认规则来决定哪个视图文件将被采用。

当操作方法返回 View 方法,如 return View()或 return View("<ViewName>")时,这个操作方法的名称被用作视图名称。例如,如果一个叫作"About"的控制器操作方法被调用,那么这个方法的名称被用于搜索名为 About.cshtml 的视图文件。当应用运行时,首先会在 Views/[ControllerName]文件夹中搜索"About"视图。如果在此处找不到对应的视图,则会在"Shared"文件夹中搜索"About"视图。

视图搜索都会按照以下顺序搜索匹配的视图文件:

```
Views/[ControllerName]/[ViewName].cshtml
Views/Shared/[ViewName].cshtml
```

操作方法除了有以上两种返回方式之外,还有以下返回方式:

```
return View("Views/Home/About.cshtml");
return View("../Manage/Index");
```

```
return View("./About");
```

视图搜索是依赖于按照文件名称来查找视图文件的。如果操作系统的文件系统区分大小写,那么视图名称可能也会区分大小写。为了跨操作系统的兼容性,应当总是保持控制器与操作名称之间,以及关联视图文件夹与文件名称之间的大小写一致。如果在处理区分大小写的文件系统中遇到无法找到视图文件的错误,则请确认请求的视图文件与实际视图文件名称之间的大小写是否一致。

通常情况下,如果控制器继承自 Controller,那么简单地使用 View 辅助方法即可返回 ViewResult。例如 HomeController 控制器中的 About 方法如下:

```
public IActionResult About()
{
    ViewData["Message"] = "Your application description page.";
    return View();
}
```

当此操作返回时,显示的 About.cshtml 视图结果如图 6.18 所示。

图 6.18 About 页面

View 方法有如下多个重载,可选择指定:

① 要返回的显式视图:

```
return View("Cargoes");
```

② 要传递给视图的实体:

```
return View(Cargo);
```

③ 返回的视图和模型:

```
return View("Cargoes", Cargo);
```

6.5.5 向视图传递数据

1. 强类型数据（ViewModel）

可以使用多种机制给视图传递数据，最可靠的方法是在视图中指定模型类型，此模型通常称为 ViewModel，然后从操作中给视图传递此类型的实例。

ViewModel 是一个用来渲染 ASP.NET MVC 视图的强类型类，可用来传递来自一个或多个视图模型（即类）或数据表的数据，因此可将其看作一座连接着模型、数据和视图的桥梁，其生命期为当前视图。

在视图中通过 @model 指令添加模型，这是控制器将强类型化的模型传递给视图的首选方式，可使视图利用强类型的检查优势，并在 Visual Studio 2017 中具备智能提示和静态检查。

例如，以下视图呈现了类型为 IEnumerable<RazorMvcDemo.Models.Cargo> 的模型：

```
@model IEnumerable<RazorMvcDemo.Models.Cargo>

@{
    ViewData["Title"] = "Index";
}

<h2>Index</h2>

<p>
    <a asp-action="Create">Create New</a>
</p>
<table class="table">
<thead>
    <tr>
        <th>
            @Html.DisplayNameFor(model => model.Name)
        </th>
        <th>
            @Html.DisplayNameFor(model => model.Code)
        </th>
        <th>
            @Html.DisplayNameFor(model => model.Country)
        </th>
        <th>
            @Html.DisplayNameFor(model => model.MinStockQty)
```

```
            </th>
            <th>
                @Html.DisplayNameFor(model => model.GrossWt)
            </th>
            <th>
                @Html.DisplayNameFor(model => model.Unit)
            </th>
            <th></th>
        </tr>
    </thead>
    <tbody>
        @foreach (var item in Model) {
        <tr>
            <td>
                @Html.DisplayFor(modelItem => item.Name)
            </td>
            <td>
                @Html.DisplayFor(modelItem => item.Code)
            </td>
            <td>
                @Html.DisplayFor(modelItem => item.Country)
            </td>
            <td>
                @Html.DisplayFor(modelItem => item.MinStockQty)
            </td>
            <td>
                @Html.DisplayFor(modelItem => item.GrossWt)
            </td>
            <td>
                @Html.DisplayFor(modelItem => item.Unit)
            </td>
            <td>
                <a asp-action="Edit" asp-route-id="@item.Id">Edit</a> |
                <a asp-action="Details" asp-route-id="@item.Id">Details</a> |
                <a asp-action="Delete" asp-route-id="@item.Id">Delete</a>
            </td>
        </tr>
        }
    </tbody>
</table>
```

为了将模型提供给视图,控制器将其作为参数进行传递,代码如下:

```
public async Task<IActionResult> Index()
{
    return View(await _context.Cargo.ToListAsync());
}
```

2. 弱类型数据(ViewData 和 ViewBag)

除了强类型数据之外,还可以使用弱类型数据的集合将少量数据传入及传出控制器和视图。所有的视图都可以访问弱类型的数据集合,这个集合可以通过控制器和视图的 ViewData 或 ViewBag 来引用。

由于 ViewData 和 ViewBag 属性不提供编译时的类型检查,而只在运行时才进行动态解析,因此使用这两者通常比使用 ViewModel 更容易出错。

(1) ViewData

ViewData 是一个 Dictionary⟨string,object⟩的字典对象,数据以键值对的形式存储在 ViewData 中。ViewData 在控制器和视图之间传递数据,也可以在视图和分部视图之间传递数据,其生存期至当前视图渲染结束。由于在使用 ViewData 时返回的是一个 object 对象,所以在应用实际类型的属性或值时需要使用强制类型转换。当传递的是字符串数据时,可以直接存储和使用,而不需要强制转换,因为 C♯ 的每个对象存在 ToString 方法,而且在 C♯ 视图中会自动调用该方法。以下代码是在 About 操作方法中使用 ViewData 的示例,结果如图 6.19 所示。

图 6.19 ViewData 的示例结果

```
public IActionResult About()
{
    ViewData["Message"] = "Your application description page.";
    ViewData["Supplier"] = "MUEY";
    ViewData["Cargo"] = new Cargo()
    {
        Name = "Steve",
```

```
            Code = "St123",
            Country = "中国",
            Unit = "个",
            MinStockQty = 10,
            GrossWt = 100
        };
        return View();
    }
```

在 About 视图中处理以下数据：

```
@{
    ViewData["Title"] = "About";
}
<h2>@ViewData["Title"]</h2>
<h3>@ViewData["Message"]</h3>
@{
    // 从 ViewData 中把对象转换成实体
    var cargo = ViewData["Cargo"] as Cargo;
}
<p>供应商：ViewData["Supplier"]，货物名称：cargo.Name,货物代码：cargo.Code,计量单位：cargo.Unit,毛重：cargo.GrossWt</p>
```

(2) ViewBag

ViewBag 是 ViewData 的动态封装器，从 ControllerBase 继承而来。ViewBag 为储存在 ViewData 里的对象提供动态访问。ViewBag 不需要强制转换，因此使用起来更加方便。下列代码演示了如何使用与上述 ViewData 有相同结果的 ViewBag：

```
public IActionResult About()
{
    ViewData["Message"] = "Your application description page.";
    ViewBag.Supplier = "MUEY";
    ViewBag.Cargo = new Cargo()
    {
        Name = "Steve",
        Code = "St123",
        Country = "中国",
        Unit = "个",
        MinStockQty = 10,
        GrossWt = 100
    };
    return View();
}
```

在 About 视图中处理以下数据：

```
@{
    ViewData["Title"] = "About";
}
<h2>@ViewData["Title"]</h2>
<h3>@ViewData["Message"]</h3>
```

<p>供应商：@ViewBag.Supplier,货物名称:@ViewBag.Cargo.Name,货物代码:@ViewBag.Cargo.Code,计量单位:@ViewBag.Cargo.Unit,毛重:@ViewBag.Cargo.GrossWt</p>

（3）ViewData 与 ViewBag 数据共享

由于 ViewBag 内部真正存储数据的还是 ViewData,也就是说 ViewData 与 ViewBag 的数据是共享的,因此,通过 ViewData 设置的数据,可以通过 ViewBag 访问;同样,通过 ViewBag 设置的数据,可以通过 ViewData 访问。下面通过以下示例代码做进一步说明。在 About.cshtml 视图顶部,使用 ViewData["Title"]设置标题,使用 ViewBag.Desc 设置说明;而在页面中,使用 ViewBag 设置标题,使用 ViewData 设置说明,结果如图 6.20 所示。

```
@{
    ViewData["Title"] = "About";
    ViewBag.Desc = "这是关于页面";
}
<h2>@ViewBag.Title</h2>
<h3>@ViewData["Desc"]</h3>
```

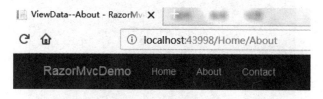

图 6.20　ViewData 与 ViewBag 数据共享

ViewData 与 ViewBag 之间的差异如表 6.6 所列。

表 6.6　ViewData 与 ViewBag 之间的差异

ViewData	ViewBag
派生自 ViewDataDictionary,因此它有可用的字典属性,如 ContainsKey、Add、Remove 和 Clear	派生自 DynamicViewData,因此它可使用点表示法(@ViewBag.SomeKey =〈value or object〉)创建动态属性,且无须强制转换。ViewBag 的语法使添加到控制器和视图的速度更快
字典中的键是字符串,因此允许有空格。示例：ViewData["Some Key With Whitespace"]	更易于检查 NULL 值。示例：@ViewBag.Person?.Name
任何非 String 类型均须在视图中进行强制转换才能使用 ViewData	在 Razor 页中不可用

6.6　ASP.NET Core 中的布局

6.6.1　什么是布局

一般 Web 应用程序都有类似的外观和操作,可在页面间切换时为用户提供一致性体验,这就是 Web 应用程序的通用布局。该布局通常包括页头、导航或菜单元素以及页脚等常见的用户界面元素,如图 6.21 所示。

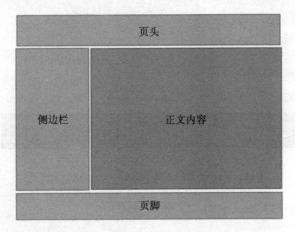

图 6.21　布　局

应用中的许多页面也经常使用脚本和样式表等常用的 HTML 结构。所有这些共享元素均可在布局文件中进行定义,随后应用内使用的任何视图均可引用此文件。布局可减少视图中的重复代码。

按照约定,ASP.NET 应用的默认布局名为 _Layout.cshtml。这个页面主要有以下两个作用：一是控制每个页面的布局（页面选择退出布局时除外）；二是导入 HTML 结构，例如 JavaScript 和样式表。

Visual Studio 的 ASP.NET Core MVC 项目模板在 Views/Shared 文件夹中包含此布局文件。

此布局为应用中的视图定义顶级模板。当应用程序需要不同的布局时，可以定义多个布局，并且为不同的视图指定不同的布局。布局页面_Layout.cshtml 的示例代码如下：

```html
<!DOCTYPE html>
<html>
<head>
    <meta charset="utf-8" />
    <meta name="viewport" content="width=device-width, initial-scale=1.0" />
    <title>@ViewData["Title"] - RazorMvcDemo</title>
    <environment include="Development">
    <link rel="stylesheet" href="~/lib/bootstrap/dist/css/bootstrap.css" />
    <link rel="stylesheet" href="~/css/site.css" />
    </environment>
    <environment exclude="Development">
    <link rel="stylesheet" href="https://ajax.aspnetcdn.com/ajax/bootstrap/3.3.7/css/bootstrap.min.css" asp-fallback-href="~/lib/bootstrap/dist/css/bootstrap.min.css" asp-fallback-test-class="sr-only" asp-fallback-test-property="position" asp-fallback-test-value="absolute" />
    <link rel="stylesheet" href="~/css/site.min.css" asp-append-version="true" />
    </environment>
</head>
<body>
    <nav class="navbar navbar-inverse navbar-fixed-top">
    <div class="container">
    <div class="navbar-header">
    <button type="button" class="navbar-toggle" data-toggle="collapse" data-target=".navbar-collapse">
        <span class="sr-only">Toggle navigation</span>
        <span class="icon-bar"></span>
        <span class="icon-bar"></span>
        <span class="icon-bar"></span>
    </button>
    <a asp-area="" asp-controller="Home" asp-action="Index" class="navbar-brand">RazorMvcDemo</a>
    </div>
    <div class="navbar-collapse collapse">
```

```html
        <ul class="nav navbar-nav">
            <li><a asp-area="" asp-controller="Home" asp-action="Index">Home</a></li>
            <li><a asp-area="" asp-controller="Home" asp-action="About">About</a></li>
            <li><a asp-area="" asp-controller="Home" asp-action="Contact">Contact</a></li>
        </ul>
        </div>
    </div>
</nav>
<div class="container body-content">
    @RenderBody()
    <hr />
    <footer>
        <p>&copy; 2018 - RazorMvcDemo</p>
    </footer>
</div>

<environment include="Development">
    <script src="~/lib/jquery/dist/jquery.js"></script>
    <script src="~/lib/bootstrap/dist/js/bootstrap.js"></script>
    <script src="~/js/site.js" asp-append-version="true"></script>
</environment>
<environment exclude="Development">
    <script src="https://ajax.aspnetcdn.com/ajax/jquery/jquery-2.2.0.min.js" asp-fallback-src="~/lib/jquery/dist/jquery.min.js" asp-fallback-test="window.jQuery" crossorigin="anonymous" integrity="sha384-K+ctZQ+LL8q6tP7I94W+qzQsfRV2a+AfHIi9k8z8l9ggpc8X+Ytst4yBo/hH+8Fk">
    </script>
    <script src="https://ajax.aspnetcdn.com/ajax/bootstrap/3.3.7/bootstrap.min.js" asp-fallback-src="~/lib/bootstrap/dist/js/bootstrap.min.js" asp-fallback-test="window.jQuery && window.jQuery.fn && window.jQuery.fn.modal" crossorigin="anonymous" integrity="sha384-Tc5IQib027qvyjSMfHjOMaLkfuWVxZxUPnCJA7l2mCWNIpG 9mGCD8wGNIcPD7Txa">
    </script>
    <script src="~/js/site.min.js" asp-append-version="true"></script>
</environment>
    @RenderSection("Scripts", required: false)
</body>
</html>
```

上面代码中需要注意的是"@RenderBody"方法的调用,这是一个占位符,用来标记使用这个布局的视图,以呈现视图中主要内容的位置。

6.6.2 指定布局

Razor视图具有Layout属性,视图可以通过设置此属性来指定是否使用布局页

面,代码如下:

```
@{
    Layout = "_Layout";
}
```

在视图中指定布局有两种方式:一是使用完整路径(例如:/Views/Shared/_Layout.cshtml),二是使用布局文件名称(例如:_Layout)。当使用布局文件名称时,Razor 视图引擎将使用它的标准发现流程来搜索布局文件。首先搜索与 Controller 相关的文件夹,然后搜索 Shared 文件夹。

在布局文件中可以通过调用 RenderSection 方法来选择引用一个或多个节。节提供了组织某些页面元素放置的方法。对于每一次 RenderSection 的调用,都可以指定该节是必需的还是可选的。例如,下面代码中的 Scripts 节是可选的,如果没有也不会抛出错误:

```
@RenderSection("Scripts", required: false)
```

而下面代码中的 Scripts 节则是必需的,如果找不到所需的节,则会引发异常:

```
@RenderSection("Scripts")
```

6.6.3 导入共享指令

视图可以使用 Razor 指令来执行许多操作,例如导入命名空间或执行依赖关系注入。_ViewImports.cshtml 文件里引入的命名空间和 TagHelper 会自动包含在所有视图里。_ViewImports.cshtml 文件通常放置在 Pages(Views)文件夹及其子文件夹的视图中。_ViewImports.cshtml 文件不支持函数和节定义等其他 Razor 功能,其支持的指令如表 6.7 所列。

表 6.7　_ViewImports.cshtml 文件支持的指令

序号	指令	序号	指令
1	@addTagHelper	5	@tagHelperPrefix
2	@removeTagHelper	6	@inherits
3	@model	7	@inject
4	@using		

_ViewImports.cshtml 文件的示例代码如下:

```
@using RazorMvcDemo
@using RazorMvcDemo.Models
@addTagHelper *, Microsoft.AspNetCore.Mvc.TagHelpers
@using Microsoft.AspNetCore.Identity
```

在 ASP.NET Core MVC 应用程序中，_ViewImports.cshtml 文件通常放置在 Views 文件夹中。实际上，_ViewImports.cshtml 文件可以放置在任何文件夹中，在这种情况下，它将只对该文件夹及其子文件夹中的视图起作用。在执行顺序上，首先执行根目录下的_ViewImports.cshtml 文件，然后再执行视图所在文件夹下的_ViewImports.cshtml 文件，所以在根目录中的_ViewImports.cshtml 文件里指定的设置可能会被覆盖掉。

例如，如果根目录中的 _ViewImports.cshtml 文件指定了@model 和@addTagHelper，另外一个 Controller 相关文件夹中的 _ViewImports.cshtml 文件指定了一个不同的 @model，并添加了另外一个@addTagHelper，则视图将有权访问这两个标记助手，并使用后者指定的 @model。

如果一个视图中有多个 _ViewImports.cshtml 文件，那么多个_ViewImports.cshtml 文件中的指令的组合行为如表 6.8 所列。

表 6.8 指令组合行为

指　　令	指令行为
@addTagHelper,@removeTagHelper	按顺序全部运行
@tagHelperPrefix	离视图最近的文件会覆盖其他任何文件
@model	离视图最近的文件会覆盖其他任何文件
@inherits	离视图最近的文件会覆盖其他任何文件
@using	全部包含；忽略重复项
@inject	对每一个属性而言（通过属性名区分），离视图最近的一个覆盖其他具体相同属性名的属性

6.6.4　在呈现每个视图之前运行代码

如果有需要在呈现每个视图之前运行公共代码，则可以在 Views 文件夹中找到_ViewStart.cshtml 文件，并在其中输入相应代码。按照约定，在_ViewStart.cshtml 文件中所写的代码语句，将在呈现每个完整的视图（不是布局，也不是分部视图）之前运行。与 ViewImports.cshtml 一样，_ViewStart.cshtml 也是分层的。例如，如果在 Views 文件夹的根目录中定义的_ ViewStart.cshtml 文件，同时又在 Views\Home 文件夹中定义了 一遍，则后者在前者之后运行。就如 6.6.2 小节的例子中，假如每一个视图都使用 Layout 属性来指定其布局，如果有许多视图都使用同一个布局，那么就会产生冗余，并且维护困难。这时就可以使用_ViewStart.cshtml 文件来消除这种冗余。在_ViewStart.cshtml 文件中设置 Layout 属性，用于指定模板页面，也就是布局页面，代码如下：

@{

```
        Layout = "_Layout";
}
```

注意：_ViewStart.cshtml 和 _ViewImports.cshtml 文件通常不会放置在 /Views/Shared 文件夹中，这些应用级别版本的文件应该直接放置在 /Views 文件夹中。

6.7　ASP.NET Core 中的标记助手

6.7.1　什么是标记助手

　　TagHelper 称为标记助手，专注于在.cshmlt 文件中辅助生成 HTML 标记，是一种利用自定义标记赋予元素功能或添加属性的方式。标记助手能够让服务器端的代码在 Razor 文件中参与创建和呈现 HTML 元素，例如，使用 LabelTagHelper 修改现有标记的行为（如〈label〉）。标记助手是基于元素名称及其属性绑定到特定元素上的，它们既提供了服务器端呈现的优势，又保留了 HTML 编辑的体验。

　　ASP.NET Core MVC 内置了多种常见的标记助手（如文本、图片、标签等），公共 GitHub 存储库和 NuGet 包中甚至还有更多可用的标记助手程序。标记助手程序使用 C# 语言来编写，并基于元素名称、属性名称或父标记创建 HTML 元素目标。

　　例如，内置 LinkTagHelper 可以用来创建指向列表页面的链接，代码如下：

```
〈div〉
    〈a asp-action="Index"〉Back to List〈/a〉
〈/div〉
```

可以使用 EnvironmentTagHelper 在视图中根据运行时环境（如开发、暂存或生产）的不同来运行不同的脚本，代码如下：

```
〈environment include="Development"〉
〈link rel="stylesheet" href="~/lib/bootstrap/dist/css/bootstrap.css" /〉
〈link rel="stylesheet" href="~/css/site.css" /〉
〈/environment〉
〈environment exclude="Development"〉
〈link rel="stylesheet" href="https://ajax.aspnetcdn.com/ajax/bootstrap/3.3.7/css/bootstrap.min.css" asp-fallback-href="~/lib/bootstrap/dist/css/bootstrap.min.css" asp-fallback-test-class="sr-only" asp-fallback-test-property="position" asp-fallback-test-value="absolute" /〉
〈link rel="stylesheet" href="~/css/site.min.css" asp-append-version="true" /〉
〈/environment〉
```

　　标记助手提供 HTML 友好的开发体验和用于创建 HTML 和 Razor 标记的丰

富的 IntelliSense 环境。大多数内置标记助手以现有 HTML 元素为目标，为该元素提供服务器端属性。

6.7.2 标记助手的功能

标记助手的功能包括以下几点：

① 提供 HTML 友好的开发体验。在大多数情况下，Razor 标记使用标记助手后看起来很像标准的 HTML，对于熟悉 HTML/CSS/JavaScript 的前端设计师来说，无须学习 C# Razor 语法即可编辑 Razor。

② 提供一个丰富的智能提示环境用于创建 HTML 和 Razor 标记。这与 HTML 助手形成了鲜明的对比，HTML 助手是在服务器端创建 Razor 视图中的标记。即使是熟悉 C# Razor 语法的开发人员，使用标记助手也比编写 C# Razor 标记更高效。

③ 使用仅在服务器上可用的信息可以提高生产力，并能生成更稳定、可靠和可维护的代码。例如，过去在更新图片时，必须在更改图片时更改图片名称。出于性能原因，需要主动缓存图片，而若不更改图片的名称，客户端就可能获得过时的副本。以前，编辑完图片后必须更改名称，同时需要更新 Web 应用中对该图片的每个引用，这不仅大费周章，而且容易出错（可能会漏掉某个引用，或者意外输入错误的字符串等）；而内置 ImageTagHelper 后可自动执行更新操作。

ImageTagHelper 可将版本号追加到图片名称中，这样每当更改图片时，服务器都会自动为该图片生成新的唯一版本，从而客户端总是能获得最新的图片。使用 ImageTagHelper 实质上是免费获得了稳健性而节省了劳动力。

大多数内置标记助手程序以标准 HTML 元素为目标，为该元素提供服务器端属性。例如，〈input〉用于包含 asp-for 特性，此特性将指定模型属性的名称提取至所呈现的 HTML 中。以下代码以一个具体的 Razor 视图为例进行说明：

```
@model RazorMvcDemo.Models.Cargo

@{
    ViewData["Title"] = "Edit";
}

<h2>Edit</h2>

<h4>Cargo</h4>
<hr/>
<div class="row">
<div class="col-md-4">
<form asp-action="Edit">
<div asp-validation-summary="ModelOnly" class="text-danger"></div>
<input type="hidden" asp-for="Id"/>
```

```html
<div class="form-group">
<label asp-for="Name" class="control-label"></label>
<input asp-for="Name" class="form-control" />
<span asp-validation-for="Name" class="text-danger"></span>
</div>
<div class="form-group">
<label asp-for="Code" class="control-label"></label>
<input asp-for="Code" class="form-control" />
<span asp-validation-for="Code" class="text-danger"></span>
</div>
<div class="form-group">
<label asp-for="Country" class="control-label"></label>
<input asp-for="Country" class="form-control" />
<span asp-validation-for="Country" class="text-danger"></span>
</div>
<div class="form-group">
<label asp-for="MinStockQty" class="control-label"></label>
<input asp-for="MinStockQty" class="form-control" />
<span asp-validation-for="MinStockQty" class="text-danger"></span>
</div>
<div class="form-group">
<label asp-for="GrossWt" class=" control-label"></label>
<input asp-for="GrossWt" class="form-control" />
<span asp-validation-for="GrossWt" class="text-danger"></span>
</div>
<div class="form-group">
<label asp-for="Unit" class=" control-label"></label>
<input asp-for="Unit" class=" form-control" />
<span asp-validation-for="Unit" class="text-danger"></span>
</div>
<div class="form-group">
<input type="submit" value="Save" class="btn btn-default" />
</div>
</form>
</div>
</div>
<div>
<a asp-action="Index">Back to List</a>
</div>
@section Scripts {
    @{await Html.RenderPartialAsync("_ValidationScriptsPartial");}
}
```

6.7.3 管理标记助手的作用域

标记助手的作用域由 @addTagHelper、@removeTagHelper 和 "!" 联合控制。

1. 使用@addTagHelper 添加标记助手

使用@addTagHelper 添加标记助手有两种方式,一种是全局性的,将@addTagHelper 指令添加到 Views/_ViewImports.cshtml 文件中,将使标记助手对 Views 目录及其子目录中的所有视图文件可用;另一种是仅对特定视图公开标记助手,可在这些视图文件中使用 @addTagHelper 指令。

@addTagHelper 指令有两个参数,第一个参数指定要加载的标记助手(或者使用通配符"*"指定程序集"Microsoft.AspNetCore.Mvc.TagHelpers"中的所有标记助手),第二个参数"Microsoft.AspNetCore.Mvc.TagHelpers"指定包含标记助手的程序集。Microsoft.AspNetCore.Mvc.TagHelpers 是内置 ASP.NET Core 标记助手的程序集。下列代码是一个_ViewImports.cshtml 文件的内容:

```
@using RazorMvcDemo
@using RazorMvcDemo.Models
@addTagHelper *, Microsoft.AspNetCore.Mvc.TagHelpers
@addTagHelper *, RazorMvcDemo
@addTagHelper RazorMvcDemo.TagHelpers.EmailTagHelper, RazorMvcDemo
@addTagHelper RazorMvcDemo.TagHelpers.E*, RazorMvcDemo
@addTagHelper RazorMvcDemo.TagHelpers.Email*, RazorMvcDemo
```

2. 使用@removeTagHelper 删除标记助手

@removeTagHelper 的作用是删除之前添加的标记助手,@removeTagHelper 与@addTagHelper 具有相同的两个参数。例如,在 Views/Folder/_ViewImports.cshtml 文件中使用 "@removeTagHelper *, Microsoft.AspNetCore.Mvc.TagHelpers"将从 Folder 的所有视图中删除指定的标记助手。

3. 禁用元素

使用感叹号("!")可在元素级别禁用标记助手。例如,使用标记助手选择感叹号在〈span〉中禁用 Email 验证,代码如下:

```
<!span asp-validation-for = "Email" class = "text-danger"></!span>
```

注意,须将标记助手选择感叹号应用于标签的开始和结束(将感叹号添加到标签开始位置时,Visual Studio 编辑器会自动为标签结束添加相应感叹号)。添加感叹号后,元素和标记助手属性不再以独特的字体显示。

4. 使用@tagHelperPrefix 阐明标记助手的用途

@tagHelperPrefix 指令可指定一个标记前缀字符串,以启用标记助手支持并阐明标记助手的用途。例如，可以将标记"@tagHelperPrefix th:"添加到 Views/_ViewImports.cshtml 文件中。

在以下代码图像中，标记助手前缀设置为"th:"，所以只有使用前缀"th:"的元素才支持标记助手(可使用标记助手的元素以独特的字体显示)，如图 6.22 所示。从图中可以看出，〈label〉和〈input〉元素具有标记助手前缀，可使用标记助手，而〈span〉元素则相反。

```
<div class="form-group">
    <th:label asp-for="Password" class="col-md-2"></th:label>
    <div class="col-md-10">
        <th:input asp-for="Password" class="form-control" />
        <span asp-validation-for="Password" class="text-danger"></span>
    </div>
</div>
```

图 6.22 标记助手前缀设置

6.7.4 标记助手的智能提示支持

当在 Visual Studio 中创建新的 ASP.NET Web 应用时，会添加 NuGet 包"Microsoft.AspNetCore.Razor.Tools"，这是添加标记助手工具的包。

现在考虑编写 HTML〈label〉元素。只要在 Visual Studio 编辑器中输入"〈l"，"智能提示"就会显示匹配的元素，如图 6.23 所示。此时，不仅会显示 HTML 帮助，而且会在元素名称旁边显示如图 6.24 所示的图标(下方带有"〈〉"的"@"字符)。然后，将该元素标识为标记助手的目标。对于纯 HTML 元素(如 fieldset)，其名称旁边只显示"〈〉"图标，如图 6.23 所示。

图 6.23 标签智能提示

输入"〈label"后，"智能提示"会列出可用的 HTML/CSS 属性和以标记助手为目标的属性，如图 6.25 所示。

"智能提示"可以帮助完成整行代码的录入，按 Tab 键即可用选择的值完成语句，如图 6.26 所示。

图 6.24 "@"字符

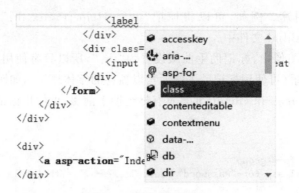

图 6.25　智能提示

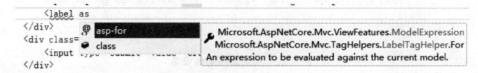

图 6.26　自动完成语句

只要输入标记助手属性,标记和属性字体就会更改,如图 6.27 所示。

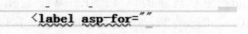

图 6.27　字体更改

图 6.28 正在编辑(Edit)视图,所以有可用的实体类 Cargo。LabelTagHelper 将 asp-for 属性值("Unit")的内容设置为"Unit"。

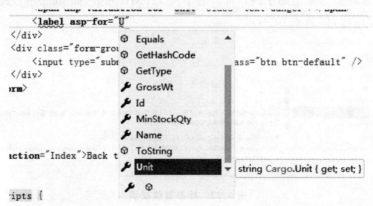

图 6.28　Unit 属性

如图 6.28 所示,"智能提示"会列出页面上模型可用的属性和方法。丰富的"智能提示"环境可帮助选择 CSS 类,如图 6.29 和图 6.30 所示。

图 6.29 样式表的类名

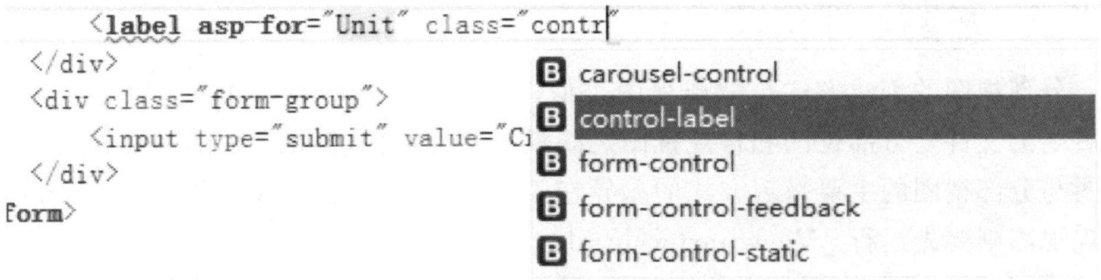

图 6.30 CSS 类名

至此就完成了一个纯 HTML〈label〉标记的录入,它以棕色字体显示 HTML 标记(当使用默认的 Visual Studio 主题颜色时),以红色字体显示属性,以蓝色字体显示属性值,如图 6.31 所示。

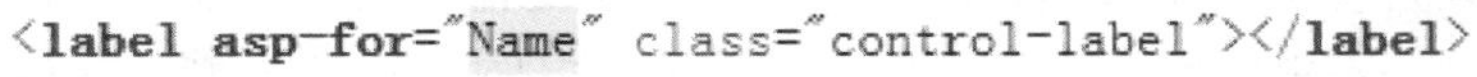

`<label asp-for="Name" class="control-label"></label>`

图 6.31 HTML 标记主题颜色

6.8 ASP.NET Core 中的分部视图

ASP.NET Core MVC 支持分部视图,用于在不同的视图中共享网页的可重用部分。分部视图能够把页面中共同的部分提取出来,在使用时简单地引入即可。分部视图除了不能指定布局外,几乎与普通视图一样。

6.8.1 什么是分部视图

在介绍分部视图之前,先来回顾一下 ASP.NET Web Form 中的 User Control。在 ASP.NET Web Form 中使用 User Control 进行组件开发可以减少重复的代码,利于页面模块化,这个概念也被引入了 ASP.NET MVC 即"分部视图"中。分部视图指在父视图内呈现的视图。通过执行分部视图,可将生成的 HTML 输出到父视图中。与视图一样,分部视图也使用.cshtml 文件扩展名。

6.8.2 何时使用分部视图

分部视图是将大型视图分解为较小组件的有效方法。分部视图可减少视图内容的重复,并使视图元素得以重复使用。常见的布局元素应在 _Layout.cshtml 中指定,非布局可重用内容可封装到分部视图中。

在由多个逻辑部分组成的复杂页面中,会有许多重用的部分,可以将这些重用的部分独立出来,进行封装,在需要使用的那些页面中进行引用,以减少代码重复,并使页面本身的视图变得更加简单,因为它仅包含整体页面结构,并且可通过调用来呈现分部视图。

6.8.3 声明分部视图

分部视图的创建方式与常规视图类似,即在 Views 文件夹内创建以.cshtml 为扩展名的文件。分部视图和常规视图之间没有语义差异,但是它们的呈现方式不同。视图与分部视图的主要呈现方式的差异在于分部视图不运行_ViewStart.cshtml,而常规视图则要先运行 _ViewStart.cshtml。

按照约定,分部视图的文件名通常以"_"开头,虽然未强制要求遵从此命名约定,但这样做有助于直观地将分部视图与常规视图区分开来。

不同的视图文件夹中可以存在具有相同文件名的不同的分部视图。当按名称(不带文件扩展名)引用视图时,每个文件夹中的视图都会使用与其位于同一文件夹中的分部视图。可以在 Shared 文件夹中创建默认(共享)的分部视图供视图调用。任何视图均可使用共享分部视图。视图中如果设置了默认分部视图(位于 Shared 中),那么该分部视图也会被与父视图位于同一文件夹并具有与默认分部视图相同名称的分部视图替代。

分部视图可以调用其他分部视图(只要未创建循环调用)。在每个视图或分部视图内,相对路径始终相对于该视图,而不相对于根视图或父视图。

注意,分部视图中定义的 Razorsection 对父视图不可见,定义的 section 仅对定义它时所在的分部视图可见。

6.8.4 分部视图访问示例

在视图页中,有多种方法可呈现分部视图,其中最佳的做法是使用异步呈现,具体方法是:

① 在 Visual Studio 2017 的"解决方案资源管理器"中找到 Views 文件夹下的 Shared 文件夹后右击,在弹出的快捷菜单中选择"添加视图",弹出"添加 MVC 视图"对话框,在"视图名称"文本框中输入"PartView",在"选项"区选中"创建为分部视

图",然后单击"添加"按钮,创建分部视图,如图 6.32 所示。

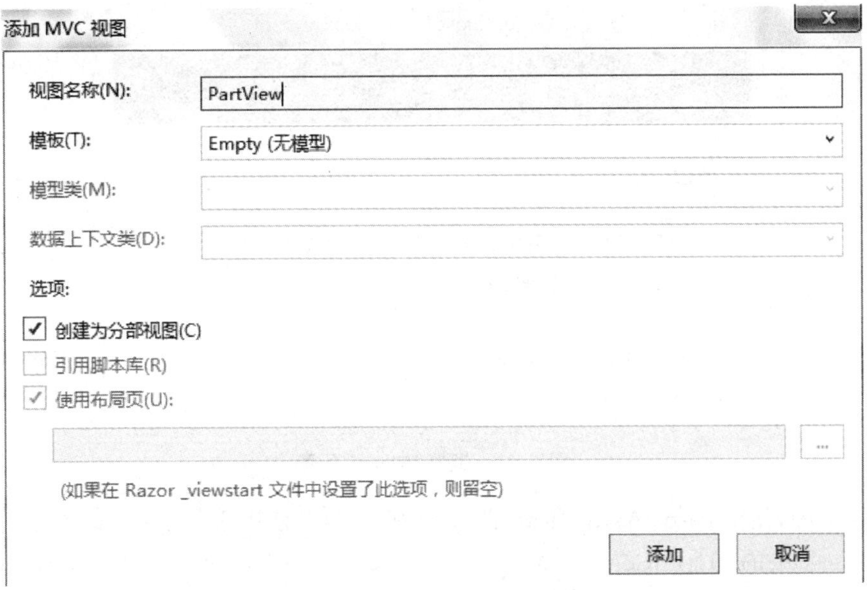

图 6.32 创建分部视图

② 在 Visual Studio 2017 中打开刚才创建的分部视图 PartView.cshmtl 文件,添加如下代码:

`<p>这是分部视图中的 P 元素文本</p>`

③ 引用分部视图的方法有两种:

第一种,使用 Partial 同步引用分部视图,但不建议大家使用这种方式。操作方法是:在 Visual Studio 2017 中打开 About.cshmtl 文件,编写如下同步引用分部视图的代码:

```
@{
    ViewData["Title"] = "About";
}
<h2>@ViewData["Title"]</h2>
<h3>@ViewData["Message"]</h3>

@Html.Partial("~/Views/Shared/PartView.cshtml");

<p>Use this area to provide additional information.</p>
<p>About 视图文件内容</p>
```

在 Visual Studio 2017 中按 F5 键运行应用程序,然后在浏览器中浏览"About"页面,效果如图 6.33 所示。

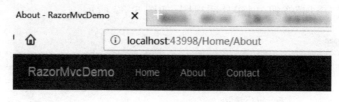

图 6.33 浏览"About"页面

第二种，使用 PartialAsync 异步引用分部视图，推荐使用这种方法。该方法返回包装在 Task 中的 IHtmlContent 类型。在 About.cshmtl 页面中，把 @Html.Partial 修改为 @Html.PartialAsync 即可，代码如下：

```
@{
    ViewData["Title"] = "About";
}
<h2>@ViewData["Title"]</h2>
<h3>@ViewData["Message"]</h3>
@Html.PartialAsync("~/Views/Shared/PartView.cshtml");
<p>Use this area to provide additional information.</p>

<p>About 视图文件内容</p>
```

在 Visual Studio 2017 中按 F5 键运行应用程序，然后在浏览器中浏览"About"页面，效果如图 6.33 所示。

④ 当实例化分部视图时，可以将父视图的 ViewData 字典的副本传递到分部视图中。在分部视图内对数据所做的更新不会保存到父视图中。分部视图中对 ViewData 的更改会在分部视图返回时丢失。在 Visual Studio 2017 中打开 About.cshtml 文件，修改后的代码如下：

```
@{
    ViewData["Title"] = "About";
}
<h2>@ViewData["Title"]</h2>
<h3>@ViewData["Message"]</h3>
```

```
@{
    var data = new ViewDataDictionary(this.ViewData);
        data.Add("name", "张三称");
        data.Add("id","10001")
}

@await Html.PartialAsync("~/Views/Shared/PartView.cshtml", data);

<p>Use this area to provide additional information.</p>
<p>About 视图文件内容</p>
```

⑤ 在 Visual Studio 2017 中打开刚才创建的分部视图 PartView.cshmtl 文件进行代码修改,修改后的分部视图中的代码如下:

```
<p>这是分部视图中的P元素文本</p>
<p>名称为:@ViewData["name"]</p>
<p>id 为:@ViewData["id"]</p>
```

⑥ 在 Visual Studio 2017 中按 F5 键运行应用程序,然后在浏览器中浏览"About"页面,效果如图 6.34 所示。

图 6.34　实例化分部视图后浏览"About"页面

⑦ 还可将模型传入分部视图,该模型可以是页面的视图模型或自定义对象。可将模型传递到 PartialAsync 或 RenderPartialAsync 中,代码如下:

```
@await Html.PartialAsync("~/Views/Shared/PartView.cshtml", viewModel);
```

也可将 ViewDataDictionary 的实例和视图模型传递到分部视图中。

6.9 ASP.NET Core 中的视图组件

6.9.1 什么是视图组件

视图组件是 ASP.NET Core MVC 中的新特性,与分部视图相似,但视图组件的功能更加强大,它可以包装并呈现逻辑于整个应用程序中,且可以重复使用它。与分部视图不同的是,视图组件不依赖于控制器。视图组件既适用于 ASP.NET Core MVC,也适用于 Razor 页面。

视图组件的特点是:
① 呈现页面响应的某一部分,而不是整个响应。
② 在控制器和视图之间同样包含了关注点分离和可测试性带来的优势。
③ 可以有参数和业务逻辑。
④ 通常在布局页调用。

视图组件可以用于任何需要重复逻辑并且对于分部视图来说相对复杂的场景,例如,动态导航菜单、登录、购物车、最近发布的文章、一个典型博客的侧边栏内容、会在所有页面显示的登录状态等。

一个视图组件由两部分组成:类(通常继承自 ViewComponent)和它返回的结果(通常为一个视图)。与控制器一样,视图组件也可以是 POCO 类型,但是建议使用继承自 ViewComponent 的方法和属性。

6.9.2 如何创建视图组件类

通过以下三种方法之一来创建视图组件类:
① 继承自 ViewComponent。
② 使用 [ViewComponent] 特性装饰一个类,或者从具有 [ViewComponent] 特性的类派生。
③ 创建一个类,并以 ViewComponent 后缀结尾。

视图组件必须是公开的、非嵌套和非抽象的类。视图组件名称是去掉了"ViewComponent"后缀的类名称。也可以使用 ViewComponentAttribute.Name 属性显式地指定。视图组件类支持构造函数的注入,不参与控制器的生命周期,不能在视图组件中使用过滤器。

视图组件一般会定义一个 Task⟨IViewComponentResult⟩ InvokeAsync 方法,并在这个方法中定义逻辑和返回 IViewComponentResult。方法参数直接来自视图组件的调用,而不是来自模型绑定。视图组件不会直接处理请求。可通过调用 View 方法来初始化视图组件。

在应用程序运行时,可在以下路径中搜索视图组件:

① "Views/〈controller_name〉/Components/〈view_component_name〉/〈view_name〉";

② "Views/Shared/Components/〈view_component_name〉/〈view_name〉"。

视图组件的默认视图名称为"Default",这意味着视图文件通常命名为"Default.cshtml"。当创建视图组件时,也可以指定不同的视图名称。

建议将视图文件命名为"Default.cshtml"并将视图文件放在"Views/Shared/Components/〈view_component_name〉/〈view_name〉"目录下。下面的示例学习如何使用CargoList视图组件。视图组件的路径为Views/Shared/Components/CargoList/Default.cshtml。

6.9.3 创建一个简单的视图组件

创建一个简单的视图组件的步骤是:

① 在Visual Studio 2017中打开RazorMvcDemo项目,在"解决方案资源管理器"中右击"RazorMvcDemo"项目名称,在弹出的快捷菜单中选择"添加"→"新建文件夹"菜单项,并把"新文件夹"重命名为"ViewComponents"。

② 在ViewComponents上右击,在弹出的快捷菜单中选择"添加"→"类"菜单项,添加一个CargoListViewComponent类,如图6.35所示。

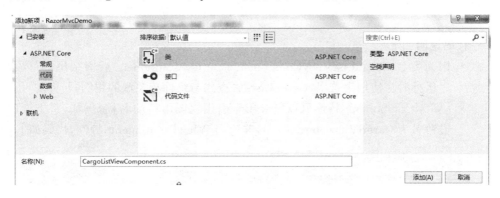

图6.35 创建视图组件类

③ 在Visual Studio 2017中打开CargoListViewComponent.cs文件,添加如下代码:

```
using System;
using System.Collections.Generic;
using System.Linq;
using System.Threading.Tasks;
using Microsoft.AspNetCore.Mvc;
using Microsoft.EntityFrameworkCore;
```

```csharp
namespace RazorMvcDemo.Models
{
    public class CargoListViewComponent:ViewComponent
    {
        private readonly EFCoreDemoContext db;
        public CargoListViewComponent(EFCoreDemoContext context)
        {
            db = context;
        }
        public async Task<IViewComponentResult> InvokeAsync(int maxPriority, bool isDone)
        {
            var items = await GetItemsAsync(maxPriority, isDone);
            return View(items);
        }
        private Task<List<Cargo>> GetItemsAsync(int minQty, bool isDone)
        {
            return db.Cargo.Where(x => x.MinStockQty > minQty).ToListAsync();
        }
    }
}
```

代码说明如下：

- 因为类名 CargoListViewComponent 以 ViewComponent 后缀结尾,所以程序运行时将使用字符串"CargoList"去查找与组件相关联的视图。
- [ViewComponent]特性可以更改用于引用视图组件的名称,例如,将上面的类命名为 CargoViewComponent,并应用[ViewComponent]特性,代码如下：

```csharp
[ViewComponent(Name = "CargoList")]
public class CargoViewComponent:ViewComponent
```

- 上面的[ViewComponent]特性通知视图组件选择器在查找与组件相关联的视图时使用名称 CargoList。
- 组件使用依赖关系注入以使数据上下文可用。
- InvokeAsync 是一个可以从视图中调用的公共方法,并且可以有任意数量的参数。
- GetItemsAsync 方法返回满足 minQty 参数条件的 Cargo 集合。

④ 创建视图组件的视图文件。在 Visual Studio 2017 的"解决方案资源管理器"中依次打开文件夹"Views/Shared",在"Shared"文件夹上右击,在弹出的快捷菜单中选择"添加"→"新建文件夹"菜单项,并把"新文件夹"重命名为"Components",注意必须是这个名称,不能修改。

⑤ 在刚才创建的"Components"文件上右击,在弹出的快捷菜单中选择"添加"→"新建文件夹"菜单项,并把"新文件夹"重命名为"CargoList"。此文件夹的名称必须与视图组件类的名称或去掉后缀(如果遵照约定并在类名中使用了"ViewComponent"后缀)的类的名称相同。

⑥ 在刚才创建的"CargoList"文件上右击,在弹出的快捷菜单中选择"添加"→"视图"菜单项,并把视图名称命名为"Default",如图6.36所示。

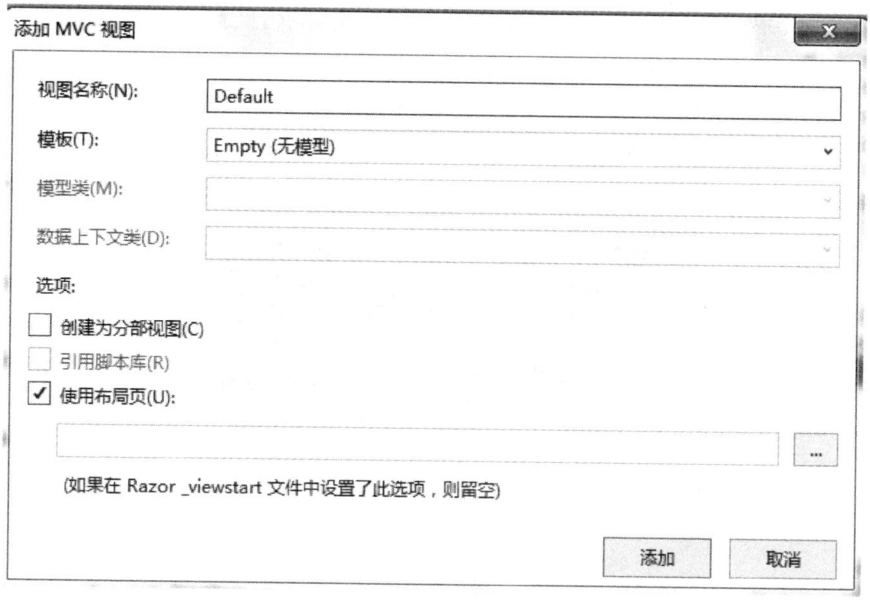

图6.36 创建视图组件视图

⑦ 在Visual Studio 2017中打开刚才创建的视图文件"Default.cshmtl",添加如下代码:

```
@{
    ViewData["Title"] = "Cargo List View";
}
<h2>@ViewData["Title"]</h2>

@model IEnumerable<RazorMvcDemo.Models.Cargo>

<ul>
@foreach (var c in Model)
{
    <li>@c.Name</li>
}
</ul>
```

⑧ 使用视图组件。在 Visual Studio 2017 中打开"Home/Index.cshmtl"文件，将调用视图组件的代码添加到文件底部，代码如下：

```
<div>
    @await Component.InvokeAsync("CargoList", new { minQty = 2 })
</div>
```

@await Component.InvokeAsync("视图组件名",〈匿名类型参数〉)是从视图中调用视图组件的语法，第一个参数是要调用的组件名称，随后是传递给组件的参数。InvokeAsync 可以采用任意数量的参数。

⑨ 在 Visual Studio 2017 中按 F5 键运行应用程序，在浏览器中显示首页，然后拉到底部，会显示视图组件中的货物名称内容，如图 6.37 所示。

图 6.37　显示视图组件中的货物名称

6.9.4　调用视图组件作为标记助手

对于 ASP.NET Core 1.1 及更高版本，可将视图组件作为标记助手注册到任何引用视图组件的文件中，具体方法是：

① 为了将视图组件用作标记助手，必须使用 @addTagHelper 指令注册包含视图组件的程序集。在 Visual Studio 2017 中打开_ViewImports.cshtml 文件，添加指令"@addTagHelper *, RazorMvcDemo"。

② 标记助手将采用 Pascal 大小写格式的类和方法参数转换为各自相应的小写短横线格式。使用〈vc〉〈/vc〉元素调用视图组件的标记助手，并按如下方式指定视图组件：

〈vc:[view-component-name]
　　parameter1 = "parameter1 value"
　　parameter2 = "parameter2 value"〉
〈/vc:[view-component-name]〉

③ 使用视图组件。在 Visual Studio 2017 中打开 Home/Index.cshmtl 文件，在文件底部输入"〈vc"，随后"智能提示"就会显示相应的视图组件，如图 6.38 所示。

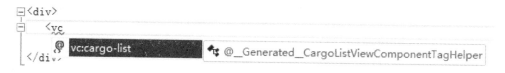

图 6.38　"智能提示"视图组件

④ 从图 6.38 可以看到，CargoList 视图组件变为 cargo-list。视图组件的参数以小写短横线格式的属性进行传递。调用视图组件的代码如下：

〈div〉
　　〈vc:cargo-list min-qty = "2"〉
　　〈/vc:cargo-list〉
〈/div〉

⑤ 运行应用程序，最后的结果如图 6.37 所示。

6.9.5　在控制器方法中直接调用视图组件

视图组件通常在视图中调用，但也可以在控制器方法中直接调用。当视图组件没有像控制器一样定义终结点时，可以简单实现一个控制器的 Action，并使用一个 ViewComponentResult 作为返回内容。在控制器方法中调用视图组件的操作方法是：

① 在 Visual Studio 2017 中打开 Controller/HomeController.cs 文件，将以下调用视图组件的代码添加到文件中：

```
public IActionResult IndexVC()
{
    return ViewComponent("CargoList", new { minQty = 3 });
}
```

② 在 Visual Studio 2017 中按 F5 键运行应用程序。在浏览器中输入 http://localhost:43998/Home/IndexVC，浏览器就会显示出视图组件中的货物名称内容，如图 6.39 所示。

图 6.39　在控制器方法中直接调用视图组件的运行结果

6.9.6　指定视图名称

在某些情况下，复杂的视图组件可能需要指定非默认视图，例如，以下代码显示了如何在 InvokeAsync 方法中指定"PVC"视图。具体操作方法如下：

① 修改 CargoListViewComponent 类中的 InvokeAsync 方法，代码如下：

```
using System;
using System.Collections.Generic;
using System.Linq;
using System.Threading.Tasks;
using Microsoft.AspNetCore.Mvc;
using Microsoft.EntityFrameworkCore;
using RazorMvcDemo.Models;

namespace RazorMvcDemo.ViewComponents
{
    public class CargoListViewComponent:ViewComponent
    {
        private readonly EFCoreDemoContext db;
        public CargoListViewComponent(EFCoreDemoContext context)
        {
            db = context;
        }
        public async Task<IViewComponentResult> InvokeAsync(int minQty, bool isDone)
        {
            string MyView = "Default";
```

```
        //If asking for all completed tasks, render with the "PVC" view.
        if(minQty>3)
        {
            MyView = "PVC";
        }
        var items = await GetItemsAsync(minQty, isDone);
        return View(MyView, items);
    }
    private Task<List<Cargo>> GetItemsAsync(int minQty, bool isDone)
    {
        return db.Cargo.Where(x => x.MinStockQty > minQty).ToListAsync();
    }
}
```

② 将 Views/Shared/Components/CargoList/Default.cshtml 文件复制到 CargoList 文件夹中,并重命名为 PVC.cshtml。添加标题以指示正在使用 PVC 视图,代码如下:

```
@{
    ViewData["Title"] = "PVC Cargo List View";
}
<h2>@ViewData["Title"]</h2>

@model IEnumerable<RazorMvcDemo.Models.Cargo>

<ul>
@foreach (var c in Model)
{
    <li>货物名称:@c.Name,货物代码:@c.Code,计量单位:@c.Unit</li>
}
</ul>
```

③ 使用视图组件。在 Visual Studio 2017 中打开 Home/Index.cshmtl 文件,将调用视图组件的代码添加到文件底部,代码如下:

```
<div>
    @await Component.InvokeAsync("CargoList", new { minQty = 4 })
</div>
```

④ 在 Visual Studio 2017 中按 F5 键运行应用程序。在浏览器中显示首页,然后拉到底部,会显示视图组件"PVC"中的内容,而不是显示默认视图组件中的内容,如图 6.40 所示。

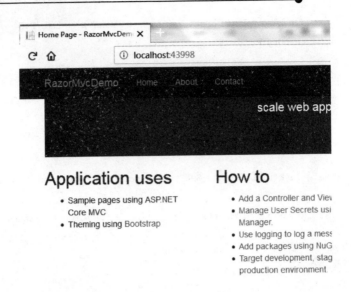

图 6.40 视图组件"PVC"中的内容

6.10 在 ASP.NET Core MVC 中使用控制器处理请求

控制器、操作和操作结果是开发人员如何使用 ASP.NET Core MVC 生成应用的一个基本组成部分。

6.10.1 什么是控制器

控制器接收用户的输入并调用模型和视图去响应用户的输入,所以当单击 Web 页面中的超链接和发送 HTML 表单时,控制器本身不输出任何东西和做任何处理,而只是接收请求并通过执行自定义的代码去处理请求,然后再确定用哪个视图来显示返回的数据。控制器关注的是应用程序中的程序流,处理用户输入的数据,以及提供输出到视图的相关数据中。

控制器类必须在项目根目录下的"Controllers"文件夹中生成,并继承自 Microsoft.Asp NetCore.Mvc.Controller。按照约定,控制器类的名称必须带有"Controller"后缀,即或者是继承自带有"Controller"后缀的类,或者是使用[Controller]特性来修饰该类。

控制器类不可含有关联的[NonController]特性。

在"模型-视图-控制器"模式中,控制器负责对请求进行初始处理和模型的实例化操作。通常情况下,应在模型中执行业务决策。

6.10.2 定义操作

控制器上的公共方法(除了那些使用[NonAction]特性修饰的方法)均是控制器操作。操作上的参数会绑定到请求数据上,并使用模型绑定进行验证。所有模型绑定的内容都会执行模型验证。ModelState.IsValid 属性值指示模型绑定和验证是否成功。

操作方法应包含将请求映射到某个业务关注点上的逻辑。操作方法的工作就是响应 URL 请求,执行通过依赖关系注入来访问的服务,并向浏览器或操作用户做出响应。

操作可以返回任何内容,但是通常情况下返回一个 IActionResult(或其异步方法返回的 Task〈IActionResult〉)实例生成的响应;操作方法负责选择响应的类型;操作结果负责响应的执行。

6.10.3 控制器响应返回的方法

控制器响应返回如下几种类型。

1. 视 图

此类型返回一个使用模型渲染 HTML 的视图。如语句"return View(customer);"将模型传递给视图以进行数据绑定。

2. 已格式化的响应

此类型返回 JSON 格式或类似以特定方式格式化的某个对象的数据交换格式。例如,语句"return Json(customer);"将提供的对象串行化为 JSON 格式。

此类型的其他常见方法包括 File、PhysicalFile 和 VirtualFile。例如,语句"return PhysicalFile(customerFilePath, "text/xml");"返回由 Content-Type 响应头值"text/xml"所描述的 XML 文件。

3. HTTP 状态代码

此类型返回 HTTP 状态代码。此类型的几种帮助程序方法是 BadRequest、NotFound 和 Ok。例如,语句"return BadRequest();"执行时生成"400"状态代码。

4. 重定向

此类型返回一个指向其他 Action 或目标的重定向(使用 Redirect、LocalRedirect、RedirectToAction 或 RedirectToRoute)。例如,语句" return RedirectToAction ("Complete", new {id = 123});"重定向到 Complete,传递一个匿名对象。

重定向类型与 HTTP 状态代码类型的不同之处主要在于是否添加了 Location HTTP 响应头。

5. 内容协商的响应

除了直接返回一个对象外，Action 还可以返回一个内容协商的响应（使用 Ok、BadRequest、CreatedAtRoute 或 CreatedAtAction）。例如，语句"return Ok();"或语句"return CreatedAtRoute("routename",values,newobject());"。

6.11　ASP.NET Core 中的过滤器

6.11.1　过滤器

ASP.NET Core MVC 中的过滤器允许在请求处理管道中的特定阶段之前或之后运行代码。过滤器以一种横切的方式应用到应用程序中，例如开发人员可以针对一个 Action，在 Action 级别使用过滤器；也可以针对一个控制器的所有操作，在控制器级别使用过滤器；还可以针对所有的控制器和所有的操作，在全局级别使用过滤器。过滤器可以合并错误处理。过滤器可以避免跨操作复制代码。过滤器允许配置为全局有效、仅对控制器有效或仅对 Action 有效，这是过滤器的三种不同级别的作用域。过滤器的作用域还决定了过滤器的执行顺序，全局过滤器涵盖类过滤器，类过滤器又涵盖方法过滤器，也可以通过重写行为或显式设置顺序来改变执行顺序。

6.11.2　过滤器的工作原理

不同的过滤器类型会在执行管道的不同阶段运行，因此它们各自有一套适用的场景，故可根据实际要解决的问题以及在请求管道中执行的位置来选择创建不同的过滤器。运行于 MVC Action 调用管道内的过滤器有时被称为过滤器管道，过滤器管道在 MVC 选择要执行的 Action 之后运行。过滤器的工作原理如图 6.41 所示。

每种过滤器的类型都在过滤器管道中的不同阶段执行，如"授权过滤器"只运行在管道的靠前位置，并且其后也不会跟随 Action。其他过滤器（如 Action 过滤器等）可以在管道的其他部分之前或之后执行，如图 6.42 所示。

所有过滤器均可通过不同的接口定义来支持同步和异步实现，具体是根据所需执行任务的不同来选择同步或异步实现。过滤器接口的

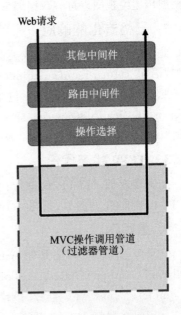

图 6.41　过滤器工作原理

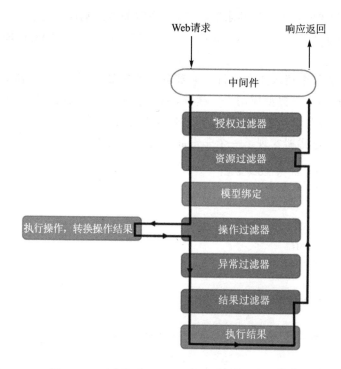

图 6.42　过滤器在过滤器管道中的不同阶段执行

同步和异步版本只能实现一个,要么是同步版本,要么是异步版本,鱼和熊掌不可兼得。如果需要执行异步工作,那么就去实现异步接口;否则应该实现同步接口。框架会先查看过滤器是否实现了异步接口,如果是,则调用该接口;如果不是,则调用同步接口的方法。如果在一个类中同时实现了这两个接口,则只调用异步方法。下面按照图 6.42 依次介绍各个过滤器。

6.11.3　授权过滤器

授权过滤器最先运行,用于过滤掉对于操作没有合适授权的请求。

授权过滤器控制对 Action 方法的访问,也是过滤器管道中第一个被执行的过滤器。它只有一个前置阶段,不像其他大多数过滤器那样既支持前置阶段方法,又支持后置阶段方法。只有在用户使用自己的授权框架时才需要定制授权过滤器。谨记,勿在授权过滤器内抛出异常,这是因为所抛出的异常不会被处理(异常过滤器也不会处理它们)。在出现异常时,可以记录该问题或寻求其他办法。

6.11.4　资源过滤器

资源过滤器是授权过滤器之后最先处理请求的过滤器。资源过滤器在模型绑定之前运行,所以可以影响模型绑定。资源过滤器可以实现缓存或以其他方式让过器管道短路,以减少后续管道请求的步骤,提高服务器的响应性能。

资源过滤器要么实现 IResourceFilter 接口,要么实现 IAsyncResourceFilter 接口。

如果需要使某个请求正在执行的大部分工作短路,则资源过滤器会很有用。例如,资源过滤器的一个典型应用——缓存,如果响应已经被缓存,则过滤器会立即将之置为结果,以绕开管道中其余阶段的多余操作过程。

下例的资源过滤器实现了一个非常简单的缓存功能,代码如下:

```
using Microsoft.AspNetCore.Mvc;
using Microsoft.AspNetCore.Mvc.Filters;
using System;
using System.Collections.Generic;
using System.Linq;
using System.Threading.Tasks;

namespace RazorMvcDemo.MyFilters
{
    public class CacheResourceFilterAttribute : Attribute,IResourceFilter
    {
        private static readonly Dictionary<string, object> _cache = new Dictionary<string, object>();
        private string _cacheKey;
        public void OnResourceExecuting(ResourceExecutingContext context)
        {
            _cacheKey = context.HttpContext.Request.Path.ToString();
            if (_cache.ContainsKey(_cacheKey))
            {
                var cachedValue = _cache[_cacheKey] as ViewResult;
                if (cachedValue != null)
                {
                    context.Result = cachedValue;
                }
            }
        }
        public void OnResourceExecuted(ResourceExecutedContext context)
        {
            if (!String.IsNullOrEmpty(_cacheKey) && !_cache.ContainsKey(_cacheKey))
            {
                var result = context.Result as ViewResult;
                if (result != null)
                {
```

```
                    _cache.Add(_cacheKey, result);
                }
            }
        }
    }
}
```

在 OnResourceExecuting 方法中,如果结果已经在缓存(静态变量_cache)中,则 Result 属性被设置到 context 上,同时 Action 被短路并返回缓存的结果。在 OnResourceExecuted 方法中,如果当前请求的键未被使用过,那么 Result 就会被保存到缓存中,用于之后的请求。

6.11.5 操作过滤器

操作过滤器允许在 Action 方法执行前后对调用进行拦截以执行一些额外操作,它们可用于处理传入某个操作的参数以及从该操作返回的结果。

操作过滤器要么实现 IActionFilter 接口,要么实现 IAsyncActionFilter 接口。操作过滤器中定义了两个方法 OnActionExecuting 和 OnActionExecuted。在调用 Action 方法之前调用 OnActionExecuting,在 Action 方法返回之后调用 OnActionExecuted。委托 ActionExecutionDelegate 用于调用操作方法或下一个操作过滤器,用户可以在调用 ActionExecutionDelegate 之前和之后执行代码,代码如下:

```
using Microsoft.AspNetCore.Mvc.Filters;
using System;
using System.Collections.Generic;
using System.Linq;
using System.Threading.Tasks;

namespace RazorMvcDemo.MyFilters
{
    public class MyAsyncActionFilter: IAsyncActionFilter
    {
        public async Task OnActionExecutionAsync(ActionExecutingContext context, ActionExecutionDelegate next)
        {
            // do something before the action executes
            var resultContext = await next();
            // do something after the action executes; resultContext.Result will be set
        }
    }
}
```

操作过滤器非常适合放置在诸如查看模型绑定结果、修改控制器或输入到操作方法的逻辑中。另外,操作过滤器还可以查看并直接修改操作方法的结果。

下面代码是一个操作过滤器的示例:

```
public class MyActionFilter : IActionFilter
{
    public void OnActionExecuting(ActionExecutingContext context)
    {
        //在这里可以编写所需要的功能
    }
    public void OnActionExecuted(ActionExecutedContext context)
    {
        //在这里可以编写所需要的功能
    }
}
```

OnActionExecuting 方法中的参数 ActionExecutingContext 有三个属性:

① ActionArguments:用于处理对操作的输入。

② Controller:用于处理控制器实例。

③ Result:使操作方法和后续操作过滤器的执行短路。虽然引发异常也会阻止操作方法和后续过滤器的执行,但会被视为失败,而不是一个成功的结果。

OnActionExecuted 方法中的参数 ActionExecutedContext 有两个属性:

① Canceled:如果操作的执行已被另一个过滤器设置为短路,则为 TRUE。

② Exception:如果操作或后续操作过滤器引发了异常,则为非 NULL 值。将此属性设置为 NULL 可有效地"处理"异常,并且会执行 Result,就像是从操作方法正常返回的一样。

下例的操作过滤器可用于验证模型状态,并在状态为无效时返回任何错误。为了显示测试结果,在状态正常时直接返回"404"。具体代码如下:

```
using Microsoft.AspNetCore.Mvc;
using Microsoft.AspNetCore.Mvc.Filters;
using System;
using System.Collections.Generic;
using System.Linq;
using System.Threading.Tasks;

namespace RazorMvcDemo.MyFilters
{
    public class ValidateModelActionFilterAttribute : ActionFilterAttribute
    {
        public override void OnActionExecuting(ActionExecutingContext context)
```

```
        {
            if(!context.ModelState.IsValid)
            {
                context.Result = new BadRequestObjectResult(context.ModelState);
            }
            else
            {
                context.Result = new BadRequestObjectResult("404");
            }
        }
    }
}
```

6.11.6 异常过滤器

异常过滤器用于在向响应正文写入任何内容之前,对未经处理的异常进行检查,并进行一些处理。

异常过滤器可实现 IExceptionFilter 或 IAsyncExceptionFilter 接口。它们可为应用实现常见的错误处理策略。异常过滤器用于处理"未处理异常",包括发生在控制器创建、模型绑定、操作过滤器或操作方法中发生的未处理异常。尽量不要捕获资源过滤器、结果过滤器或 MVC 结果执行中发生的异常。

只在管道内发生异常时才会调用异常过滤器,它们提供了一个单一的位置来实现应用程序内的公共异常处理策略。框架提供了抽象的 ExceptionFilterAttribute,可根据自己的需要继承这个类。异常过滤器适用于捕获 MVC Action 内出现的异常,但异常过滤器并不像错误处理中间件那么灵活,一般来讲优先使用中间件,只有在需要做一些基于所选 MVC Action 的、有别于错误处理的工作时才选择使用异常过滤器。

异常过滤器没有前置和后置两个事件,它只实现 OnException 或 OnExceptionAsync。

若要处理异常,则将 ExceptionContext.ExceptionHandled 属性设置为 TRUE,或者编写响应,这可以停止传播异常。异常过滤器无法将异常转变为"成功",只有操作过滤器才能执行该转变。

下列的异常过滤器示例使用开发人员自定义的错误视图来显示在开发应用时发生的与异常相关的详细信息,代码如下:

```
using Microsoft.AspNetCore.Hosting;
using Microsoft.AspNetCore.Mvc;
using Microsoft.AspNetCore.Mvc.Filters;
using Microsoft.AspNetCore.Mvc.ModelBinding;
```

```csharp
using Microsoft.AspNetCore.Mvc.ViewFeatures;
using System;
using System.Collections.Generic;
using System.Linq;
using System.Threading.Tasks;

namespace RazorMvcDemo.MyFilters
{
    public class MyExceptionFilterAttribute : ExceptionFilterAttribute
    {
        private readonly IHostingEnvironment _hostingEnvironment;
        private readonly IModelMetadataProvider _modelMetadataProvider;

        public MyExceptionFilterAttribute (IHostingEnvironment hostingEnvironment, IModelMetadataProvider modelMetadataProvider)
        {
            _hostingEnvironment = hostingEnvironment;
            _modelMetadataProvider = modelMetadataProvider;
        }
        public override void OnException(ExceptionContext context)
        {
            if (!_hostingEnvironment.IsDevelopment())
            {
                // do nothing
                return;
            }
            Exception exception = context.Exception;
            if (context.ExceptionHandled)
            {
                return;
            }
            var result = new ViewResult { ViewName = "CustomError"};
            /*
             * context.Exception.Message  错误信息
             */
            string messager = context.Exception.Message;
            result.ViewData = new ViewDataDictionary ( _modelMetadataProvider, context.ModelState);
            result.ViewData.Add("Exception", context.Exception);
            result.ViewData.Add("errMsg", messager);
            // TODO: Pass additional detailed data via ViewData
            context.Result = result;
```

        }
    }
}

6.11.7 结果过滤器

结果过滤器的作用与操作过滤器的非常相似。结果过滤器可以在操作即将把结果返回到客户端之前和之后立即运行代码,且仅当操作方法成功执行时,它们才会被运行。

结果过滤器实现了 IResultFilter 或 IAsyncResultFilter 接口。当操作或操作过滤器执行成功并将生成操作结果时,才会执行结果过滤器。

下例的结果过滤器会向响应添加 HTTP 标头,代码如下:

```
using Microsoft.AspNetCore.Mvc.Filters;
using Microsoft.Extensions.Logging;
using System;
using System.Collections.Generic;
using System.Linq;
using System.Threading.Tasks;

namespace RazorMvcDemo.MyFilters
{
    public class AddHeaderAttribute: IResultFilter
    {
        private readonly string _name;
        private readonly string _value;
        public AddHeaderAttribute(string name, string value)
        {
            _name = name;
            _value = value;
        }
        public override void OnResultExecuting(ResultExecutingContext context)
        {
            context.HttpContext.Response.Headers.Add(_name, new string[]{ _value });
            base.OnResultExecuting(context);
        }
        public void OnResultExecuted(ResultExecutedContext context)
        {
            // Can't add to headers here because response has already begun.
        }
    }
}
```

当操作或操作过滤器生成操作结果时,结果过滤器针对成功的结果可以替换或更改操作结果。当异常过滤器处理异常时，不执行结果过滤器。

由于 OnResultExecuting 方法运行于操作结果执行之前，因此其可通过 ResultExecutingContext.Result 来操作操作结果。如果将 ResultExecutingContext.Cancel 设置为 TRUE，则 OnResultExecuting 方法可短路操作结果和后续结果过滤器的执行。如果发生了短路，MVC 将不会修改响应，所以当发生短路时，为了避免生成空响应，一般应该直接修改响应对象。如果在 OnResultExecuting 方法内抛出异常，那么将阻止操作结果以及后续过滤器的执行，但会被当作失败的结果(而非成功的结果)。

OnResultExecuted 方法运行于操作结果执行之后。也就是说，如果没有抛出异常，则响应可能就会被发送到客户端而且不能再修改。如果操作结果在执行中被其他过滤器短路，则 ResultExecutedContext.Canceled 将被设置为 TRUE。如果操作结果或后续结果过滤器抛出异常，则 ResultExecutedContext.Exception 将被置为非 NULL 值。把 ResultExecutedContext.Exception 设置为 NULL 可有效地"处理"异常，并且防止异常在之后的管道内被 MVC 重新抛出。在处理结果过滤器内的异常时，可能无法将某些数据写入响应中。如果操作结果在执行中途抛出异常，并且标头已经刷新到客户端，那么将没有任何可靠的机制来发送失败代码。

6.11.8　内置过滤器特性

ASP.NET Core 框架中包含许多内置过滤器，这些内置过滤器可用作自定义实现的基类，也可作为特性使用。这些特性包括：

- ActionFilterAttribute；
- ExceptionFilterAttribute；
- ResultFilterAttribute；
- FormatFilterAttribute；
- ServiceFilterAttribute；
- TypeFilterAttribute。

例如，下例的结果过滤器会向响应添加标头，代码如下：

```
using Microsoft.AspNetCore.Mvc.Filters;
using System;
using System.Collections.Generic;
using System.Linq;
using System.Threading.Tasks;

namespace RazorMvcDemo.MyFilters
{
    public class AddHeaderAttribute : ResultFilterAttribute
    {
        private readonly string _name;
```

```
        private readonly string _value;

        public AddHeaderAttribute(string name, string value)
        {
            _name = name;
            _value = value;
        }
        public override void OnResultExecuting(ResultExecutingContext context)
        {
            context.HttpContext.Response.Headers.Add(_name, newstring[]{ _value });
            base.OnResultExecuting(context);
        }
    }
}
```

允许过滤器特性采用参数,如上面的示例所示。可将特性添加到控制器或操作方法中,并指定 HTTP 标头的名称和值。

6.11.9 取消和设置短路

通过设置传入过滤器方法的上下文参数中的 Result 属性,可以在过滤器管道的任意位置设置短路。例如,下面的资源过滤器将阻止管道内所有在它之后的过滤器,包括所有的 Action 过滤器,代码如下:

```
using Microsoft.AspNetCore.Mvc;
using Microsoft.AspNetCore.Mvc.Filters;
using System;
using System.Collections.Generic;
using System.Linq;
using System.Threading.Tasks;

namespace RazorMvcDemo.MyFilters
{
    public class ShortResourceFilterAttribute : Attribute, IResourceFilter
    {
        public void OnResourceExecuting(ResourceExecutingContext context)
        {
            context.Result = new ContentResult()
            {
                Content = "Resource unavailable - 资源过滤器短路"
            };
        }
        public void OnResourceExecuted(ResourceExecutedContext context)
```

```
        {
        }
    }
}
```

6.11.10 依赖关系注入

过滤器的添加方式有两种：按类型或按实例。如果添加实例，则该实例将用于每个请求；如果添加类型，则将激活该类型，这意味着将为每个请求创建一个实例，并且依赖关系注入（DI）将填充所有构造函数的依赖项。

如果将过滤器以特性形式实现并直接添加到控制器类或操作方法中，则该过滤器不能由依赖关系注入（DI）提供构造函数的依赖项，这是因为特性在应用时所需的构造函数参数必须由使用处直接提供，这是特性工作原理上的限制。

如果过滤器需要从 DI 中获得依赖项，那么可以使用 TypeFilterAttribute 和 ServiceFilterAttribute 两个特性，将过滤器应用于类（Class）或操作（Action）方法。

使用 ServiceFilterAttribute 特性之前需要在 Startup.cs 文件的 ConfigureServices 方法中进行过滤器注册，如果没有注册，则会抛出异常。ServiceFilter 的使用示例代码如下：

```
[ServiceFilter(typeof(AddHeaderFilterWithDi))]
public IActionResult Index()
{
    return View();
}
```

在使用 ServiceFilter 之前必须在 ConfigureServices 中为 AddHeaderFilterWithDI 类型注册，代码如下：

```
public void ConfigureServices(IServiceCollection services)
{
    services.AddMvc();
    services.AddScoped<AddHeaderFilterWithDi>();
}
```

TypeFilterAttribute 与 ServiceFilterAttribute 很像，但 TypeFilterAttribute 在使用之前不需要先进行注册，因为它使用 Microsoft.Extensions.DependencyInjection.ObjectFactory 来实例化类型。

同样，TypeFilterAttribute 能可选地接受该类型的构造函数参数，代码如下：

```
[TypeFilter(typeof(CacheResourceFilterAttribute))]//资源过滤器
public class HomeController : Controller
{
}
```

6.11.11 过滤器示例

根据前面的介绍,下面通过一个项目进行实践,具体步骤是:

① 在 Visual Studio 2017 中打开 RazorMvcDemo 项目,并在"解决方案资源管理器"中新建一个文件夹"MyFilters"。

② 在"解决方案资源管理器"中右击刚才创建的文件夹"MyFilters",添加一个类"CacheResourceFilterAttribute.cs"文件,这个类用来实现资源过滤器,是一个非常简单的缓存功能,代码见 6.11.4 小节。

③ 继续在"MyFilters"文件夹中添加一个类"ValidateModelActionFilterAttribute.cs"文件,这个类用来实现操作过滤器,示例代码见 6.11.5 小节。

④ 继续在"MyFilters"文件夹中添加一个类"MyExceptionFilterAttribute.cs"文件,这个类用来实现异常过滤器,示例是使用开发人员自定义的错误视图来显示在开发应用时发生的与异常相关的详细信息,代码见 6.11.6 小节。

⑤ 继续在"MyFilters"文件夹中添加一个类"AddHeaderFilterWithDI.cs"文件,该类用来实现一个添加 HTTP 标头的结果过滤器,代码如下:

```
using Microsoft.AspNetCore.Mvc.Filters;
using Microsoft.Extensions.Logging;
using System;
using System.Collections.Generic;
using System.Linq;
using System.Threading.Tasks;

namespace RazorMvcDemo.MyFilters
{
    public class AddHeaderFilterWithDI : IResultFilter
    {
        private ILogger _logger;
        public AddHeaderFilterWithDI(ILoggerFactory loggerFactory)
        {
            _logger = loggerFactory.CreateLogger<AddHeaderFilterWithDI>();
        }
        public void OnResultExecuting(ResultExecutingContext context)
        {
            var headerName = "OnResultExecuting";
            context.HttpContext.Response.Headers.Add(headerName, new string[] { "ResultExecutingSuccessfully" });
            _logger.LogInformation($"Header added: {headerName}");
        }
        public void OnResultExecuted(ResultExecutedContext context)
```

```
            {
                // Can't add to headers here because response has already begun.
            }
        }
    }
```

⑥ 继续在"MyFilters"文件夹中添加一个类"AddHeaderAttribute.cs"文件,该类用来实现一个添加 HTTP 标头的结果过滤器,代码如下:

```
using Microsoft.AspNetCore.Mvc.Filters;
using System;
using System.Collections.Generic;
using System.Linq;
using System.Threading.Tasks;

namespace RazorMvcDemo.MyFilters
{
    public class AddHeaderAttribute : ResultFilterAttribute
    {
        private readonly string _name;
        private readonly string _value;

        public AddHeaderAttribute(string name, string value)
        {
            _name = name;
            _value = value;
        }
        public override void OnResultExecuting(ResultExecutingContext context)
        {
            context.HttpContext.Response.Headers.Add(_name, new string[] { _value });
            base.OnResultExecuting(context);
        }
    }
}
```

⑦ 在解决方案资源管理器中打开"Controllers\HomeController.cs"文件,添加各种过滤器。在下面的代码中,ShortResourceFilter 和 AddHeader 过滤器都指向了名为 HomeController 的操作(Action)方法 About。

```
using System;
using System.Collections.Generic;
using System.Diagnostics;
using System.Linq;
using System.Threading.Tasks;
```

```csharp
using Microsoft.AspNetCore.Mvc;
using RazorMvcDemo.Models;
using RazorMvcDemo.MyFilters;

namespace RazorMvcDemo.Controllers
{
    [AddHeader("Author", "Chill@address.com")]          //结果过滤器
    [TypeFilter(typeof(CacheResourceFilterAttribute))]  //资源过滤器
    public class HomeController : Controller
    {
        [ValidateModelActionFilterAttribute]            //操作过滤器
        public IActionResult Index()
        {
            return View();
        }
        [ShortResourceFilter]                            //短路过滤器
        public IActionResult About()
        {
            ViewData["Message"] = "Successful access to resource - header should be set. About View";
            ViewBag.Supplier = "MUEY";
            ViewBag.Cargo = new Cargo()
            {
                Name = "Steve",
                Code = "St123",
                Country = "中国",
                Unit = "个",
                MinStockQty = 10,
                GrossWt = 100
            };
            return View();
        }
        public IActionResult Contact()
        {
            ViewData["Message"] = "Your contact page. 时间为:" + DateTime.Now.ToString();
            return View();
        }
    }
}
```

⑧ 在 Visual Studio 2017 中按 F5 键运行应用程序,并在浏览器中浏览首页(Index)。由于在 Index 方法上添加了 ValidateModelActionFilterAttribute 过滤器,因此会看到如图 6.43 所示的结果。从图中看到并没有显示首页,而是直接显示了"404",说明操作过滤器起作用了。

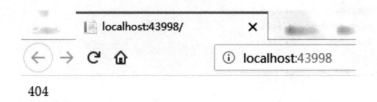

图 6.43 操作过滤器输出"404"

⑨ 现在对操作过滤器"ValidateModelActionFilterAttribute.cs"进行修改,代码如下:

```
using Microsoft.AspNetCore.Mvc;
using Microsoft.AspNetCore.Mvc.Filters;
using System;
using System.Collections.Generic;
using System.Linq;
using System.Threading.Tasks;

namespace RazorMvcDemo.MyFilters
{
    public class ValidateModelActionFilterAttribute : ActionFilterAttribute
    {
        public override void OnActionExecuting(ActionExecutingContext context)
        {
            if (!context.ModelState.IsValid)
            {
                context.Result = new BadRequestObjectResult(context.ModelState);
            }
        }
    }
}
```

⑩ 在 Visual Studio 2017 中按 F5 键运行应用程序。浏览器首先显示首页,然后单击 About 菜单,浏览 About 页面。由于在 About 方法上添加了 ShortResourceFilter 过滤器,因此,首先运行的是 ShortResourceFilter,该过滤器对管道中的其余部分进行了短路处理,故控制器(HomeController)上的 AddHeader 过滤器并不会运行。如果这两个过滤器都应用于操作方法级别,则只要 ShortResourceFilter 先运行,其运行结果也是一样的,如图 6.44 所示。

下面来看看全局过滤器是如何起作用的。

图 6.44 过滤器短路操作

① 修改 Index 方法中的代码，直接抛出异常，代码如下：

```
public IActionResult Index()
{
    throw (new Exception("异常过滤器,一个错误"));
}
```

② 在 startup.cs 文件的 ConfigureServices 方法中注册异常过滤器，以便使用。此例中把异常过滤器设为全局过滤器，代码如下：

```
public void ConfigureServices(IServiceCollection services)
{
    services.AddDbContext<EFCoreDemoContext>(options => options.UseSqlServer(Configuration.GetConnectionString("EFCoreDemoContext")));
    services.AddMvc(opt =>
    {
        opt.Filters.Add<MyFilters.MyExceptionFilterAttribute>();//异步过滤器
    });        //.AddXmlSerializerFormatters();
}
```

③ 在 Visual Studio 2017 中按 F5 键运行应用程序。浏览器首先显示首页，然后应用程序直接抛出了一个异常，如图 6.45 所示。

图 6.45 异常过滤器

④ 在浏览器中单击 Contact 菜单，浏览 Contact 页面，接着看一下结果过滤器与资源过滤器的作用。在打开 Contact 页面后，AddHeader 过滤器输出的结果是：响应

头显示在左下角,如图 6.46 所示。

图 6.46　添加响应头

⑤ 再次单击 Contact 菜单,由于在第一次单击时资源过滤器 CacheResourceFilter-Attribute 已经将结果保存到缓存中,故在 OnResourceExecuting 方法中会将缓存结果设置到 context 上,同时 Action 被短路并返回缓存的结果。注意页面中的时间和右下角的时间,如图 6.47 所示。

图 6.47　缓存过滤器

⑥ 在添加头时不能使用中文,只能使用 ASCII 码。例如使用过滤器代码"[AddHeader("Author","张小三")]"则会报错,如图 6.48 所示。

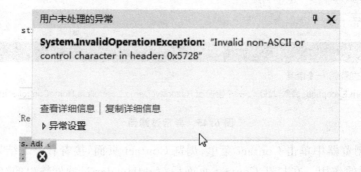

图 6.48　异常信息

6.12　ASP.NET Core 中的区域

区域（Areas）是 ASP.NET MVC 用来将相关功能组织成一组单独的命名空间（用于路由）和文件夹结构（用于视图）的功能。使用区域创建层次结构的路由，是通过添加另一个路由参数 Area 到 Controller 和 Action 中实现的。

区域是应用程序内的一个 MVC 结构，它允许将模型、视图和控制器分成一组组单独的功能。在 MVC 项目中，模型、控制器和视图等逻辑组件保存在不同的文件夹中，MVC 使用命名约定来创建这些组件之间的关系。对于大型 ASP.NET Core MVC Web 应用，将应用分为一个个独立的高级功能区域不失为一个良好的解决方案，例如，一个多业务单元的电子商务应用，如结账、计费和搜索等。每个单元都有自己的逻辑组件视图、控制器和模型，在这种情况下，可使用区域对同一项目中的业务组件进行物理分区。

区域具有如下特性：
- 一个 ASP.NET Core MVC 应用可以有任意数量的区域；
- 每个区域都有自己的控制器、模型和视图；
- 可以将大型 MVC 项目分为可以独立工作的多个高级组件；
- 在不同的区域可以创建相同名称的控制器。

下面的示例演示了如何创建和使用区域。假设有一个信息管理系统应用，它有两组不同的控制器和视图：后台和库区。操作步骤如下：

① 在 Visual Studio 2017 的"解决方案资源管理器"中右击项目名称，在弹出的快捷菜单中选择"添加"→"区域"菜单项，如图 6.49 所示。

图 6.49　添加区域

② 在弹出的"添加 MVC 区域"对话框中输入区域名称"Areas",创建该文件夹,如图 6.50 所示。

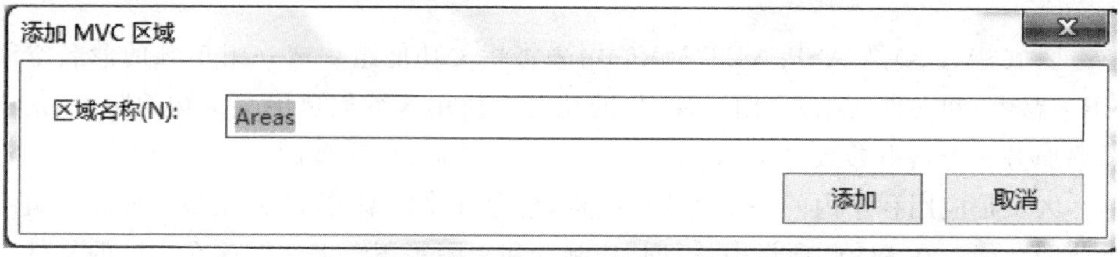

图 6.50 输入区域名称

③ 在 Visual Studio 2017 的"解决方案资源管理器"中右击刚才创建的"Areas"文件夹,在弹出的快捷菜单中选择"添加"→"区域"菜单项。在弹出的"添加 MVC 区域"对话框中输入区域名称"Account"。则创建的 MVC 区域的文件夹结构如图 6.51 所示。

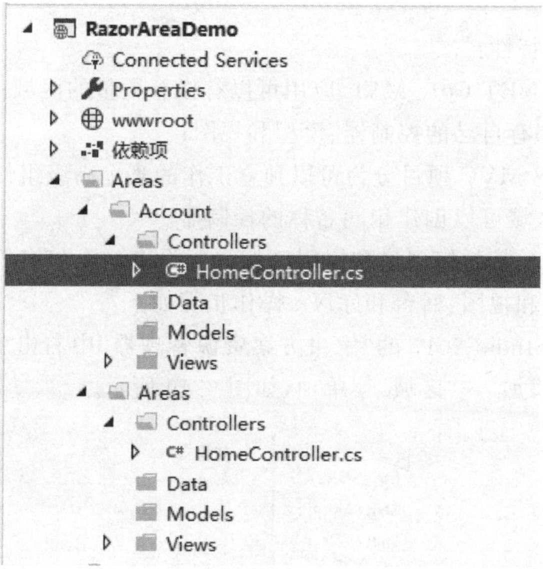

图 6.51 MVC 区域的文件夹结构

④ 在定义了文件夹层次结构之后,需要告知 MVC 每一个相关区域的控制器的关联关系。现在分别在 Account\Controlls 和 Areas\Controlls 目录下创建 HomeController.cs 文件,并分别在类名上面使用[Area]特性进行修饰,代码如下:

```
using System;
using System.Collections.Generic;
using System.Diagnostics;
using System.Linq;
using System.Threading.Tasks;
using Microsoft.AspNetCore.Mvc;
```

```csharp
using RazorAreaDemo.Models;

namespace RazorAreaDemo.Areas.Account.Controllers
{
    [Area("Account")]
    public class HomeController : Controller
    {
        public IActionResult Index()
        {
            return View();
        }
        public IActionResult About()
        {
            ViewData["Message"] = "Your application description page.";
            return View();
        }
        public IActionResult Contact()
        {
            ViewData["Message"] = "Your contact page.";
            return View();
        }
        public IActionResult Privacy()
        {
            return View();
        }
        [ResponseCache(Duration = 0, Location = ResponseCacheLocation.None, NoStore = true)]
        public IActionResult Error()
        {
            return View(new ErrorViewModel { RequestId = Activity.Current?.Id ?? HttpContext.TraceIdentifier });
        }
    }
}
```

⑤ 在 Views 文件夹中添加相应的视图文件和共享文件。区域内部的 Razor 视图可以使用外面的布局页面(也就是根目录下的/Views/Shared)。当然可以为每个区域定义不同的布局页面。这里，针对 Account 区域使用了根目录下的布局，而对 Areas 区域则没有使用布局，如图 6.52 所示。

⑥ 在 Visual Studio 2017 中打开 Startup.cs 文件，修改 Configure 方法，在这个方法中添加一个名为 areas 的路由定义，为新创建的区域设置路由定义，代码如下：

```csharp
public void Configure(IApplicationBuilder app, IHostingEnvironment env)
{
    if (env.IsDevelopment())
```

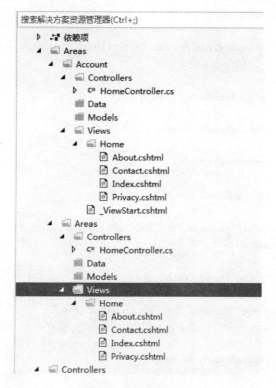

图 6.52　添加视图文件

```
{
    app.UseDeveloperExceptionPage();
}
else
{
    app.UseExceptionHandler("/Home/Error");
}
app.UseStaticFiles();
app.UseCookiePolicy();
app.UseMvc(routes =>
{
    routes.MapRoute(
        name: "areas",
        template: "{area:exists}/{controller=Home}/{action=Index}/{id?}");
    routes.MapRoute(
        name: "default",
        template: "{controller=Home}/{action=Index}/{id?}");
});
}
```

⑦ 在 Visual Studio 2017 中按 F5 键运行应用程序。在浏览器的地址栏中输入

http://localhost:5000/Account，将调用 Account 区域中 HomeController 的 Index 操作方法，如图 6.53 所示。

图 6.53　Account 区域首页

⑧ 在浏览器的地址栏中输入 http://localhost:5000/Areas，将调用 Areas 区域中 HomeController 的 Index 操作方法。由于 Areas 区域中没有页面布局，所以页面中的图片显示与图 6.53 不一样，会多出 1、2、3 三个序号，同时文字也不在图片上显示，如图 6.54 所示。

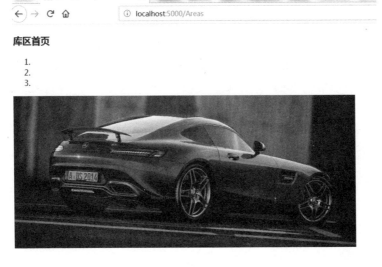

图 6.54　Areas 区域首页

最后讲一下，当 MVC 尝试呈现 Area 中的视图时，首先默认尝试搜索以下位置：

/Areas/⟨Area-Name⟩/Views/⟨Controller-Name⟩/⟨Action-Name⟩.cshtml
/Areas/⟨Area-Name⟩/Views/Shared/⟨Action-Name⟩.cshtml
/Views/Shared/⟨Action-Name⟩.cshtml

当然以上的默认位置可以通过 Microsoft.AspNetCore.Mvc.Razor.RazorViewEngineOptions 上的 AreaViewLocationFormats 进行修改。例如，在下面的代码中，将区域文件夹名称从"Areas"改为了"Stocks"。

```
services.Configure<RazorViewEngineOptions>(options =>
{
    options.AreaViewLocationFormats.Clear();
    options.AreaViewLocationFormats.Add("/Stocks/{2}/Views/{1}/{0}.cshtml");
    options.AreaViewLocationFormats.Add("/Stocks/{2}/Views/Shared/{0}.cshtml");
    options.AreaViewLocationFormats.Add("/Views/Shared/{0}.cshtml");
});
```

需要注意的是，这里唯独看重的是 Views 文件夹的结构，而 Controllers 和 Models 等文件夹则无关紧要。

第 7 章

依赖注入

7.1 什么是依赖注入

依赖注入（Dependency Injection，DI）是一种用于在对象与被调用者或依赖项之间实现松散耦合的技术。该技术不是直接实例化被调用者或使用静态引用，而是以某种方式向类提供该类需要的被调用者对象。大多数情况下，类通过其构造函数声明它们的依赖项，从而允许它们遵循显式依赖关系原则。此方法称为"构造函数注入"。

在遵循 DI 原则来设计类时，由于它们与被调用者没有直接的、硬编码的依赖关系，所以它们之间的耦合更为松散。这遵循了依赖倒置原则，其中指出"高级模块不应依赖于低级模块；同时两者都应依赖于抽象"。类通过该类的构造函数提供被调用者的抽象（通常是接口），这些接口的实现作为参数提供。

当系统被设计为使用 DI 时，许多类通过其构造函数（或属性）来请求它们的依赖项，所以要有一个专门用来创建这些类以及它们相关的依赖项的类，这些类统称为容器，或者更具体地说，叫控制反转（IoC）容器或依赖注入（DI）容器。

依赖注入是什么？为什么要使用它？

7.1.1 什么是依赖

首先来讲一个生活的例子。老王是一个电工，现在接到派单去做维修任务，他要先去申领个螺丝刀。

老王对库管老赵说："请给我一把可以拧三角螺丝的三角螺丝刀。"库管老赵就从仓库里拿了一把牛头牌三角螺丝刀给老王。

在这个例子中，电工老王只需告诉库管老赵要一把"可以拧三角螺丝的三角螺丝刀"即可，他不用关心螺丝刀的品牌，也不用采购螺丝刀，更不用关心这把螺丝刀是怎么来的；而对于库管老赵，他只需提供满足这个要求的一把螺丝刀即可，不用去关心老王拿着这把螺丝刀之后去干什么。所以老王和老赵都只关心"可以拧三角螺丝"这个要求即可，也就是说，如果后期仓库里不再提供牛头牌螺丝刀，而是提供了同样规格的马头牌螺丝刀，无论换了什么牌子和样式，只要仍满足这个要求，老王仍然可以正常工作。他们定义了一个规则（比如接口 IScrewdriver），二者都依赖于这个规则，然后仓库无论提供牛头牌（ScrewdriverNiu：IScrewdriver）还是马头牌（Screwdriver-

Ma:IScrewdriver),都不影响正常工作。

在程序中也类似,当一个类需要另一个类协作来完成工作的时候就产生了依赖,比如AccountController控制器需要完成与用户相关的注册、登录等事情,其中的登录由EF结合Idnetity来完成,所以就封装了一个EFLoginService。这里AccountController就有一个ILoginService的依赖,如图7.1所示。

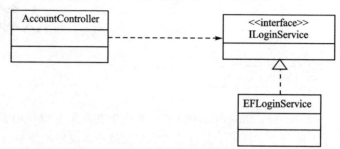

图7.1 登 录

这里有一个设计原则:依赖于抽象,而不是具体的实现。所以就给EFLoginService定义了一个接口,抽象了LoginService的行为。

7.1.2 什么是注入

注入体现的是一个IoC(控制反转的思想)。在反转之前,先看看正转。AccountController自己来实例化需要的依赖,代码如下:

```
private ILoginService<AppUser> _loginService;
public AccountController()
{
    _loginService = new EFLoginService()
}
```

计算机业的大师们说,这样不好,不应该自己创建它,而应该由调用者给自己。于是就通过构造函数让外界把这个依赖传给自己,代码如下:

```
public AccountController(ILoginService<AppUser> loginService)
{
    _loginService = loginService;
}
```

把依赖的创建丢给其他人,其他人在依赖创建成功之后,把依赖丢给你,而你自己只负责使用依赖,而不负责创建依赖的过程就可理解为注入。

7.1.3 为什么要反转

首先来看一下机械表齿轮组。打开机械式手表的后盖,会看到各个齿轮分别带动时针、分针和秒针顺时针旋转,从而在表盘上产生正确的时间。手表中的齿轮组拥有多个独立的齿轮,这些齿轮相互啮合在一起,协同工作,共同完成某项任务,如图 7.2 所示。从图中可以看到,在这样的齿轮组中,如果有一个齿轮出了问题,就可能影响到整个齿轮组的正常运转。

图 7.2 机械表齿轮组

齿轮组中齿轮之间的啮合关系,与软件系统中对象之间的耦合关系非常相似,对象之间的耦合关系是无法避免的,也是必要的,这是协同工作的基础。现在,伴随着工业级应用的规模越来越庞大,对象之间的依赖关系也越来越复杂,经常会出现对象之间的多重依赖关系。

耦合关系不仅会出现在对象与对象之间,也会出现在软件系统的各模块之间,以及软件系统和硬件系统之间。如何降低系统之间、模块之间和对象之间的耦合度,是软件工程永远追求的目标之一。为了解决对象之间耦合度过高的问题,软件专家 Michael Mattson 提出了 IoC 理论,用来实现对象之间的"解耦"。

尽量避免在业务变化时因改动代码而带来的问题。比如现在要把在 EF 中去验证登录改为在 Redis 中去验证,这样就增加了一个 RedisLoginService,如图 7.3 所示。这时只需在原来注入的地方修改一下即可,修改后的代码如下:

```
public AccountController(ILoginService<AppUser> loginService)
{
    _loginService = loginService;
}
//用 Redis 来替换原来的 EF 登录
```

```
var controller = new AccountController(new RedisLoginService());
controller.Login(userName, password);
```

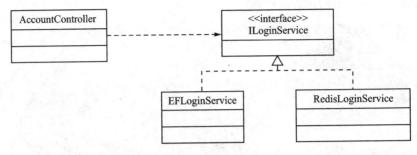

图 7.3　登录接口继承

7.1.4　何为容器

继续 7.1.1 小节中关于维修任务的例子。库管老赵为什么会提供给老王牛头牌而不是马头牌的螺丝刀呢？那是因为领导给了他一份如下的构建仓库的物品购置及发放清单：

A. 当有人要三角螺丝刀时，给他一个牛头牌的螺丝刀，当再有人来要时再给另一把。

B. 但对于冲击钻，每个小组只能给一台，小组内所有人共用这一台。

C. 卡车更是全单位只有一辆，谁申请都是同一辆。

在使用 AccountController 时，通过代码创建了一个 ILoggingServce 的实例。想象一下，一个系统中如果有 100 个这样的地方，是不是要在 100 个地方都做这样的事情？控制是反转了，依赖的创建也移交到了外部。但现在的问题是依赖太多，因此就需要一个地方来统一管理系统中所有的依赖，故而容器诞生了。

容器本质上是一个工厂，它负责提供请求类型的实例。如果给定的类型已声明它具有依赖项，并且容器已经配置提供了这些依赖类型，那么容器将在创建请求实例时创建依赖项。通过这种方式，可以为类提供复杂的依赖关系图，而无须任何硬编码对象的构造。除了创建包含依赖项的对象外，容器通常还可以管理应用程序中的对象生命周期。

创建请求的对象（父对象）和它需要的所有对象（子对象），以及这些对象（父子对象）需要的所有对象被称为对象图。同样，必须被解析的依赖关系的集合通常被称为依赖关系树或依赖项关系图。容器负责解决图中所有的依赖关系，并返回完全解析后的服务。

简单来说，容器负责以下两件事：

① 绑定服务与实例之间的关系；

② 获取实例，并对实例进行管理（创建与销毁）。

7.2 .NET Core DI

7.2.1 构造函数注入行为

　　构造函数是一种特殊的方法，主要用来在创建对象时初始化对象，即为对象成员变量赋初始值。构造函数注入要求写在公共的构造函数内；否则，应用程序会引发 InvalidOperationException 异常——构造函数注入请确保该类型有具体的实现类，并且为公共构造函数的所有参数注册服务。

　　构造函数注入要求只存在一个适用的构造函数，虽然它支持构造函数重载，但其参数能够全部通过依赖注入来实现的构造函数重载只能存在一个。如果存在多个，应用程序将引发 InvalidOperationException 异常——构造函数可以接受依赖注入没有提供的参数，但这些参数必须给定默认值。请看如下代码：

```
public class AboutModel: PageModel
{
    private readonly ILogger m_logger;
    private readonly RazorMvcBooks.Models.EFCoreDemoContext _context;

    public AboutModel(ILogger<AboutModel> logger, RazorMvcBooks.Models.EFCoreDemoContext context, ICar transientCar, IJeep scopedJeep, IBus singletonBus)
    {
        m_logger = logger;
        _context = context;
        TransientCar = transientCar;
        ScopedJeep = scopedJeep;
        SingletonBus = singletonBus;
    }
}
```

7.2.2 实例的注册

　　在.NET Core 中 DI 的核心分为两个组件 IServiceCollection 和 IServiceProvider，它们各自的功能如下：

　　① IServiceCollection 负责注册；

　　② IServiceProvider 负责提供实例。

　　默认的 ServiceCollection 有三个方法（AddTransient、AddScoped 和 AddSingleton），这三个扩展方法放置在 Microsoft.Extensions.DependencyInjection 命名空间中，代码如下：

```
public void ConfigureServices(IServiceCollection services)
{
    services.AddDbContext<EFCoreDemoContext>(options => options.UseSqlServer(Configuration.GetConnectionString("EFCoreDemoContext")));

    services.AddMvc();
    services.AddTransient<ICar, Vehicle>();
    services.AddScoped<IJeep, Vehicle>();
    services.AddSingleton<IBus, Vehicle>();
}
```

这三个方法都是将实例注册进去，只是实例的生命周期不一样。这三个注册方法都有两个泛型参数，第一个泛型参数的类型表示将从容器请求的类型（通常是一个接口）；第二个泛型参数的类型表示将由容器实例化并用于满足此类请求的具体类型。

7.2.3 实例的生命周期

从上面的示例代码中可以看出，.NET Core DI 有三个注册的方法 AddSingleton、AddScoped 和 AddTransient，这三个方法对应的三个选项 Singleton、Scoped 和 Transient 体现出三种对服务对象生命周期的控制形式：

① Singleton：在当前请求（Request）的应用程序生命周期内返回相同的一个实例。

② Scoped：在同一个作用域内只初始化一个实例，可以理解为如果同一个请求获取多次的话，会得到相同的实例。

③ Transient：每次访问时都创建一个新的实例。这种生存期适合轻量级和无状态的服务。

这三个选项对应了 Microsoft.Extensions.DependencyInjection.ServiceLifetime 的三个枚举值，代码如下：

```
public enum ServiceLifetime
{
    Singleton,
    Scoped,
    Transient
}
```

三个方法 AddSingleton、AddScoped 和 AddTransient 用于将抽象类型映射到为每个需要它的对象分别实例化的具体服务，这称为服务的生存期。为注册的每个服务选择适当的生存期非常重要。选择生存期时需要考虑如下问题：

① 是否应该向每个请求它的类提供一个新的服务实例？

② 是否在整个给定的 Web 请求中使用同一个实例？

③ 是否应该在应用程序生存期内使用单例？

　　服务可以通过多种方式注册到容器中。通过前面的学习已经知道了如何通过指定要使用的具体类型来注册一个给定类型的服务实现。此外，可以指定一个工厂，然后将其用于按需创建实例。另一种方法是直接指定要使用的类型的实例，在这种情况下，容器将永远不会尝试创建实例（也不会释放实例）。

　　为了能够更好地理解生命周期的概念，下面通过一个简单的接口，将一个或多个任务表示为具有唯一标识符 Id 的操作。根据此服务的生存期配置方式，容器将向请求的类提供相同或不同的实例。为了明确正在请求哪个生存期，按照如下步骤可以为每个生命周期选项创建一个类型：

① 在 Visual Studio 2017 的"解决方案资源管理器"中右击"RazorMvcBooks"项目，在弹出的快捷菜单中选择"添加"→"新文件夹"菜单项创建一个新文件夹，并命名为"DI"。

② 在 Visual Studio 2017 的"解决方案资源管理器"中右击"DI"文件夹，在弹出的快捷菜单中选择"添加"→"类"菜单项创建一个新文件，并命名为"IVehicle"，代码如下：

```
using System;
using System.Collections.Generic;
using System.Linq;
using System.Threading.Tasks;

namespace RazorMvcBooks.DI
{
    public interface IVehicle
    {
        Guid VehicleId { get; }
        string CreateTime { get; }
    }
}
```

③ 重复第②步，创建"ICar""IJeep""IBus"三个接口文件，代码如下：

```
using System;
using System.Collections.Generic;
using System.Linq;
using System.Threading.Tasks;

namespace RazorMvcBooks.DI
{
```

```csharp
    public interface ICar:IVehicle
    {
    }
}
using System;
using System.Collections.Generic;
using System.Linq;
using System.Threading.Tasks;

namespace RazorMvcBooks.DI
{
    public interface IBus:IVehicle
    {
    }
}
using System;
using System.Collections.Generic;
using System.Linq;
using System.Threading.Tasks;

namespace RazorMvcBooks.DI
{
    public interface IJeep : IVehicle
    {
    }
}
```

④ 重复第②步,创建"Vehicle"类文件来实现以上创建的接口。Vehicle 类在构造函数中接受一个 Guid,如果没有提供,则使用一个新的 Guid,代码如下:

```csharp
using System;
using System.Collections.Generic;
using System.Linq;
using System.Threading.Tasks;

namespace RazorMvcBooks.DI
{
    public class Vehicle:ICar,IJeep,IBus
    {
        public Vehicle():this(Guid.NewGuid())
        {
        }
```

```
        public Vehicle(Guid id)
        {
            VehicleId = id;
        }
        public Guid VehicleId { get; privateset; }
        public string CreateTime { get { return DateTime.Now.ToString("yyyy-MM-dd HH:mm:ss"); } }
    }
}
```

⑤ 在 Visual Studio 2017 的"解决方案资源管理器"中打开 Startup.cs 文件,并在 ConfigureServices 方法中,将每个类型用不同的生命周期添加到容器中,代码如下:

```
public void ConfigureServices(IServiceCollection services)
{
    services.AddDbContext<EFCoreDemoContext>(options => options.UseSqlServer(Configuration.GetConnectionString("EFCoreDemoContext")));

    services.AddMvc();
    services.AddTransient<ICar, Vehicle>();
    services.AddScoped<IJeep, Vehicle>();
    services.AddSingleton<IBus, Vehicle>();
}
```

请注意,由于已经注册了依赖于每个其他 IVehicle 类型的 Vehicle,因此对于每个 Vehicle 类型来说,可以在请求中明确该服务是获得与控制器相同的实例,还是获得一个新实例。此服务的全部作用就是将其依赖项作为属性公开,以便它们可以显示在视图中。

⑥ 重复第②步,创建"VehicleServicecs"类文件,代码如下:

```
using System;
using System.Collections.Generic;
using System.Linq;
using System.Threading.Tasks;

namespace RazorMvcBooks.DI
{
    public class VehicleServicecs
    {
        public VehicleServicecs(ICar transientCar, IJeep scopedJeep, IBus singletonBus)
        {
            TransientCar = transientCar;
```

ASP.NET Core 应用开发入门教程

```
            ScopedJeep = scopedJeep;
            SingletonBus = singletonBus;
        }
        public ICar TransientCar { get; }
        public IJeep ScopedJeep { get; }
        public IBus SingletonBus { get; }
    }
}
```

⑦ 为了演示对应用程序的各个独立请求之内和之间的对象的生存期,该示例在 About 页面中请求了每一种 IVehicle 类型以及一个 VehicleService。在刷新页面操作后显示所有控制器和服务的 Id 值,代码如下:

```
using System;
using System.IO;
using System.Threading.Tasks;
using Microsoft.AspNetCore.Mvc;
using Microsoft.AspNetCore.Mvc.RazorPages;
using Microsoft.EntityFrameworkCore;
using Microsoft.Extensions.Logging;
using Microsoft.Extensions.DependencyInjection;
using RazorMvcBooks.DI;

namespace RazorMvcBooks.Pages
{
    public class AboutModel: PageModel
    {
        private readonly ILogger m_logger;
        private readonly RazorMvcBooks.Models.EFCoreDemoContext _context;

        public AboutModel(ILogger<AboutModel> logger, RazorMvcBooks.Models.EFCoreDemoContext context, ICar transientCar, IJeep scopedJeep, IBus singletonBus, VehicleServicecs vehicleService)
        {
            m_logger = logger;
            _context = context;
            TransientCar = transientCar;
            ScopedJeep = scopedJeep;
            SingletonBus = singletonBus;
            VehicleService = vehicleService;
```

```csharp
}
public string Message { get; set; }
public string Name { get; set; }
public string Code { get; set; }
public void OnGet()
{
    Message = string.Format("TransientCar ID:{0} -- Time :{1} \r\n", TransientCar.VehicleId, TransientCar.CreateTime) + string.Format("ScopedJeep ID:{0} -- Time:{1} \r\n", ScopedJeep.VehicleId, ScopedJeep.CreateTime) + string.Format("SingletonBus ID:{0} -- Time:{1} \r\n ", SingletonBus.VehicleId, SingletonBus.CreateTime);
    Console.WriteLine("");
    Console.WriteLine("         --------About 页面请求--------         ");
    Console.WriteLine(" TransientCar ID:{0} -- Time:{1} ", TransientCar.VehicleId, TransientCar.CreateTime);
    Console.WriteLine(" ScopedJeep ID:{0} -- Time:{1} ", ScopedJeep.VehicleId, ScopedJeep.CreateTime);
    Console.WriteLine("SingletonBus ID:{0} -- Time:{1} ", SingletonBus.VehicleId, SingletonBus.CreateTime);

    Console.WriteLine("");
    Console.WriteLine("         --------VehicleService 操作--------         ");
    Console.WriteLine("TransientCar ID:{0} -- Time:{1} ", VehicleService.TransientCar.VehicleId, VehicleService.TransientCar.CreateTime);
    Console.WriteLine(" ScopedJeep ID:{0} -- Time:{1} ", VehicleService.ScopedJeep.VehicleId, VehicleService.ScopedJeep.CreateTime);
    Console.WriteLine("SingletonBus ID:{0} -- Time:{1} ", VehicleService.SingletonBus.VehicleId, VehicleService.SingletonBus.CreateTime);
}
public VehicleServicecs VehicleService { get; }
public ICar TransientCar { get; }
public IJeep ScopedJeep { get; }
public IBus SingletonBus { get; }
}
}
```

⑧ 在 Visual Studio 2017 中按 F5 键运行应用程序。在浏览器中浏览 About 页面。这个操作发起了两个单独的请求,如图 7.4 和图 7.5 所示。

大家注意到,在图 7.4 和图 7.5 中一共得到了 4 个 Transient 实例,2 个 Scoped 实例,1 个 Singleton 实例,如图 7.6 所示。

图 7.4　第一次浏览 About 页面

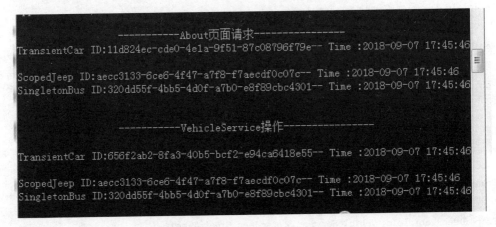

图 7.5　第二次浏览 About 页面

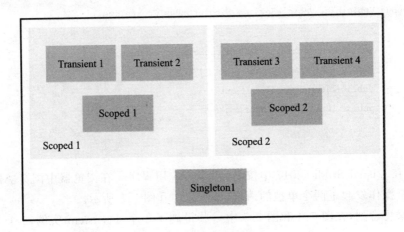

图 7.6　实例图解

下面观察一下在请求内和请求之间的哪些 Id 值不同：

① Tranisent 生命周期的对象始终不同，且向每个控制器和每项服务都提供了一个新的实例。

② Scoped 生命周期的对象在请求内是相同的，但在不同的请求中是不同的 Scope 注册的对象，在同一个请求内 Scope 相当于是单例。

③ Singleton 生命周期的对象对于每个对象和每个请求都是相同的（不管 ConfigureServices 是否提供了一个实例）。

单例生存期服务在第一次被请求时或者 ConfigureServices 运行时（如果在其中指定了实例）被创建，然后每个后续请求使用同一实例。如果应用程序需要单例行为，则建议允许服务容器管理服务的生存期，而不是实现单例设计模式并在类中自行管理对象的生存期。

7.3　DI 在 ASP.NET Core 中的应用

ASP.NET Core 的设计从头至尾以支持和利用依赖注入为目标。ASP.NET Core 包含一个简单的内置容器（表示为 IServiceProvider 接口），默认情况下，该容器支持构造函数注入。

ASP.NET Core 的容器指其作为服务管理的类型。本书后面，服务指的是由 ASP.NET Core 的 IoC 容器管理的类型。ASP.NET Core 提供的默认服务容器提供了一个最小功能的集合。开发人员可以在应用程序的 Startup 类的 ConfigureServices 方法中配置内置容器的服务。

下面通过一些简单的示例来展示.Net Core 框架的依赖注入是如何在 ASP.NET Core 中工作的。.Net Core 框架的依赖注入虽然使用简单，但功能强大，足以完成大部分的依赖注入工作。操作步骤如下：

① 在 Visual Studio 2017 的"解决方案资源管理器"中右击"DI"文件夹，在弹出的快捷菜单中选择"添加"→"类"菜单项创建一个接口文件，并命名为"IUser"，代码如下：

```
using System;
using System.Collections.Generic;
using System.Linq;
using System.Threading.Tasks;

namespace RazorMvcBooks.DI
{
    public interface IUser
    {
        string Name { get; }
```

```
        string UserId { get; }
    }
}
```

② 重复第①步,创建一个用户信息类"UserInfo"类文件,实现 IUser 接口,UserInfo 类在构造函数中接受一个默认的用户账号"Admin",代码如下:

```
using System;
using System.Collections.Generic;
using System.Linq;
using System.Threading.Tasks;

namespace RazorMvcBooks.DI
{
    public class UserInfo:IUser
    {
        public UserInfo() : this("Admin")
        {
        }
        public UserInfo(string id)
        {
            UserId = id;
        }
        public string UserId { get; privateset; }
        public string Name { get { return"张三石"; } }
    }
}
```

③ 重复第①步,创建一个自定义页面信息类"PageInfo"类文件,为页面提供当前用户信息,代码如下:

```
using System;
using System.Collections.Generic;
using System.Linq;
using System.Threading.Tasks;

namespace RazorMvcBooks.DI
{
    public class PageInfo
    {
        private readonly IUser _user;

        public PageInfo(IUser user)
```

```
        {
            _user = user;
        }
        public string PageTitle;
        public string UserTitle
        {
            get
            {
                var title = (PageTitle ?? "").Trim();
                if (!string.IsNullOrWhiteSpace(title) && !string.IsNullOrWhiteSpace(_user.UserId))
                {
                    title += " | ";
                }
                title += _user.Name.Trim();
                return title;
            }
        }
        public string UserID
        {
            get
            {
                return _user.UserId;
            }
        }
    }
}
```

7.3.1 在 Startup 类中初始化

在 Razor 页面中,在把上面示例中创建的类作为服务使用之前,需要在应用程序启动时注册这些服务,ASP.NET Core 可以在 Startup 类的 ConfigureServices()方法中完成注册。每个 services.Add〈ServiceName〉扩展方法都可以添加服务,ASP.NET Core 的一些组件已经提供了一些实例的绑定。例如,services.AddMvc()是在 IServiceCollection 上定义的扩展方法,主要功能是添加 MVC 中间件。下面介绍操作方法。

在 Visual Studio 2017 的"解决方案资源管理器"中打开 Startup.cs 文件,修改 ConfigureServices()方法,代码如下:

```
public void ConfigureServices(IServiceCollection services)
{
```

```
        services.AddDbContext<EFCoreDemoContext>(options => options.UseSqlServer
(Configuration.GetConnectionString("EFCoreDemoContext")));

        services.AddMvc();
        services.AddTransient<ICar, Vehicle>();
        services.AddScoped<IJeep, Vehicle>();
        services.AddSingleton<IBus, Vehicle>();
        services.AddTransient<VehicleServicecs, VehicleServicecs>();

        services.AddSingleton<IUser, UserInfo>();
        services.AddScoped<PageInfo, PageInfo>();
    }
```

现在即可将这些类注入到支持依赖注入的控制器和其他 UI 组件中了。

7.3.2　在控制类中使用

一般可以通过构造函数或属性来实现注入，但官方推荐通过构造函数，这也是所谓的显式依赖。只要在相应的类文件的构造函数中写出了这个参数，ServiceProvider 就会帮助实现注入。操作方法如下：

① 在 Visual Studio 2017 的"解决方案资源管理器"中打开 Index.cshtml.cs 文件，添加如下代码：

```
using System;
using System.Collections.Generic;
using System.Linq;
using System.Text;
using System.Threading.Tasks;
using Microsoft.AspNetCore.Mvc;
using Microsoft.AspNetCore.Mvc.RazorPages;
using Microsoft.AspNetCore.Http;
using RazorMvcBooks.DI;

namespace RazorMvcBooks.Pages
{
    public class IndexModel: PageModel
    {
        public IndexModel(DI.PageInfo pageContext)
        {
            PageInfo = pageContext;
        }
        public DI.PageInfo PageInfo { get; set; }
```

```
        public string CacheDate { get; set; }

        public void OnGet()
        {
        }
    }
}
```

② 在 Visual Studio 2017 的"解决方案资源管理器"中打开 Index.cshtml 文件，添加如下代码：

```
@page
@model IndexModel
@{
    ViewData["Title"] = "Home page";
}

<div class = "row">
<div class = "col-md-3">
<h3>用户账号:@Model.PageInfo.UserID</h3>
</div>
<div class = "col-md-3">
<h3>用户名:@Model.PageInfo.UserTitle</h3>
</div>
</div>
```

③ 在 Visual Studio 2017 中按 F5 键运行应用程序，结果如图 7.7 所示。

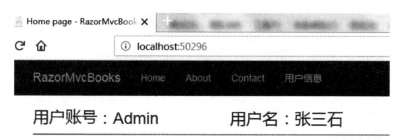

图 7.7　显示账号信息

7.3.3　通过 HttpContext 来获取实例

HttpContext 类下有一个 RequestedService 属性同样可以用来获取实例对象，不过这种方法一般不推荐。同时要注意 GetService 方法默认返回的是 object 对象。代码如下：

```csharp
using System;
using System.Collections.Generic;
using System.Linq;
using System.Text;
using System.Threading.Tasks;
using Microsoft.AspNetCore.Mvc;
using Microsoft.AspNetCore.Mvc.RazorPages;
using Microsoft.AspNetCore.Http;
using RazorMvcBooks.DI;

namespace RazorMvcBooks.Pages
{
    public class IndexModel : PageModel
    {
        public DI.PageInfo PageInfo { get; set; }
        public string CacheDate { get; set; }

        public void OnGet()
        {
            var provider1 = HttpContext.RequestServices;
            PageInfo = (PageInfo) provider1.GetService(typeof(PageInfo));
        }
    }
}
```

7.4 在 ASP.NET Core 中将依赖项注入到视图中

ASP.NET Core 支持将依赖关系注入视图,即视图注入,这对于视图特定的服务很有用,例如仅为填充视图元素所需的本地化或数据应尽量少使用视图注入,视图中显示的大部分数据应该从控制器传入。

7.4.1 简单示例

在 View 中使用@inject 指令将服务注入到视图中,并给服务起一个别名。可将@inject 看作向视图添加属性,并用 DI 填充该属性。@inject 的语法是:

@inject ⟨type⟩⟨name⟩

继续使用 7.3 节中的示例,具体操作如下:

① 在 Visual Studio 2017 的"解决方案资源管理器"中打开 Index.cshtml 文件,显示用户信息,添加如下代码:

```
@page
@model IndexModel
@inject DI.PageInfo pinfo
@{
    ViewData["Title"] = "Home page";
}

<div class="row">
<div class="col-md-3">
<h3>用户账号:@pinfo.UserID</h3>
</div>
<div class="col-md-3">
<h3>用户名:@pinfo.UserTitle</h3>
</div>
</div>
```

② 在 Visual Studio 2017 的"解决方案资源管理器"中打开 Index.cshtml.cs 文件,把依赖注入代码删除,代码如下:

```
using System;
using System.Collections.Generic;
using System.Linq;
using System.Text;
using System.Threading.Tasks;
using Microsoft.AspNetCore.Mvc;
using Microsoft.AspNetCore.Mvc.RazorPages;
using Microsoft.AspNetCore.Http;
using RazorMvcBooks.DI;

namespace RazorMvcBooks.Pages
{
    public class IndexModel: PageModel
    {
        public DI.PageInfo PageInfo { get; set; }
        public string CacheDate { get; set; }

        public void OnGet()
        {
        }
    }
}
```

③ 在 Visual Studio 2017 中按 F5 键运行应用程序。Index 页面显示绑定到视图

的模型数据以及注入到视图中的服务,如图 7.7 所示。

7.4.2 填充查找数据

视图注入可用于填充 UI 元素(如下拉列表)中的选项。请考虑一个用户资料填写页面的情景,其中包含用于指定性别的下拉选项。使用标准的 MVC 方法呈现这样的页面,需让控制器为每组选项请求数据访问服务,然后将选项绑定到下拉列表框上。

另一种方法是将服务直接注入视图以获取选项,这样做能最大限度地减少控制器所需的代码量,同时将此视图元素的构造逻辑移入视图本身。显示个人资料编辑页面的控制器操作只需要把个人资料实例传递给页面,再通过实例来实现以上功能,具体操作如下:

① 在 Visual Studio 2017 的"解决方案资源管理器"中打开 PageInfo.cs 文件,添加一个性别列表方法 ListGenders(),代码如下:

```
using System;
using System.Collections.Generic;
using System.Linq;
using System.Threading.Tasks;

namespace RazorMvcBooks.DI
{
    public class PageInfo
    {
        private readonly IUser _user;
        public PageInfo(IUser user)
        {
            _user = user;
        }
        public string PageTitle;

        public string UserTitle
        {
            get
            {
                var title = (PageTitle ?? "").Trim();
                if (!string.IsNullOrWhiteSpace(title) &&
                    !string.IsNullOrWhiteSpace(_user.UserId))
                {
                    title += " | ";
                }
```

```
            title += _user.Name.Trim();
            return title;
        }
    }
    public string UserID
    {
        get
        {
            return _user.UserId;
        }
    }
    public List<String> ListGenders()
    {
        return new List<string>() { "男", "女" };
    }
}
```

② 在 Visual Studio 2017 的"解决方案资源管理器"中打开 Index.cshtml 文件，更新 HTML 窗体，添加性别下拉列表，这个列表由已注入视图的服务填充，代码如下：

```
@page
@model IndexModel
@inject DI.PageInfo pinfo
@{
    ViewData["Title"] = "Home page";
}

<div class="row">
<div class="col-md-3">
<h3>用户账号:@pinfo.UserID</h3>
</div>
<div class="col-md-3">
<h3>用户名:@pinfo.UserTitle</h3>
</div>
<div class="col-md-3">
<h3>
    性别:@Html.DropDownList("Gender",pinfo.ListGenders().Select(g =>new SelectListItem() { Text = g, Value = g }))
</h3>
</div>
</div>
```

③ 在 Visual Studio 2017 中按 F5 键运行应用程序。Index 页面显示绑定到视图的模型数据以及注入到视图中的服务，结果如图 7.8 所示。

图 7.8　账号信息中显示性别

使用依赖注入时，请记住以下**建议**：
- DI 适用于具有复杂的依赖关系的对象，控制器、服务、适配器和仓储都是可能添加到 DI 中的对象示例。
- 避免在 DI 中直接存储数据和配置，例如，用户的购物车通常不应添加到服务容器中。配置应使用选项模型。同样，避免"数据持有者"对象，也就是仅仅为实现对某些其他对象的访问而存在的对象。如果可能，最好通过 DI 请求所需的实际项目。
- 避免静态访问服务。
- 在应用程序代码中避免使用服务定位器模式。
- 避免静态访问 HttpContext。

7.5　如何替换其他的 IoC 容器

　　.NET Core 内置的容器服务框架对于一些小型的项目来说完全够用，甚至大型项目也能用，只是会有些麻烦，原因在于它只提供了最基本的 AddXXXX 方法来绑定实例关系，操作中需要一个一个地添加这些方法，如果是大型项目，可能要添加好几百行这样的方法。为了提高效率，下面来学习第三方的可用于 .NET 的 IoC 容器，其中 Autofac 是 .NET 领域最为流行的 IoC 框架之一，其优点是：

　　① 它与 C# 语言联系很紧密，也就是说 C# 里的很多编程方式都可以在 Autofac 中使用，例如可以用 Lambda 表达式注册组件。

　　② 较低的学习曲线，学习它非常简单，只要理解了 IoC 和 DI 的概念以及在何时使用它们即可。

　　③ 支持 XML 配置。

　　④ 可进行自动装配。

　　⑤ 可与 ASP.NET MVC 集成。

　　⑥ 微软的 Orchad 开源程序使用的就是 Autofac，从该源码可以看出它的方便和强大。

7.5.1 Autofac 的基本使用

1. 使用实例

Autofac 是一款轻量级的 IoC 框架，有较高的使用率。下面先通过一个例子来学习一下 Autofac 的基本使用方法：

① 在 Visual Studio 2017 中选择"文件"→"新建"→"项目"菜单项，在"新建项目"对话框的"名称"文本框中输入"AutofacDemoApp"，然后单击"确定"按钮，如图 7.9 所示。

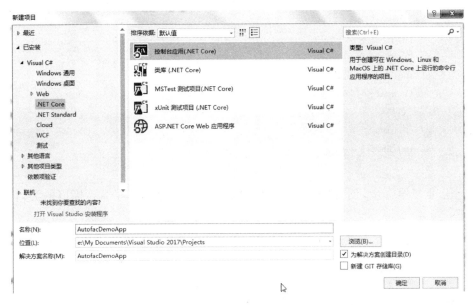

图 7.9 创建项目

② 在 Visual Studio 2017 中选择"工具"→"NuGet 包管理器"→"管理解决方案的 NuGet 程序包"菜单项，在管理界面中安装"Autofac"和"Autofac.Extensions.DependencyInjection"两个包，如图 7.10 所示。

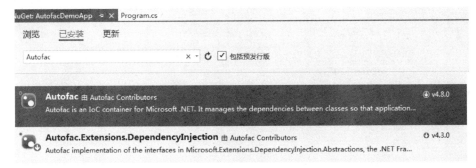

图 7.10 已安装 NuGet 包

③ 在 Visual Studio 2017 的"解决方案资源管理器"中右击"AutofacDemoApp"项目,在弹出的快捷菜单中选择"添加"→"类"菜单项创建一个新接口文件,并命名为"IPerson",代码如下：

```
public interface IPerson
{
    ///<summary>
    ///年龄
    ///</summary>
    string Age { get; }
    ///<summary>
    ///性别
    ///</summary>
    string Sex { get; }
    ///<summary>
    /// 名称
    ///</summary>
    string Name { get; }
}
```

④ 重复第③步,创建"Student"和"Worker"两个实现 IPerson 接口的类,代码如下：

```
///<summary>
///学生
///</summary>
public class Student: IPerson
{
    ///<summary>
    ///年龄
    ///</summary>
    public string Age
    {
        get
        {
            return "18";
        }
    }
    ///<summary>
    /// 名称
    ///</summary>
    public string Name
```

```csharp
        {
            get
            {
                return "学生黄";
            }
        }
        ///<summary>
        ///性别
        ///</summary>
        public string Sex
        {
            get
            {
                return "女";
            }
        }
    }
    ///<summary>
    ///职工
    ///</summary>
    public class Worker：IPerson
    {
        ///<summary>
        ///年龄
        ///</summary>
        public string Age
        {
            get
            {
                return "35";
            }
        }
        ///<summary>
        /// 名称
        ///</summary>
        public string Name
        {
            get
            {
                return "工人张";
            }
        }
```

```csharp
        ///<summary>
        ///性别
        ///</summary>
        public string Sex
        {
            get
            {
                return "男";
            }
        }
    }
```

⑤ 在 Visual Studio 2017 的"解决方案资源管理器"中打开 Program.cs 文件,添加注册类型方法 Register(),代码如下:

```csharp
using Autofac;
using System;
using System.Collections.Generic;

namespace AutofacDemoApp
{
    class Program
    {
        static void Main(string[] args)
        {
            Console.WriteLine("Hello World!");
            Register();
        }
        public static void Register()
        {
            var builder = new ContainerBuilder();
            //注册 Student 指定为 IPerson 实现
            builder.RegisterType<Student>().As<IPerson>();
            builder.RegisterType<Worker>().As<IPerson>();
            using (var container = builder.Build())
            {
                var persons = container.Resolve<IEnumerable<IPerson>>();
                foreach (var person in persons)
                {
                    Console.WriteLine($"名称:{person.Name},年龄:{person.Age},性别:{person.Sex}");
                }
```

            }
          }
        }
      }

⑥ 在 Visual Studio 2017 中按 F5 键运行应用程序,结果如图 7.11 所示。

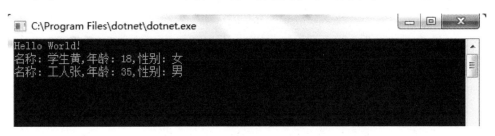

图 7.11 Autofac 注入的执行结果

2. 注册程序集

可以使用 AsImplementedInterfaces 直接注册程序集下的所有类型,注入这些类的所有公共接口作为服务,代码如下:

```
public static void RegisterAssemblyTypes()
{
    var builder = new ContainerBuilder();
    //注册程序集下的所有类型
    builder.RegisterAssemblyTypes(typeof(Program).Assembly).AsImplementedInterfaces();
    using (var container = builder.Build())
    {
        var persons = container.Resolve<IEnumerable<IPerson>>();
        foreach (var person in persons)
        {
            Console.WriteLine($"名称:{person.Name},年龄:{person.Age},性别:{person.Sex}");
        }
    }
}
```

3. 配置文件注册

下面介绍一种适用于 Aufofac 4+版本的配置文件注册方式,操作如下:

① 在 Visual Studio 2017 的"解决方案资源管理器"中新建一个"AutofacApp.xml"文件,在文件中添加注册类型配置,代码如下:

```
<?xml version="1.0" encoding="utf-8"?>
<autofac defaultAssembly="AutofacDemoApp">
```

```xml
<components name="0">
<type>AutofacDemoApp.Worker, AutofacDemoApp</type>
<services name="0" type="AutofacDemoApp.IPerson" />
<injectProperties>true</injectProperties>
</components>
<components name="1">
<type>AutofacDemoApp.Student, AutofacDemoApp</type>
<services name="0" type="AutofacDemoApp.IPerson" />
<injectProperties>true</injectProperties>
</components>
</autofac>
```

② 在 Visual Studio 2017 中选择"工具"→"NuGet 包管理器"→"管理解决方案的 NuGet 程序包"菜单项,在管理界面中安装"Autofac.Configuration"和"Microsoft.Extensions.Configuration.xml"两个包。

③ 在 Visual Studio 2017 中打开 Program.cs 文件,然后添加读取配置文件中的信息并进行注册的代码如下:

```
public static void RegisterAppConfig()
{
    var config = new ConfigurationBuilder();
    config.AddXmlFile("AutofacApp.xml");

    var module = new ConfigurationModule(config.Build());
    var builder = new ContainerBuilder();
    builder.RegisterModule(module);
    using (var container = builder.Build())
    {
        var persons = container.Resolve<IEnumerable<IPerson>>();
        foreach (var person in persons)
        {
            Console.WriteLine($"名称:{person.Name},年龄:{person.Age},性别:{person.Sex}");
        }
    }
}
```

7.5.2 用 Autofac 代替原来的 IoC

除了 ASP.NET Core 自带的 IoC 容器外,还可以使用其他成熟的 DI 框架,这可以给初始化带来便利性。下面的例子实现了如何用 Autofac 替换 ASP.NET Core 的默认 DI,操作如下:

① 在 Visual Studio 2017 中打开"RazorMvcBooks"项目,可以看到之前示例中用于 DI 测试的代码,其结构如图 7.12 所示。

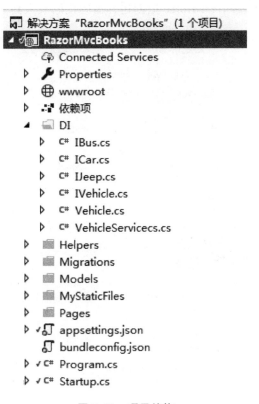

图 7.12　项目结构

② 安装 Autofac。在 Visual Studio 2017 中选择"工具"→"NuGet 包管理器"→"管理解决方案的 NuGet 程序包"菜单项,在管理界面中安装"Autofac"和"Autofac.Extensions.DependencyInjection"两个包,如图 7.13 所示。

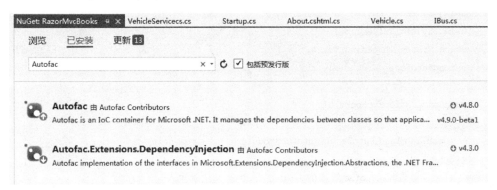

图 7.13　安装 Autofac NuGet 包

③ 替换内置的 DI 框架。分以下 4 步进行：

ⓐ 把 Startup 类中的 ConfigureService 的返回值从 void 改为 IServiceProvider，然后新建一个方法 RegisterAutofac 把创建容器的代码放入其中，代码如下：

```
public IServiceProvider ConfigureServices(IServiceCollection services)
{
    services.AddDbContext<EFCoreDemoContext>(options => options.UseSqlServer(Configuration.GetConnectionString("EFCoreDemoContext")));
    services.AddMvc();
    return RegisterAutofac(services);//注册 Autofac
}
private IServiceProvider RegisterAutofac(IServiceCollection services)
{
    //实例化 Autofac 容器
    var builder = new ContainerBuilder();
    //将 Services 中的服务填充到 Autofac 中
    builder.Populate(services);
    //新模块组件注册
    builder.RegisterModule<AutofacDI>();
    //创建容器
    var Container = builder.Build();
    //第三方 IoC 接管 Core 内置 DI 容器
    return new AutofacServiceProvider(Container);
}
```

ⓑ 新建一个 AutofacDI 类继承 Autofac 的 Module，然后重写 Module 的 Load 方法来存放新组件的注入代码，代码如下：

```
using System;
using System.Collections.Generic;
using System.Linq;
using System.Reflection;
using System.Threading.Tasks;
using Autofac;

namespace RazorMvcBooks.DI
{
    public class AutofacDI:Autofac.Module
    {
        //重写 Autofac 管道的 Load 方法，在这里注册注入
        protected override void Load(ContainerBuilder builder)
        {
```

```csharp
            //注册服务的对象,这里以命名空间名称中含有"DI"字符串为要求,否则注册失败
            builder.RegisterAssemblyTypes(GetAssemblyByName("RazorMvcBooks")).Where(a =>
                a.Namespace.EndsWith("DI")).AsImplementedInterfaces();
        }
        ///<summary>
        /// 根据程序集的名称获取程序集
        ///</summary>
        ///<param name = "AssemblyName">程序集名称</param>
        ///<returns></returns>
        public static Assembly GetAssemblyByName(String AssemblyName)
        {
            return Assembly.Load(AssemblyName);
        }
    }
}
```

ⓒ 至此,Autofac 的基本使用已经配置完毕,下面在 Visual Studio 2017 中打开 About.cshtml.cs 文件,并修改代码如下:

```csharp
using System;
using System.IO;
using System.Threading.Tasks;
using Microsoft.AspNetCore.Mvc;
using Microsoft.AspNetCore.Mvc.RazorPages;
using Microsoft.EntityFrameworkCore;
using Microsoft.Extensions.Configuration;
using Microsoft.Extensions.DependencyInjection;
using Microsoft.Extensions.Logging;
using Microsoft.Extensions.Options;
using RazorMvcBooks.DI;
using RazorMvcBooks.Models;

namespace RazorMvcBooks.Pages
{
    public class AboutModel : PageModel
    {
        private readonly ILogger m_logger;
        private readonly RazorMvcBooks.Models.EFCoreDemoContext _context;
        public AboutModel(ILogger<AboutModel> logger, RazorMvcBooks.Models.EFCoreDemoContext context, ICar transientCar, IJeep scopedJeep, IBus singletonBus)
        {
            m_logger = logger;
            _context = context;
            TransientCar = transientCar;
```

```csharp
                ScopedJeep = scopedJeep;
                SingletonBus = singletonBus;
        }
            public string Message { get; set; }
            public string Name { get; set; }
            public string Code { get; set; }
            public void OnGet()
            {
                Message = string.Format("TransientCar ID:{0} -- Time:{1} \r\n", TransientCar.VehicleId, TransientCar.CreateTime) +
                    string.Format("ScopedJeep ID:{0} -- Time:{1} \r\n", ScopedJeep.VehicleId, ScopedJeep.CreateTime) +
                    string.Format("SingletonBus ID:{0} -- Time:{1} \r\n ", SingletonBus.VehicleId, SingletonBus.CreateTime);
                Console.WriteLine("");
                Console.WriteLine("        --------About 页面请求-------------");

                Console.WriteLine("TransientCar ID:{0} -- Time:{1} ", TransientCar.VehicleId, TransientCar.CreateTime);
                Console.WriteLine(" ScopedJeep ID:{0} -- Time:{1} ", ScopedJeep.VehicleId, ScopedJeep.CreateTime);
                Console.WriteLine("SingletonBus ID:{0} -- Time:{1} ", SingletonBus.VehicleId, SingletonBus.CreateTime);
            }
        public VehicleServicecs VehicleService { get; }
        public ICar TransientCar { get; }
        public IJeep ScopedJeep { get; }
        public IBus SingletonBus { get; }
    }
}
```

④ 在 Visual Studio 2017 中按 F5 键运行应用程序，结果如图 7.14 所示。

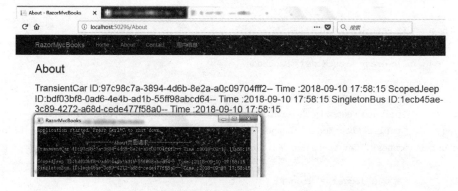

图 7.14　Autofac 替换内置注入框架的执行结果

7.5.3　一个接口对应多个实现的情况

下面给出一个新的示例,用来展示一个接口对应多个实现的情况,操作如下:

① 在 Visual Studio 2017 的"解决方案资源管理器"中右击"DI"文件夹,在弹出的快捷菜单中选择"添加"→"类"菜单项创建两个类,实现接口"IVehicle",代码如下:

```
using System;
using System.Collections.Generic;
using System.Linq;
using System.Threading.Tasks;

namespace RazorMvcBooks.DI
{
    public class Car:IVehicle
    {
        public Car() : this(Guid.NewGuid())
        {
        }
        public Car(Guid id)
        {
            VehicleId = id;
        }
        public Guid VehicleId { get; private set; }
        public string CreateTime { get { return DateTime.Now.ToString("yyyy-MM-dd HH:mm:ss"); } }
        public string Name { get { return "Car"; } }
    }
}

using System;
using System.Collections.Generic;
using System.Linq;
using System.Threading.Tasks;

namespace RazorMvcBooks.DI
{
    public class Jeep:IVehicle
    {
        public Jeep() : this(Guid.NewGuid())
        {
        }
```

```csharp
        public Jeep(Guid id)
        {
            VehicleId = id;
        }
        public Guid VehicleId { get; private set; }
        public string CreateTime { get { return DateTime.Now.ToString("yyyy-MM-dd HH:mm:ss"); } }
        public string Name { get { return "Jeep"; } }
    }
}
```

② 在 Visual Studio 2017 中打开 Index.cshtml.cs 文件,输入以下代码:

```csharp
using System;
using System.Collections.Generic;
using System.Linq;
using System.Text;
using System.Threading.Tasks;
using Microsoft.AspNetCore.Mvc;
using Microsoft.AspNetCore.Mvc.RazorPages;
using Microsoft.AspNetCore.Http;
using RazorMvcBooks.DI;
using Autofac;

namespace RazorMvcBooks.Pages
{
    public class IndexModel : PageModel
    {
        private IComponentContext _componentContext;
        public IndexModel(DI.IVehicle vehicle, IComponentContext componentContext)
        {
            Vehicles = vehicle;
            _componentContext = componentContext;
            _car = _componentContext.ResolveNamed<IVehicle>(typeof(Car).Name);
            _jeep = _componentContext.ResolveNamed<IVehicle>(typeof(Jeep).Name);
        }
        private IVehicle _car;
        private IVehicle _jeep;
        public DI.IVehicle Vehicles { get; set; }
        public string VehicleName { get { returnstring.Format("小轿车:{0} 吉普:{1}", _car.Name, _jeep.Name); } }
        public void OnGet()
        {
```

            }
        }
}

③ 在 Visual Studio 2017 中打开 Index.cshtml 文件,输入以下代码:

```
@page
@model IndexModel
@{
    ViewData["Title"] = "Home page";
}
<div class = "row">
<div class = "col-md-12">
<h3>
    车:@Model.VehicleName --|-- 构造注入车辆信息:@Model.Vehicles.Name
</h3>
</div>
</div>
```

④ 在 Visual Studio 2017 中按 F5 键运行应用程序,此时会报错,如图 7.15 所示。

图 7.15　异常信息

⑤ 在 Visual Studio 2017 中打开 AutofacDI.cs 文件,使用 RegisterType 进行注册。使用 Named 来区分两个组件,后面的 typeof(Car).Name 可以是任意字符串,这里直接使用该类的名称作为标识,以便取出时仍是此名称,不易忘记,代码如下:

```csharp
using System;
using System.Collections.Generic;
using System.Linq;
using System.Reflection;
using System.Threading.Tasks;
using Autofac;

namespace RazorMvcBooks.DI
{
    public class AutofacDI:Autofac.Module
    {
        //重写 Autofac 管道的 Load 方法,在这里注册注入
        protected override void Load(ContainerBuilder builder)
        {
            builder.RegisterAssemblyTypes(GetAssemblyByName("RazorMvcBooks")).Where(a => a.Namespace.EndsWith("DI")).AsImplementedInterfaces();
            builder.RegisterType<Car>().Named<IVehicle>(typeof(Car).Name);
            builder.RegisterType<Jeep>().Named<IVehicle>(typeof(Jeep).Name);
        }
        ///<summary>
        /// 根据程序集名称获取程序集
        ///</summary>
        ///<param name="AssemblyName">程序集名称</param>
        ///<returns></returns>
        public static Assembly GetAssemblyByName(String AssemblyName)
        {
            return Assembly.Load(AssemblyName);
        }
    }
}
```

⑥ 在 Visual Studio 2017 中按 F5 键运行应用程序,结果如图 7.16 所示。

图 7.16 注入执行结果

第 8 章

Razor 视图

8.1 什么是 Razor

Razor 不是编程语言,而是服务器端的标记语言,是一种允许向网页中嵌入基于服务器代码的标记语言。

当网页被写入浏览器时,基于服务器的代码能够创建动态内容。在网页加载时,服务器在向浏览器返回页面之前,会执行页面内基于服务器的代码。由于是在服务器上运行,所以这种代码能够执行复杂的任务,比如访问数据库。

Razor 是基于 ASP.NET,并为 Web 应用程序的创建而设计的。Razor 的语法由 Razor 标记、C♯和 HTML 组成。包含 Razor 的文件通常具有 .cshtml 文件扩展名。

Razor 中的核心转换字符是@符号。从 HTML 转换为 C♯,或者 Razor 计算 C♯的表达式,并将它们呈现在 HTML 输出中都是使用这个符号作为转换字符。当@符号后跟 Razor 保留关键字时,会将它们转换为 Razor 的特定标记;否则会转换为纯 C♯。

首先创建一个示例项目。本章中的示例都将在此项目中实现。操作如下:

① 启动 Visual Studio 2017,选择"文件"→"新建"→"项目"菜单项。

② 在"新建项目"对话框中会显示几个项目模板。模板包含给定项目类型所需的基本文件和设置。在"新建项目"对话框的左侧窗格中展开"Visual C♯",然后选择".NET Core"。在中间窗格中选择"ASP.NET Core Web 应用程序",在"名称"文本框中输入"RazorDemo",最后单击"确定"按钮。

③ 在 Visual Studio 2017 中打开 About.cshtml 文件,并输入以下代码:

〈p〉如果要在 Razor 标记中显示@@符号,则需要使用两个@@符号,其中一个是转义符,例如:〈/p〉

〈p〉@@DateTime.Now〈/p〉

〈p〉如果在 Razor 标记中使用单个@@符号,则表示是代码或变量,例如:〈/p〉

〈p〉@DateTime.Now〈/p〉

〈p〉包含电子邮件地址的 HTML 属性和内容不将 @@符号视为转换字符。Razor 会自动分析,不会处理。例如:〈/p〉

```
<a href = "mailto:Support@163.com">Support@163.com</a>
```

④ 在 Visual Studio 2017 中按 F5 键运行应用程序,结果如图 8.1 所示。

```
localhost:37832/About

RazorDemo    Home    About    Contact

About
Your application description page.

如果要在Razor标记中显示@符号,则需要使用两个@符号,其中一个是转义符,例如:
@DateTime.Now

如果在Razor标记中使用单个@符号,则表示是代码或变量,例如:
2018-09-13 17:52:41

包含电子邮件地址的 HTML 属性和内容不将 @ 符号视为转换字符。Razor会自动分析,不会处理。例如:
Support@163.com
```

图 8.1　浏览 About 页面

8.2　Razor 保留关键字

8.2.1　Razor 关键字

Razor 关键字(见表 8.1)使用符号@(Razor Keyword)进行转义(例如,@(functions))。

表 8.1　Razor 关键字

序　号	关键字	序　号	关键字
1	page	4	model
2	functions	5	section
3	inherits	6	namespace

8.2.2　C♯ Razor 关键字

C♯ Razor 关键字(见表 8.2)必须使用符号@(@C♯ Razor Keyword)进行双转义(例如,@(@case))。第一个 @对 Razor 分析器转义,第二个 @对 C♯分析器转义。

表 8.2　C# Razor 关键字

序 号	关键字	序 号	关键字
1	case	8	lock
2	do	9	switch
3	default	10	try
4	for	11	catch
5	foreach	12	finally
6	if	13	using
7	else	14	while

8.3　使用 Razor 语法编写表达式

8.3.1　隐式 Razor 表达式

隐式 Razor 表达式就是以@开头，后跟 C#代码。示例操作如下：

① 在 Visual Studio 2017 中打开 About.cshtml.cs 文件，输入以下代码，为下面的操作做准备：

```
using System;
using System.Collections.Generic;
using System.Linq;
using System.Threading.Tasks;
using Microsoft.AspNetCore.Mvc.RazorPages;

namespace RazorDemo.Pages
{
    public class AboutModel : PageModel
    {
        public string Message { get; set; }
        public void OnGet()
        {
            Message = "Your application description page.";
        }
        public static Task<string> Hello(string name, string msg)
        {
            return Task.FromResult<string>(string.Format("欢迎 {0} 登录。{1}。", name, msg));
        }
    }
```

```
    public string Add<T>(int a,int b)
    {
        int sum = a + b;
        return sum.ToString();
    }
}
```

② 在 Visual Studio 2017 中打开 About.cshtml 文件，输入以下代码：

<p>隐式 Razor 表达式以@@开头，后跟 C#代码：</p>

<p>@DateTime.Now</p>

<p>@DateTime.DaysInMonth(2018,6)</p>

<p>隐式表达式不能包含空格，但 C# await 关键字除外。如果该 C#语句具有明确的结束标记，则可以混用空格：</p>

<p>@await AboutModel.Hello("张三","管理员")</p>

③ 在 Visual Studio 2017 中按 F5 键运行应用程序，结果如图 8.2 所示。

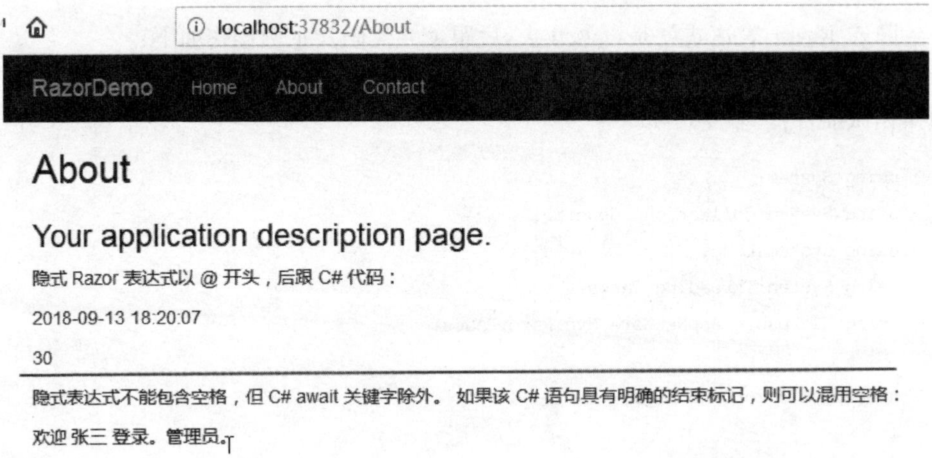

图 8.2　隐藏表达式的结果

④ 在 Visual Studio 2017 中打开 About.cshtml 文件，输入以下代码：

<p>隐式表达式不能包含 C#泛型，因为括号(<>)内的字符会被解释为 HTML 标记。以下代码无效：</p>

<p>@Model.Add<int>(1,3)</p>

⑤ 按 F5 键运行应用程序，此时会发生编译错误。上述代码生成了与以下错误之一类似的编译器错误，如图 8.3 和图 8.4 所示。

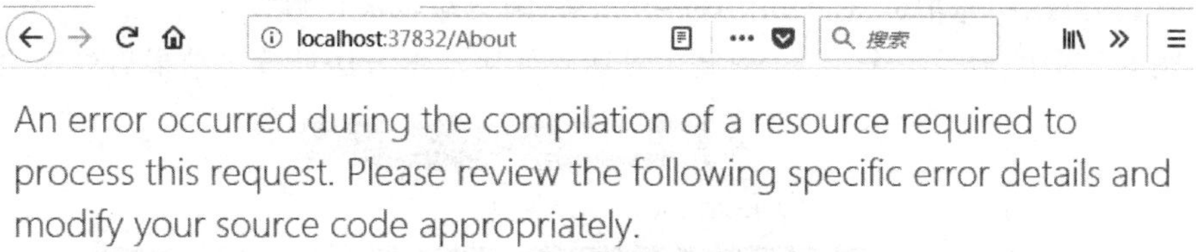
localhost:37832/About
搜索
An error occurred during the compilation of a resource required to process this request. Please review the following specific error details and modify your source code appropriately.

图 8.3 错误提示信息

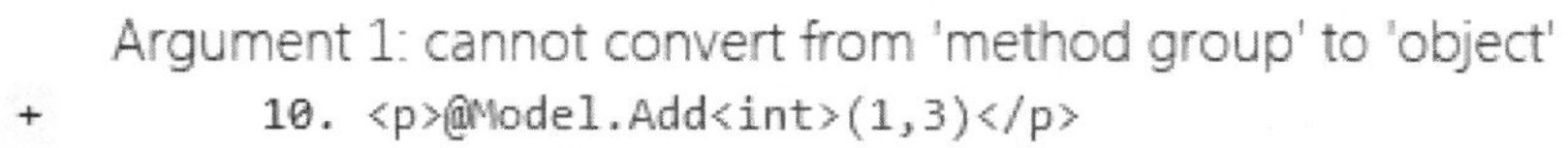
Argument 1: cannot convert from 'method group' to 'object'
+
10. <p>@Model.Add<int>(1,3)</p>

图 8.4 错误代码

8.3.2 显式 Razor 表达式

显式 Razor 表达式由符号@和一对圆括号组成。示例操作如下：

① 在 Visual Studio 2017 中打开 About.cshtml 文件，输入以下代码：

```
@page
@model AboutModel
@{
    ViewData["Title"] = "About";
}
〈h2〉@ViewData["Title"]〈/h2〉
〈h3〉@Model.Message〈/h3〉
〈p〉如果要显示上一周的时间，可使用以下 Razor 标记：〈/p〉
〈p〉上周当前时间：@(DateTime.Now - TimeSpan.FromDays(7))〈/p〉
〈p〉如果不加括号，则不会进行计算，在下面的代码中，不会从当前时间减去一周：〈/p〉
〈p〉上周当前时间：@DateTime.Now - TimeSpan.FromDays(7)〈/p〉
〈p〉可以使用显式表达式将文本与表达式结果串联起来：〈/p〉

@{
    var student = new Models.Person("Joe", 33);
}
〈p〉年龄@(student.Age)〈/p〉
〈p〉年龄@student.Age〈/p〉
〈p〉显示表达式中使用 C# 泛型：〈/p〉
〈p〉@(@Model.Add〈int〉(2,3))〈/p〉
```

② 按 F5 键运行应用程序，结果如图 8.5 所示。注意，在结果图的第二个方框中，如果不用括号把@student.Age 括起来，则会被视为电子邮件地址，而不会显示

成年龄；如果用括号把@student.Age括起来，则会显示成年龄"33"。

图8.5中的最后一个结果，是显式表达式使用泛型方法计算出结果后的输出。

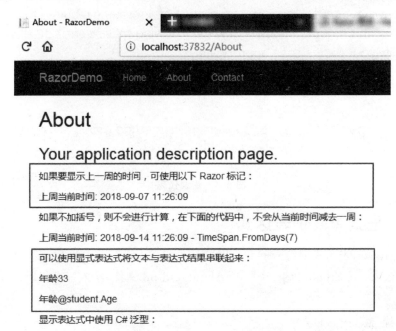

图 8.5　显示表达式的计算结果

8.3.3　表达式的编码

计算结果为字符串的 C# 表达式采用 HTML 编码。计算结果为 IHtmlContent 的 C# 表达式直接通过 IHtmlContent.WriteTo 呈现。计算结果不为 IHtmlContent 的 C# 表达式通过 ToString 方法转换为字符串，并在呈现前进行编码。示例操作是：

① 在 Visual Studio 2017 中打开 About.cshtml 文件，输入以下代码：

```
@page
@model AboutModel
@{
    ViewData["Title"] = "About";
}
<h2>@ViewData["Title"]</h2>
<h3>@Model.Message</h3>
<p>表达式计算结果为字符串：@("<span>Hello World</span>")<p>
<p>HtmlHelper.Raw输出不进行编码，但呈现为 HTML 标记：@Html.Raw("<span>Hello World
```

</span>")<p>

② 按 F5 键运行应用程序,显示结果如图 8.6 所示。

图 8.6 表达式编码的结果

8.4 Razor 代码块

Razor 代码块以 @ 开头,并括在大括号{}中。代码块内的默认语言为 C#,C# 代码不会呈现,这一点与表达式不同。一个视图中的代码块和表达式共享相同的作用域,并按顺序进行定义。示例操作如下:

① 在 Visual Studio 2017 中打开 About.cshtml 文件,输入以下代码:

```
@page
@model AboutModel
@{
    ViewData["Title"] = "About";
}
<h2>@ViewData["Title"]</h2>
<h3>@Model.Message</h3>
@{
    var quote = "真理惟一可靠的标准就是永远自相符合。—— 欧文";
}
<p>@quote</p>
@{
    quote = "学习知识要善于思考,思考,再思考。—— 爱因斯坦";
}
<p>@quote</p>
```

② 按 F5 键运行应用程序,显示结果如图 8.7 所示。

图 8.7 Razor 代码块的结果

③ 在 Visual Studio 2017 中打开 About.cshtml 文件,输入如下隐式转换的代码:

```
<p>Razor 页面可以转换回 HTML:</p>
@{
    var inCSharp = true;
    <p>当前内容在 HTML 中,这个值在 C# 中是@inCSharp</p>
}
```

④ 在 Visual Studio 2017 中打开 About.cshtml.cs 文件,输入如下代码创建学生列表。

```
public List<Models.Person> Students {
    get {
        List<Models.Person> list = new List<Models.Person>();
        list.Add(new Models.Person("大刀王五", 45));
        list.Add(new Models.Person("小李飞刀", 35));
        list.Add(new Models.Person("杀神白起", 25));
        return list;
    }
}
```

⑤ 接续上面的代码,输入如下显式转换的代码:

`<p>若要显示 HTML 的代码块,请使用 Razor <text>标记将要呈现的字符括起来:</p>`

```
@for (var i = 0; i < @Model.Students.Count; i++)
{
    var person = @Model.Students[i];
    <text>学生名称：@person.Name | </text>
}
```

上段代码中的<text>标签是一个 Razor 特殊处理的元素。Razor 将<text>块的内部内容视为内容块，显示时不呈现包含这些内容的<text>标签（这意味着只呈现<text>内部的内容，而不呈现标签本身），这使得呈现那些没有被 HTML 元素包装的多行内容块变得很方便。

⑥ 接续上面的代码，输入如下显式转换的代码：

```
<p>若要在代码块内以 HTML 的形式显示内容：</p>
@for (var i = 0; i < @Model.Students.Count; i++)
{
    var person = @Model.Students[i];
    <p>学生名称：@person.Name</p>
}
```

⑦ 按 F5 键运行应用程序，显示结果如图 8.8 所示。

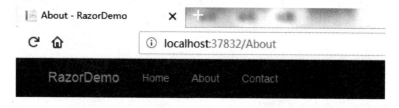

图 8.8　隐式与显式转换结果

8.5 Razor 逻辑条件控制

有如下几种逻辑条件控制语句。

8.5.1 if 和 switch 条件语句

C#允许执行基于条件的代码。

1. if 条件语句

若需测试某个条件,则可以使用 if 语句。if 语句会基于测试的结果来返回 true 或 false。if 语句的特征是:

① if 语句启动代码块;

② 条件位于括号中;

③ 如果条件为真,则执行大括号{}中的代码;

④ if 语句能够包含 else 条件。else 条件定义了当条件为 false 时要执行的代码。

⑤ 可通过 else if 条件来测试多个条件。

if 语句的示例操作是:在 Visual Studio 2017 中打开 About.cshtml 文件,输入 if 条件测试代码,其中 if 前面要加符号@,而 else 和 else if 前面不需要加符号@,代码如下:

```
@if (@Model.value % 2 == 0)
{
    <p>if 测试,这个数字是偶数.</p>
}
else if (@Model.value >= 1500)
{
    <p>if 测试,这个数字大于或等于1500.</p>
}
else
{
    <p>if 测试,这个数据比 1500 小.</p>
}
```

2. switch 条件语句

switch 条件语句是先计算表达式的值,再逐个与 case 后面的常量表达式比较,如果没有相等的,则执行 default 后面的语句;如果等于某一个常量表达式,则执行相应 case 后面的语句。

switch 语句的示例操作是：在 Visual Studio 2017 中打开 About.cshtml 文件，输入如下测试一系列条件代码的 switch 语句：

```
@switch (@Model.value)
{
    case 1:
        <p>switch 测试，数字是 1!</p>
        break;
    case 1450:
        <p>switch 测试，数字是 1450!</p>
        break;
    default:
        <p>switch 测试，请输入 1 或 1450 两个数字.</p>
        break;
}
```

在 Visual Studio 2017 中按 F5 键运行应用程序，结果如图 8.9 所示。

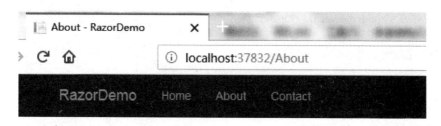

图 8.9　if 和 switch 语句的结果

8.5.2　循环语句

语句可以在循环中重复执行。有如下三种循环语句。

1. for 循环

如果需要重复运行相同的语句，则可以编写一个循环。

如果能够确定循环的次数，则可以使用 for 循环。这种循环类型是专门为计数或反向计数设计的。可以使用循环控制语句呈现模板化 HTML，如要呈现一组人员。示例操作是：在 Visual Studio 2017 中打开 About.cshtml 文件，输入如下 for 测试代码：

```
<p>For 循环:</p>
@for (var i = 0; i < @Model.Students.Count; i++)
{
    var person = @Model.Students[i];
    <p>学生名称：@person.Name</p>
}
```

2. foreach 循环

如果需要处理集合或数组，则通常要用到 foreach 循环。

集合是一组相似的对象，foreach 循环允许在每个项目上执行一次任务，foreach 循环会遍历集合直到完成为止。示例操作是：在 Visual Studio 2017 中打开 About.cshtml 文件，输入如下 foreach 测试代码：

```
<p>Foreach 循环:</p>
@foreach (var person in @Model.Students)
{
    <p>学生名称：@person.Name</p>
}
```

3. while 循环

while 循环是一种通用的循环。

while 循环以关键词 while 开始，后面跟的小括号定义了循环持续的长度；跟的大括号包含了要循环的代码块。while 循环通常会对用于计数的变量进行增减。示例操作是：在 Visual Studio 2017 中打开 About.cshtml 文件，输入如下 while 测试代码：

```
<p>While 循环:</p>
@{ var j = 0; }
@while (j < @Model.Students.Count)
{
    var person = @Model.Students[j];
    <p>学生名称：@person.Name    年龄：@person.Age</p>
    j++;
}
```

在 Visual Studio 2017 中按 F5 键运行应用程序，结果如图 8.10 所示。

Razor 视图 8

图 8.10　循环语句的结果

8.5.3　复合语句 @using

在 C# 中，using 语句用于确保释放对象。在 Razor 中，可使用相同的机制来创建包含附加内容的 HTML。在下面的代码中，HTML 帮助助手使用 @using 语句呈现表单标记。示例操作是：在 Visual Studio 2017 中打开 About.cshtml 文件，输入如下 using 测试代码：

```
<p>表单：</p>
@using (Html.BeginForm())
{
    <div>
        姓名：
        <input type="text" id="Name" value="" />
        <button>查询</button>
    </div>
}
```

8.5.4 异常处理语句@try、catch、finally

Razor 中的异常处理与 C# 中的类似，操作是：在 Visual Studio 2017 中打开 About.cshtml 文件，输入如下异常捕获测试代码：

```
<p>异常捕获：</p>
@try
{
    throw new InvalidOperationException("这是一个异步捕获测试代码.");
}
catch (Exception ex)
{
    <p>The exception message：@ex.Message</p>
}
finally
{
    <p>执行 finally 中的代码.</p>
}
```

在 Visual Studio 2017 中按 F5 键运行应用程序，结果如图 8.11 所示。

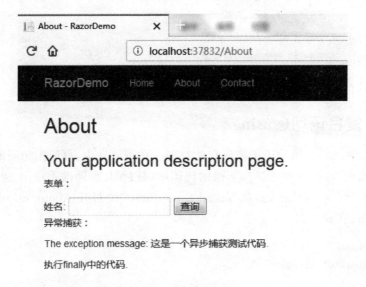

图 8.11　异常捕获的结果

8.5.5　加锁语句@lock

Razor 可以使用@lock 语句来保护关键节，用法如下：

```
@lock (IsLock)
{
    // 要加锁的代码
}
```

8.5.6 注　释

Razor 支持 C# 和 HTML 两种注释方式。在 Razor 作用域中编写的代码就是服务器代码，因此也可以使用“//”和“/ * * /”进行单行注释和多行注释。当然，使用“〈! --注释内容--〉”形式的注释也是可以的。而“@ * 注释内容 * @”是 Razor 注释特有的形式。图 8.12 中的代码已被注释禁止，因此服务器不呈现任何标记。

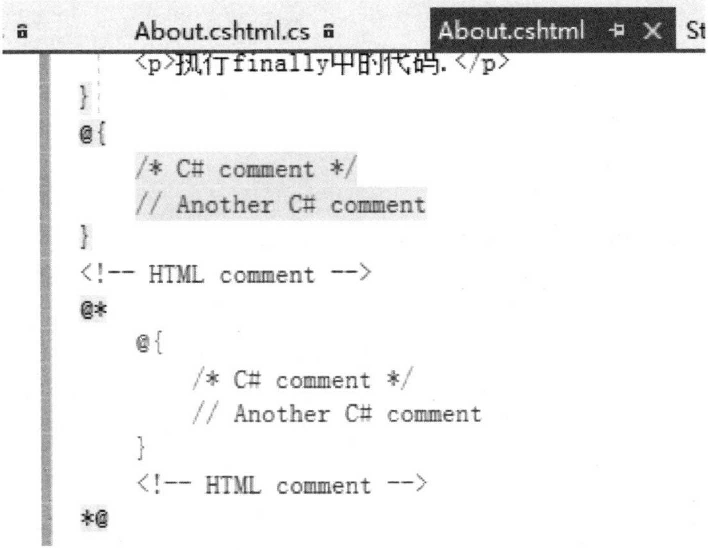
About.cshtml.cs
About.cshtml
<p>执行finally中的代码.</p>
}
@{
/* C# comment */
// Another C# comment
}
<!-- HTML comment -->
@*
@{
/* C# comment */
// Another C# comment
}
<!-- HTML comment -->
*@

图 8.12　注释方式

8.6 指　令

Razor 指令是形如“@符号+保留关键字”的隐式表达式。指令通常能改变页面的解析或为 Razor 页面启用不同的功能。理解了 Razor 如何为视图生成代码后，就能轻松理解指令是如何工作的。

1. @using 指令

@using 指令用于向生成的视图添加 C#　using 指令。示例操作如下。

在 Visual Studio 2017 中打开 About. cshtml 文件，输入如下测试代码：

```
〈p〉using 指令:〈/p〉
```

```
@using System.IO
@{
    var dir = Directory.GetCurrentDirectory();
}
<p>@dir</p>
```

2. @model 指令

@model 指令指定传递到视图的模型类型。示例操作如下。

在 Visual Studio 2017 中打开 About.cshtml 文件,输入如下测试代码:

```
@page
@model AboutModel
@{
    ViewData["Title"] = "About";
}
<h2>@ViewData["Title"]</h2>
<h3>@Model.Message</h3>
<p>model 指令:</p>
<p>AboutModel 的属性 Message:@Model.Message</p>
```

3. @functions 指令

@functions 指令允许 Razor 页面将 C#代码块添加到视图中。示例操作如下。

在 Visual Studio 2017 中打开 About.cshtml 文件,输入如下测试代码:

```
<p>functions 指令:</p>
@functions{
    public string GetHello()
    {
        return "Hello World";
    }
}
<div>functions 指令测试:@GetHello()</div>
```

此时,在 Visual Studio 2017 中按 F5 键运行应用程序,结果如图 8.13 所示。

4. @inject 指令

@inject 指令允许 Razor 页面将服务从服务容器注入到视图中。有关的详细信息可参阅 7.4 节。

5. @section 指令

@section 指令与布局结合使用,允许视图将内容呈现在 HTML 页面的不同部分。

8 Razor 视图

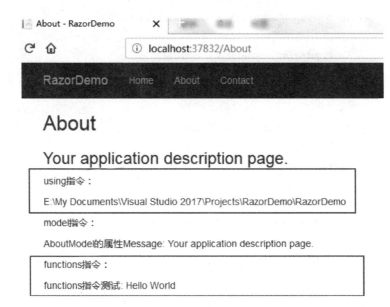

图 8.13 指令的结果

8.7 ASP.NET Core 中的 Razor 页面介绍

Razor 页面是 ASP.NET Core MVC 的一个新特性，可以使基于页面的编码方式更加简单高效。

8.7.1 启用 Razor 页面

在 Startup.cs 文件中调用以下两个用斜体字符表示的方法可以启用 Razor 页面：

```
using System;
using System.Collections.Generic;
using System.Linq;
using System.Threading.Tasks;
using Microsoft.AspNetCore.Builder;
using Microsoft.AspNetCore.Hosting;
using Microsoft.Extensions.Configuration;
using Microsoft.Extensions.DependencyInjection;

namespace RazorDemo
{
    public class Startup
    {
```

```
    public Startup(IConfiguration configuration)
    {
        Configuration = configuration;
    }
    public IConfiguration Configuration { get; }

    public void ConfigureServices(IServiceCollection services)
    {
        services.AddMvc();

    }
    public void Configure(IApplicationBuilder app, IHostingEnvironment env)
    {
        if (env.IsDevelopment())
        {
            app.UseBrowserLink();
            app.UseDeveloperExceptionPage();
        }
        else
        {
            app.UseExceptionHandler("/Error");
        }
        app.UseStaticFiles();
        app.UseMvc();
    }
}
```

8.7.2 Razor 页面介绍

下面来看如下一个基本的 Razor 页面的代码：

```
@page
@model AboutModel
@{
    ViewData["Title"] = "About";
}
<h2>@ViewData["Title"]</h2>
<h3>@Model.Message</h3>
```

上述代码是一个 Razor 视图文件，它与第 6 章中的 MVC 视图很类似，不同之处在于 @page 指令必须是页面上的第一个 Razor 指令。@page 命令使文件转换为一个 MVC 操作,这意味着 Razor 视图将直接处理请求,而不是通过控制器来处理。@page 命令将影响其他 Razor 构造的行为。

下述代码是上面视图文件的 PageModel 类:

```
using System;
using System.Collections.Generic;
using System.Linq;
using System.Threading.Tasks;
using Microsoft.AspNetCore.Mvc.RazorPages;

namespace RazorDemo.Pages
{
    public class AboutModel: PageModel
    {
        public string Message { get; set; }
        public void OnGet()
        {
            Message = "Your application description page.";
        }
    }
}
```

按照惯例,PageModel 类文件的名称与追加了".cs"的 Razor 页面文件的名称相同。例如,前面的 Razor 页面的名称为 Pages/About.cshtml,包含 PageModel 类的文件的名称为 Pages/About.cshtml.cs。

页面中 URL 路径的关联由页面所在文件系统中的位置决定,表 8.3 显示了 Razor 页面路径及匹配的 URL。

表 8.3　Razor 页面路径及匹配的 URL

文件名与路径	匹配的 URL
/Pages/Index.cshtml	/ 或 /Index
/Pages/About.cshtml	/About
/Pages/StoreIn/Delete.cshtml	/StoreIn/Delete
/Pages/StoreIn/Index.cshtml	/StoreIn 或 /StoreIn/Index

注意:
- 默认情况下,运行时在"Pages"文件夹中查找 Razor 页面文件。
- 当 URL 未包含页面时,Index 为默认页面。

8.7.3　编写基本窗体

对于 Razor 页面的设计,在构建应用时可以轻松使用模板来创建页面。Razor

页面类中定义的属性可使用模型绑定、标记帮助助手和 HTML 帮助助手。有关 Entity Framework Core 的知识请参考第 5 章。下面通过一个实例来了解如何编写以及增加、删除、修改、查询基本页面。具体操作如下：

① 在 Visual Studio 2017 中选择"工具"→"NuGet 包管理器"→"程序包管理器控制台"菜单项。

② 在程序包管理器控制台中依次输入以下三条指令，安装 NuGet 包：

```
Install-Package Microsoft.EntityFrameworkCore -version 2.0.3
Install-Package Microsoft.EntityFrameworkCore.SqlServer -version 2.0.3
Install-Package Microsoft.EntityFrameworkCore.Tools -version 2.0.3
```

③ 在 Models 文件夹中添加"Supplier"类，代码如下：

```csharp
using System;
using System.Collections.Generic;
using System.ComponentModel.DataAnnotations;
using System.ComponentModel.DataAnnotations.Schema;

namespace RazorDemo.Models
{
    public class Supplier
    {
        [Key]
        [DatabaseGeneratedAttribute(DatabaseGeneratedOption.Identity)]
        public int Id { get; set; }
        public string Name { get; set; }
        public string Code { get; set; }
        public string Contact { get; set; }
    }
}
```

④ 在 Visual Studio 2017 的"解决方案资源管理器"的 Models 目录中添加"EFCoreDemoContext"类，该类必须继承于"System.Data.Entity.DbContext"类以赋予其数据操作能力，代码如下：

```csharp
using System;
using Microsoft.EntityFrameworkCore;
using Microsoft.EntityFrameworkCore.Metadata;

namespace RazorDemo.Models
{
    public partial class EFCoreDemoContext: DbContext
    {
```

```
        public virtual DbSet<Supplier>Supplier { get; set; }

        public EFCoreDemoContext (DbContextOptions<EFCoreDemoContext> options):base(options)
        {
        }
    }
}
```

⑤ 在 Visual Studio 2017 的资源管理器中找到 appsettings.json 文件并双击打开,在文件中添加一个连接字符串,代码如下:

```
{
    "Logging": {
        "IncludeScopes": false,
        "LogLevel": {
            "Default": "Warning",
            "Microsoft": "Warning"
        }
    },
    "ConnectionStrings": {
        "EFCoreDemoContext": "Server = .\\sqlexpress;Database = EFCoreDemo;Trusted_Connection = True;MultipleActiveResultSets = true"
    }
}
```

⑥ 在 Visual Studio 2017 的资源管理器中找到 startup.cs 文件并双击打开,在 startup.cs 文件的 ConfigureServices 方法中写入依赖注入容器注册数据库上下文的代码,具体代码如下:

```
public void ConfigureServices(IServiceCollection services)
{
    services.AddDbContext<EFCoreDemoContext>(options => options.UseSqlServer(Configuration.GetConnectionString("EFCoreDemoContext")));
    services.AddMvc();
}
```

⑦ 在 Visual Studio 2017 的程序包管理器控制台上依次执行以下命令:

Add - MigrationRazorDemoInit
Update - Database

⑧ 在 Visual Studio 2017 的"解决方案资源管理器"中右击 Pages 文件夹,在弹出的快捷菜单中选择"添加"→"新建文件夹"菜单项,创建一个新文件夹,并把文件夹

命名为 Suppliers。

⑨ 在 Visual Studio 2017 的"解决方案资源管理器"中右击 Suppliers 文件夹,在弹出的快捷菜单中选择"添加"→"Razor 页面"菜单项,在弹出的对话框中选择"使用实体框架生成 Razor 页面(CURD)",然后单击"添加"按钮,弹出如图 8.14 所示对话框。

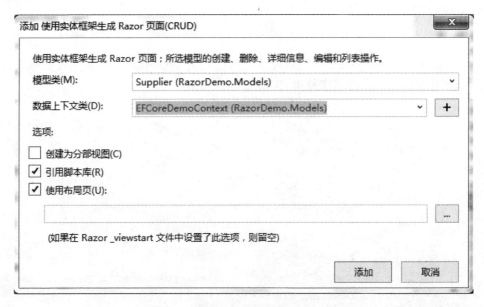

图 8.14　使用实体框架生成 Razor 页面

⑩ Visual Studio 2017 会自动生成创建、删除、详细信息、编辑和列表页面,如图 8.15 所示。

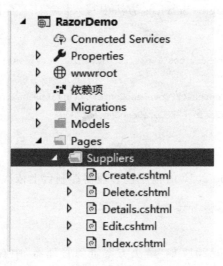

图 8.15　框架生成的 Razor 页面

⑪ 在 Visual Studio 2017 的"解决方案资源管理器"中打开 Suppliers\Create.cshtml 视图文件,代码如下:

```razor
@page
@model RazorDemo.Pages.Suppliers.CreateModel

@{
    ViewData["Title"] = "Create";
}
<h2>Create</h2>

<h4>Supplier</h4>
<hr />
<div class="row">
<div class="col-md-4">
<form method="post">
<div asp-validation-summary="ModelOnly" class="text-danger"></div>
<div class="form-group">
<label asp-for="Supplier.Name" class="control-label"></label>
<input asp-for="Supplier.Name" class="form-control" />
<span asp-validation-for="Supplier.Name" class="text-danger"></span>
</div>
<div class="form-group">
<label asp-for="Supplier.Code" class="control-label"></label>
<input asp-for="Supplier.Code" class="form-control" />
<span asp-validation-for="Supplier.Code" class="text-danger"></span>
</div>
<div class="form-group">
<label asp-for="Supplier.Contact" class="control-label"></label>
<input asp-for="Supplier.Contact" class="form-control" />
<span asp-validation-for="Supplier.Contact" class="text-danger"></span>
</div>
<div class="form-group">
<input type="submit" value="Create" class="btn btn-default" />
</div>
</form>
</div>
</div>
<div>
<a asp-page="Index">Back to List</a>
</div>
@section Scripts {
```

```
@{await Html.RenderPartialAsync("_ValidationScriptsPartial");}
}
```

⑫ 在 Visual Studio 2017 的"解决方案资源管理器"中打开 Suppliers\Create.cshtml.cs 页面模型，代码如下：

```
using System;
using System.Collections.Generic;
using System.Linq;
using System.Threading.Tasks;
using Microsoft.AspNetCore.Mvc;
using Microsoft.AspNetCore.Mvc.RazorPages;
using Microsoft.AspNetCore.Mvc.Rendering;
using RazorDemo.Models;

namespace RazorDemo.Pages.Suppliers
{
    public class CreateModel:PageModel
    {
        private readonly RazorDemo.Models.EFCoreDemoContext _context;

        public CreateModel(RazorDemo.Models.EFCoreDemoContext context)
        {
            _context = context;
        }
        public IActionResult OnGet()
        {
            return Page();
        }
        [BindProperty]
        public Supplier Supplier { get; set; }
        public async Task<IActionResult> OnPostAsync()
        {
            if (!ModelState.IsValid)
            {
                return Page();
            }
            _context.Supplier.Add(Supplier);
            await _context.SaveChangesAsync();
            return RedirectToPage("./Index");
        }
    }
}
```

默认情况下，PageModel 类的命名规则为〈PageName〉Model，并且它与页面位于同一个命名空间。

使用 PageModel 类可以将页面的逻辑与其展示分离开来。在 PageModel 类中编写页面处理程序，用于处理发送到页面的请求和呈现页面的数据。借助这种页面与代码的分离，可以通过依赖关系注入来管理页面的依赖关系，并对页面执行单元测试。

PageModel 类包含 OnPostAsync 处理程序方法，该方法在 POST 请求上运行（当用户提交窗体时）。可以为任何 HTTP 谓词添加处理程序方法。最常见的处理程序是：

- OnGet，用于初始化页面所需的状态。
- OnPost，用于处理窗体提交。

Async 的命名后缀为可选，可以为 OnGet 和 OnPost 添加 Async 后缀。如果添加了 Async 后缀，则一般会将其用于异步函数。上面示例中的 OnPostAsync 方法就是 OnPost 方法的异步执行方式。

现在介绍上面代码中 OnPostAsync 方法的执行过程：

1) 检查验证错误。如果没有错误，则保存数据并重定向；如果有错误，则再次显示页面并附带验证消息。客户端验证与传统的 ASP.NET Core MVC 应用程序相同。很多情况下，都会在客户端上检测到验证错误，并且从不将错误提交给服务器。

2) 成功保存数据后，OnPostAsync 处理程序方法调用 RedirectToPage 方法来返回 RedirectToPageResult 的实例。RedirectToPage 是新的操作结果，类似于 RedirectToAction 或 RedirectToRoute。在前面的示例中，它将重定向到根索引页(/Index)。

3) 当提交的窗体存在(已传递到服务器上)验证错误时，OnPostAsync 处理程序方法调用 Page 方法。Page 方法返回 PageResult 的实例。PageResult 是处理程序方法的默认返回类型。返回 void 类型的处理程序方法将显示页面。

上例中的 Supplier 属性使用 [BindProperty] 特性来选择加入模型绑定，代码如下：

```
[BindProperty]
public Supplier Supplier { get; set; }
```

可以使用 [BindProperty] 特性将 PageModel 属性直接绑定到 Razor 页面上。绑定属性可以减少需要编写的代码量。绑定通过使用相同的属性来显示窗体字段（<input asp-for="Supplier.Name" />）以减少代码，并接受输入。

若要将属性绑定在 GET 请求上，则将 [BindProperty] 特性的 SupportsGet 属性设置为 true，代码如下：

```
[BindProperty(SupportsGet = true)]
```

⑬ 在 Visual Studio 2017 的"解决方案资源管理器"中打开 Suppliers\Index.cshtml 视图文件，代码如下：

```
@page
@model RazorDemo.Pages.Suppliers.IndexModel

@{
    ViewData["Title"] = "Index";
}
<h2>Index</h2>
<p>
<a asp-page="Create">Create New</a>
</p>
<table class="table">
<thead>
<tr>
<th>
@Html.DisplayNameFor(model => model.Supplier[0].Name)
</th>
<th>
@Html.DisplayNameFor(model => model.Supplier[0].Code)
</th>
<th>
@Html.DisplayNameFor(model => model.Supplier[0].Contact)
</th>
<th></th>
</tr>
</thead>
<tbody>
@foreach (var item in Model.Supplier) {
<tr>
<td>
            @Html.DisplayFor(modelItem => item.Name)
</td>
<td>
            @Html.DisplayFor(modelItem => item.Code)
</td>
<td>
            @Html.DisplayFor(modelItem => item.Contact)
</td>
<td>
<a asp-page="./Edit" asp-route-id="@item.Id">Edit</a> |
<a asp-page="./Details" asp-route-id="@item.Id">Details</a> |
<a asp-page="./Delete" asp-route-id="@item.Id">Delete</a>
</td>
```

```
            </tr>
        }
    </tbody>
</table>
```

⑭ 在 Visual Studio 2017 的"解决方案资源管理器"中打开 Suppliers\Index.cshtml 视图文件(Index.cshtml.cs)关联的 PageModel 类,代码如下:

```
using System;
using System.Collections.Generic;
using System.Linq;
using System.Threading.Tasks;
using Microsoft.AspNetCore.Mvc;
using Microsoft.AspNetCore.Mvc.RazorPages;
using Microsoft.EntityFrameworkCore;
using RazorDemo.Models;

namespace RazorDemo.Pages.Suppliers
{
    public class IndexModel : PageModel
    {
        private readonly RazorDemo.Models.EFCoreDemoContext _context;

        public IndexModel(RazorDemo.Models.EFCoreDemoContext context)
        {
            _context = context;
        }

        public IList<Supplier> Supplier { get;set; }
        public async Task OnGetAsync()
        {
            Supplier = await _context.Supplier.ToListAsync();
        }
    }
}
```

Index.cshtml 文件包含以下标记来创建每个联系人项的编辑链接:

```
<a asp-page="./Edit" asp-route-id="@item.Id">edit</a>
```

定位点标记帮助程序使用"asp-route-{value}"属性生成"编辑"页面的链接,此链接包含路由数据及供应商 ID,例如 http://localhost:37832/Edit/1。

⑮ 在 Visual Studio 2017 的"解决方案资源管理器"中打开 Suppliers\Edit.cshtml 视图文件,代码如下:

```
@page "{id:int}"
@model RazorDemo.Pages.Suppliers.EditModel

@{
    ViewData["Title"] = "Edit";
}

<h2>Edit</h2>
<h4>Supplier</h4>
<hr />
<div class="row">
<div class="col-md-4">
<form method="post">
<div asp-validation-summary="ModelOnly" class="text-danger"></div>
<input type="hidden" asp-for="Supplier.Id" />
<div class="form-group">
<label asp-for="Supplier.Name" class="control-label"></label>
<input asp-for="Supplier.Name" class="form-control" />
<span asp-validation-for="Supplier.Name" class="text-danger"></span>
</div>
<div class="form-group">
<label asp-for="Supplier.Code" class="control-label"></label>
<input asp-for="Supplier.Code" class="form-control" />
<span asp-validation-for="Supplier.Code" class="text-danger"></span>
</div>
<div class="form-group">
<label asp-for="Supplier.Contact" class="control-label"></label>
<input asp-for="Supplier.Contact" class="form-control" />
<span asp-validation-for="Supplier.Contact" class="text-danger"></span>
</div>
<div class="form-group">
<input type="submit" value="Save" class="btn btn-default" />
</div>
</form>
</div>
</div>
<div>
<a asp-page="./Index">Back to List</a>
</div>
@section Scripts {
    @{await Html.RenderPartialAsync("_ValidationScriptsPartial");}
}
```

代码的第一行包含"@page "{id:int}""指令。路由约束"{id:int}"告诉页面接受包含 int 路由数据的页面请求。如果页面请求未包含可转换为 int 的路由数据,则运行时返回 HTTP404(未找到)错误。若要使 id 可选,则将"?"追加到路由约束中,代码如下:

@page "{id:int?}"

⑯ 在 Visual Studio 2017 的"解决方案资源管理器"中打开 Suppliers\Edit.cshtml.cs 文件,代码如下:

```
using System;
using System.Collections.Generic;
using System.Linq;
using System.Threading.Tasks;
using Microsoft.AspNetCore.Mvc;
using Microsoft.AspNetCore.Mvc.RazorPages;
using Microsoft.AspNetCore.Mvc.Rendering;
using Microsoft.EntityFrameworkCore;
using RazorDemo.Models;

namespace RazorDemo.Pages.Suppliers
{
    public class EditModel : PageModel
    {
        private readonly RazorDemo.Models.EFCoreDemoContext _context;

        public EditModel(RazorDemo.Models.EFCoreDemoContext context)
        {
            _context = context;
        }
        [BindProperty]
        public Supplier Supplier { get; set; }
        public async Task<IActionResult> OnGetAsync(int? id)
        {
            if (id == null)
            {
                return NotFound();
            }
            Supplier = await _context.Supplier.SingleOrDefaultAsync(m => m.Id == id);
            if (Supplier == null)
            {
                return NotFound();
```

```csharp
            }
            return Page();
        }
        public async Task<IActionResult> OnPostAsync()
        {
            if (!ModelState.IsValid)
            {
                return Page();
            }
            _context.Attach(Supplier).State = EntityState.Modified;
            try
            {
                await _context.SaveChangesAsync();
            }
            catch (DbUpdateConcurrencyException)
            {
                if (!SupplierExists(Supplier.Id))
                {
                    return NotFound();
                }
                else
                {
                    throw;
                }
            }
            return RedirectToPage("./Index");
        }
        private bool SupplierExists(int id)
        {
            return _context.Supplier.Any(e => e.Id == id);
        }
    }
}
```

至此完成了编辑页面的介绍,下面介绍删除页面。

⑰ 在 Visual Studio 2017 的"解决方案资源管理器"中打开 Suppliers\Index.cshtml 文件,此文件的代码中包括了删除每个供应商的链接,代码如下:

```html
<a asp-page="./Delete" asp-route-id="@item.Id">Delete</a>
```

⑱ 在 Visual Studio 2017 的"解决方案资源管理器"中打开 Suppliers\Delete.cshtml 文件,代码如下:

```
@page
@model RazorDemo.Pages.Suppliers.DeleteModel

@{
    ViewData["Title"] = "Delete";
}
<h2>Delete</h2>
<h3>Are you sure you want to delete this?</h3>
<div>
<h4>Supplier</h4>
<hr/>
<dl class = "dl-horizontal">
<dt>
            @Html.DisplayNameFor(model => model.Supplier.Name)
</dt>
<dd>
            @Html.DisplayFor(model => model.Supplier.Name)
</dd>
<dt>
            @Html.DisplayNameFor(model => model.Supplier.Code)
</dt>
<dd>
            @Html.DisplayFor(model => model.Supplier.Code)
</dd>
<dt>
            @Html.DisplayNameFor(model => model.Supplier.Contact)
</dt>
<dd>
            @Html.DisplayFor(model => model.Supplier.Contact)
</dd>
</dl>
<form method = "post">
<input type = "hidden" asp-for = "Supplier.Id" />
<input type = "submit" value = "Delete" class = "btn btn-default" /> |
<a asp-page = "./Index">Back to List</a>
</form>
</div>
```

删除按钮采用 HTML 呈现,其可以包括以下两个参数:
1) asp-route-id 属性指定的供应商 ID。
2) asp-page-handler 属性指定的 handler。
当用户单击删除按钮时,向服务器发送窗体 POST 请求。默认情况下是调用

OnPost 或 OnPostAnsyc 这两个方法之一。如果添加了 asp-page-handler 参数,则会根据方案 OnPost[handler]Async 并基于 handler 参数的值来选择处理程序方法的名称。

⑲ 在 Visual Studio 2017 的"解决方案资源管理器"中打开 Suppliers\Delete.cshtml.cs 文件,代码如下:

```
using System;
using System.Collections.Generic;
using System.Linq;
using System.Threading.Tasks;
using Microsoft.AspNetCore.Mvc;
using Microsoft.AspNetCore.Mvc.RazorPages;
using Microsoft.EntityFrameworkCore;
using RazorDemo.Models;

namespace RazorDemo.Pages.Suppliers
{
    public class DeleteModel : PageModel
    {
        private readonly RazorDemo.Models.EFCoreDemoContext _context;

        public DeleteModel(RazorDemo.Models.EFCoreDemoContext context)
        {
            _context = context;
        }

        [BindProperty]
        public Supplier Supplier { get; set; }
        public async Task<IActionResult> OnGetAsync(int? id)
        {
            if (id == null)
            {
                return NotFound();
            }
            Supplier = await _context.Supplier.SingleOrDefaultAsync(m => m.Id == id);
            if (Supplier == null)
            {
                return NotFound();
            }
            return Page();
        }
        public async Task<IActionResult> OnPostAsync(int? id)
        {
            if (id == null)
```

```
            {
                return NotFound();
            }
            Supplier = await _context.Supplier.FindAsync(id);
            if (Supplier != null)
            {
                _context.Supplier.Remove(Supplier);
                await _context.SaveChangesAsync();
            }
            return RedirectToPage("./Index");
        }
    }
}
```

本示例中没有指定 asp-page-handler，因此调用了默认处理程序方法 OnPostAsync 来处理 POST 请求。如果添加了 asp-page-handler 参数，并设置为不同值（如 delete），则选择名称为 OnPostDeleteAsync 的页面处理程序方法。

OnPostAsync 方法的执行过程是：

1）接受来自查询字符串的 id。

2）使用 FindAsync 从数据库中查询供应商数据。

3）如果找到供应商，则从供应商列表中将其删除，并同时从数据库中删除。

4）调用 RedirectToPage 方法重定向到根索引页（/Index）。

8.7.4 页面的 URL 生成

在 PageModel 类中一般使用 RedirectToPage 方法进行页面跳转，例如在 Pages/Suppliers/Create.cshtml 和 Pages/Suppliers/Edit.cshtml 页面中的如下代码：

```
return RedirectToPage("/Index");
```

跳转成功后将重定向到 Suppliers/Index.cshtml 页面。

在 8.7.3 小节的示例项目中有如图 8.16 中的文件/文件夹结构。

通过给 RedirectToPage 方法赋予不同的参数，会跳转到不同的页面，如表 8.4 所列。

表 8.4 RedirectToPage 方法的参数

RedirectToPage(x)	URL
RedirectToPage("/Index")	Pages/Index
RedirectToPage("./Index")	Pages/Suppliers/Index
RedirectToPage("../Index")	Pages/Index
RedirectToPage("Index")	Pages/Suppliers/Index

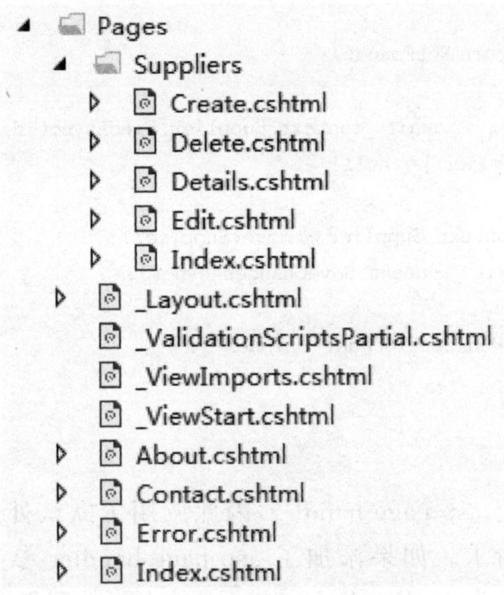

```
▲ 📁 Pages
    ▲ 📁 Suppliers
        ▷ 📄 Create.cshtml
        ▷ 📄 Delete.cshtml
        ▷ 📄 Details.cshtml
        ▷ 📄 Edit.cshtml
        ▷ 📄 Index.cshtml
    ▷ 📄 _Layout.cshtml
       📄 _ValidationScriptsPartial.cshtml
       📄 _ViewImports.cshtml
       📄 _ViewStart.cshtml
    ▷ 📄 About.cshtml
    ▷ 📄 Contact.cshtml
    ▷ 📄 Error.cshtml
    ▷ 📄 Index.cshtml
```

图 8.16 项目文件/文件夹结构

表 8.4 中的页面路径是从根文件夹"/Pages"到页面的路径(包含前导字符"/"，例如"/Index")。与硬编码 URL 相比，表 8.4 中的 URL 生成示例提供了改进的选项和功能。URL 的生成使用路由，并且可以根据用目标路径定义路由的方式来生成参数，并对参数进行编码。

RedirectToPage ("Index")、RedirectToPage ("./Index") 和 RedirectToPage ("../Index")使用的是相对路径，需结合 RedirectToPage 的参数与当前页的路径来计算目标页面的路径。

当构建结构复杂的站点时，相对路径的链接很有用。如果使用相对路径来链接文件夹中的页面，则可以重命名该文件夹，并且所有链接仍然有效(因为这些链接并未包含此文件夹的名称)。

8.7.5 针对一个页面的多个处理程序

在一个页面中有多个处理方法是很平常的事，例如，在同一页面中有"保存""删除""申请""打印"等按钮，因此，asp-page-handler 标记就是一个非常有用的标记，使用该标记可以在同一页面中实现对多个方法的处理。下面就来看一个实例，具体操作如下：

① 在 Visual Studio 2017 中打开 Suppliers\Edit.cshtml 文件，输入以下代码：

@page
@model RazorDemo.Pages.Suppliers.EditModel

@{

```html
    ViewData["Title"] = "Edit";
}
<h2>Edit</h2>
<h4>Supplier</h4>
<hr/>
<div class="row">
<div class="col-md-4">
<form method="post">
<div asp-validation-summary="ModelOnly" class="text-danger"></div>
<input type="hidden" asp-for="Supplier.Id"/>
<div class="form-group">
<label asp-for="Supplier.Name" class="control-label"></label>
<input asp-for="Supplier.Name" class="form-control"/>
<span asp-validation-for="Supplier.Name" class="text-danger"></span>
</div>
<div class="form-group">
<label asp-for="Supplier.Code" class="control-label"></label>
<input asp-for="Supplier.Code" class="form-control"/>
<span asp-validation-for="Supplier.Code" class="text-danger"></span>
</div>
<div class="form-group">
<label asp-for="Supplier.Contact" class="control-label"></label>
<input asp-for="Supplier.Contact" class="form-control"/>
<span asp-validation-for="Supplier.Contact" class="text-danger"></span>
</div>
<div class="form-group">
<input type="submit" value="Save" asp-page-handler="Save" class="btn btn-default"/>
<input type="submit" value="测试方法" asp-page-handler="New" class="btn btn-default"/>
</div>
</form>
</div>
</div>
<div>
<a asp-page="./Index">Back to List</a>
</div>
@section Scripts {
    @{await Html.RenderPartialAsync("_ValidationScriptsPartial");}
}
```

代码中包含两个提交按钮，每个按钮均使用 FormActionTagHelper 提交到不同的 URL,其中 asp-page-handler 是 asp-page 的配套属性。asp-page-handler 生成提交到页面的各个处理程序方法的 URL。在上面的代码中,两个提交按钮已经指向了各自的处理方法。

由 asp-page-handler 指定的处理方法的命名规则是:

1) OnPost〈方法名〉Ansyc。

2) OnPost〈方法名〉。

例如下面的代码中,页面方法是 OnPostSaveAsync 和 OnPostNewAsync。删除 OnPost 和 Async 后,处理方法的名称变为 Save 和 New。

② 在 Visual Studio 2017 中打开 Suppliers\Edit.cshtml.cs 文件,输入以下代码:

```
using System;
using System.Collections.Generic;
using System.Linq;
using System.Threading.Tasks;
using Microsoft.AspNetCore.Mvc;
using Microsoft.AspNetCore.Mvc.RazorPages;
using Microsoft.AspNetCore.Mvc.Rendering;
using Microsoft.EntityFrameworkCore;
using RazorDemo.Models;

namespace RazorDemo.Pages.Suppliers
{
    public class EditModel : PageModel
    {
        private readonly RazorDemo.Models.EFCoreDemoContext _context;
        public EditModel(RazorDemo.Models.EFCoreDemoContext context)
        {
            _context = context;
        }
        [BindProperty]
        public Supplier Supplier { get; set; }
        public async Task<IActionResult> OnGetAsync(int? id)
        {
            if (id == null)
            {
                return NotFound();
            }
            Supplier = await _context.Supplier.SingleOrDefaultAsync(m => m.Id == id);
            if (Supplier == null)
            {
```

```csharp
        return NotFound();
    }
    return Page();
}
public Task<IActionResult> OnPostAsync()
{
    return OnPostSaveAsync();
}
public async Task<IActionResult> OnPostSaveAsync()
{
    if (!ModelState.IsValid)
    {
        return Page();
    }
    _context.Attach(Supplier).State = EntityState.Modified;
    try
    {
        await _context.SaveChangesAsync();
    }
    catch (DbUpdateConcurrencyException)
    {
        if (!SupplierExists(Supplier.Id))
        {
            return NotFound();
        }
        else
        {
            throw;
        }
    }
    ModelState.AddModelError(string.Empty, "保存成功!");
    return Page();
}
private bool SupplierExists(int id)
{
    return _context.Supplier.Any(e => e.Id == id);
}
public async Task<IActionResult> OnPostNewAsync()
{
    try
    {
        Supplier = await _context.Supplier.SingleOrDefaultAsync();
```

```
                    ModelState.AddModelError(string.Empty,"测试一个页面多个按钮,多个
处理方法!");
                }
                catch (DbUpdateConcurrencyException)
                {
                    if (!SupplierExists(Supplier.Id))
                    {
                        return NotFound();
                    }
                    else
                    {
                        throw;
                    }
                }
                return Page();
            }
        }
```

③ 在 Visual Studio 2017 中按 F5 键运行应用程序。在浏览器中浏览 http://localhost:37832/Suppliers/Edit?id=1 页面,然后修改页面中某个字段的数据,最后单击"Save"按钮,结果如图 8.17 所示。提交到 OnPostSaveAsync 的 URL 为 http://localhost:37832/Suppliers/Edit?handler=Save。

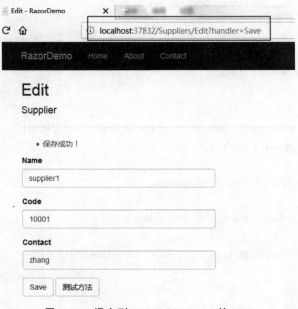

图 8.17　提交到 OnPostSaveAsync 的 URL

④ 在浏览器中单击"测试方法"按钮,结果如图 8.18 所示。提交到 OnPostNewAsync 的 URL 为 http://localhost:37832/Suppliers/Edit?handler=New。

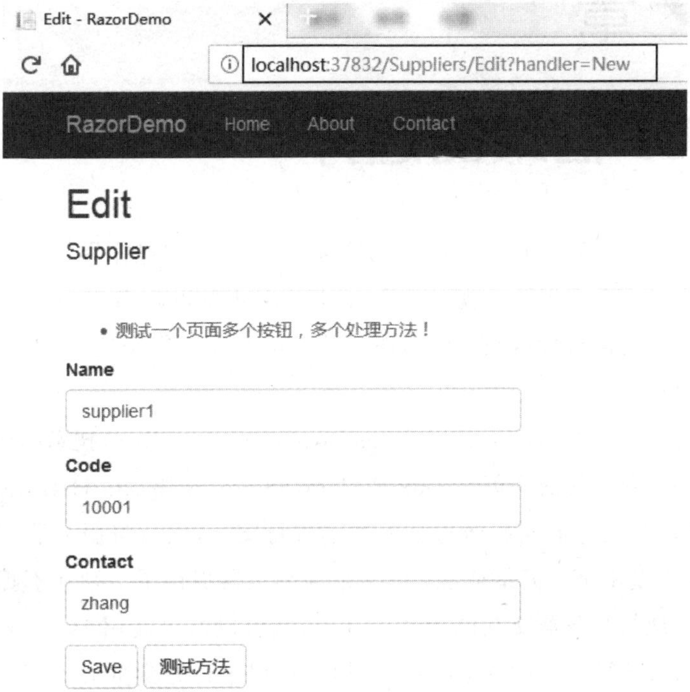

图 8.18 提交到 OnPostNewAsync 的 URL

第 9 章

Web UI 框架的选择

UI 即 User Interface（用户界面）的简称，泛指用户的操作界面。一个友好美观的用户界面会给人带来舒适的视觉享受，拉近人与电脑的距离；一个友好美观的用户界面可以让软件变得有特点、有品位，还让软件的操作变得舒适、简单、自由，充分体现软件的价值。

进入 21 世纪后，Ajax(Asynchronous JavaScript And Xml)技术的崛起绝对是互联网应用的一个划时代的变革。Ajax 通过 JavaScript 使用 XmlHttpRequest 对象将从 UI 界面获取的用户输入数据与服务器端进行交互，然后将服务器端的处理结果通过 UI 界面展现出来。简单地说，Ajax 是为浏览器提供了一种在不提交整个页面的情况下动态地与服务器交互的能力。这样，就可以通过使用 JavaScript 脚本来提交数据和刷新或渲染页面中的某些部分，用户可以不再看到页面提交与显示之间的空白状态，使得 Web 应用的用户体验得到了巨大的增强。

基于 JavaScript 的 UI 的出现，将 Web UI 的控制权从界面设计人员的手中递交给了程序员，即可以直接在 Web 的前端使用 JavaScript 脚本来描述一个 UI 组件模型，然后在运行时，由浏览器的脚本解释器调用 UI 技术框架的核心库将该模型转换成 HTML 的 UI 界面。

此类 UI 技术框架与服务器端 UI 技术的思路一致，只是在客户端的浏览器中封装了一套 UI 模型。这样，界面设计不需要服务器端的支持，在开发期间能更好地展示和测试界面效果。同时，由于 UI 界面的构建和控制都在客户端，因此，只需要与服务器端传递请求参数和数据，就能比服务器端的 UI 技术大大降低服务器端的压力和网络数据的传递量。

对于日益增加的 Web UI 控件需求，市场上也出现了很多可供选择的 Web UI 控件，以满足用户各种复杂的需求。随着 Web UI 控件的增加，又出现了一些整合 Web UI 控件的 Web UI 框架。本书介绍一些以三种核心框架（JQuery、Bootstrap、ExtJS）为基础的 Web UI 框架。

9.1 以 JQuery 为核心的前端框架

JQuery 是一个快速、简洁的 JavaScript 框架,可用于封装 JavaScript 常用的功能代码,还可以优化 HTML 文档操作、事件处理、动画设计和 Ajax 交互。下面简单介绍一些基于 JQuery 的 Web UI 框架。

9.1.1 EasyUI

第一个介绍 EasyUI,其显示界面如图 9.1 所示,这个前端框架在国内的名气不小,对于开发者来说,即使没用过,多少也听说过,而且也比较适合在.Net 环境中使用。

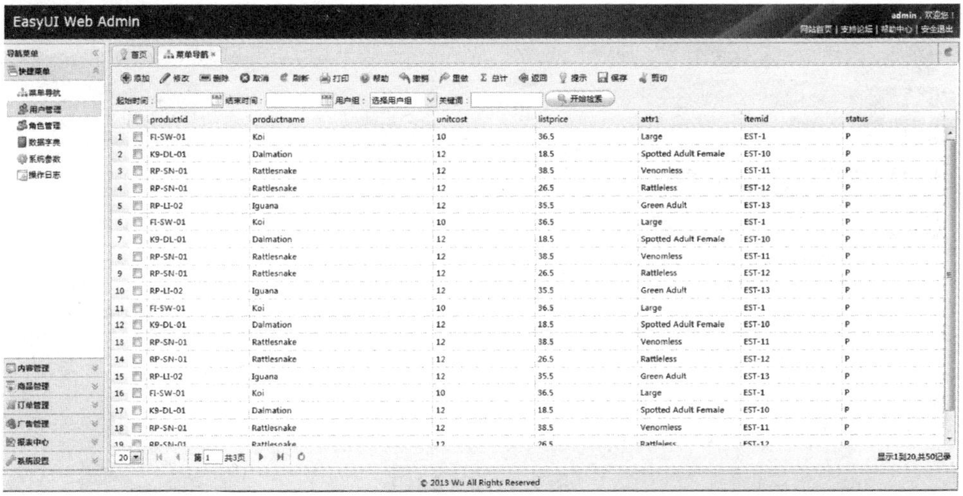

图 9.1 EasyUI

EasyUI 一开始是一种基于 JQuery 的用户界面插件集合,现在它可以基于 Vue、Angular、React 这些最新的脚本库实现用户界面。EasyUI 的目标就是帮助 Web 开发者更轻松地打造出功能丰富且美观的 UI 界面。EasyUI 支持各种皮肤以满足使用者对于页面不同风格的喜好。开发者不需要编写复杂的 JavaScript,也不需要对 CSS 样式有深入的了解,而只需了解一些简单的 HTML 标签即可定义用户界面。

EasyUI 为用户提供了大多数 UI 控件的使用,如:表单、下拉列表框、菜单、对话框、标签、窗体、按钮、数据网格、树形表格、面板等。用户可以组合使用这些组件,也可以单独使用其中一个。

EasyUI 有以下优点:

① 是基于 JQuery 用户界面插件的集合;

② 为一些当前用于交互的 JavaScript 应用提供必要的功能;

③ 支持两种渲染方式；
④ 支持 HTML 5（通过 data-options 属性）；
⑤ 开发产品时可节省时间和资源；
⑥ 轻量、免费、兼容性好，功能很强大；
⑦ 支持扩展，可根据自己的需求扩展控件；
⑧ 帮助详细，使用的人多，资源多。

9.1.2 DWZ JUI

DWZ 富客户端框架（JQuery RIA framework）如图 9.2 所示，是中国人自己开发的基于 JQuery 实现的 Ajax RIA 开源框架。DWZ 富客户端框架的设计目标是简单实用、扩展方便、快速开发、RIA 思路、轻量级，比较亲和 Java 环境。

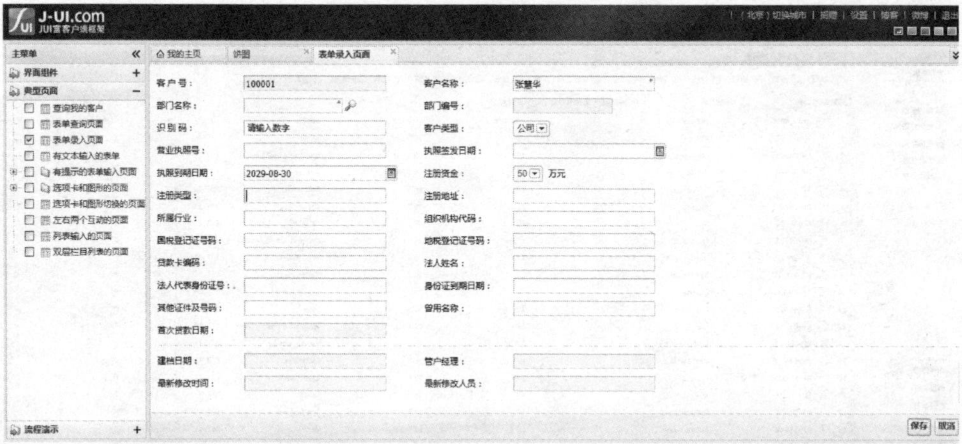

图 9.2 DWZ JUI

在做 Ajax 项目时需要编写大量的 JavaScript 才能达到满意的效果，开发人员如果对 JavaScript 不熟悉则会大大影响开发速度。DWZ 框架支持用 HTML 扩展的方式来代替 JavaScript 代码，只要懂 HTML 语法，再参考 DWZ 的使用手册就可以进行 Ajax 开发。DWZ 简单扩展了 HTML 标准，给 HTML 定义了一些特别的 Class 和 Attribute。DWZ 框架会找到当前请求结果中的那些特别的 Class 和 Attribute，并自动关联相应的 JavaScript 处理事件和效果。

DWZ 基于 JQuery，可以非常方便地定制有特定需求的 UI 组件，并以 JQuery 插件的形式发布出来，若有需要也可做定制化开发。

DWZ JUI 有以下优点：
① 完全开源，方便扩展；
② CSS 和 JavaScript 代码彻底分离，方便修改样式；
③ 简单实用，轻量级框架，快速开发；

④ 仍然保留了 HTML 的页面布局方式；

⑤ 支持用 HTML 扩展方式调用 UI 组件，开发人员不需要编写 JavaScript 代码；

⑥ 只要懂 HTML 语法，不需要精通 JavaScript，就可以使用 Ajax 开发后台程序；

⑦ 基于 JQuery，UI 组件以 JQuery 插件的形式发布。

9.1.3　LigerUI

LigerUI 是基于 JQuery 的 UI 框架，其核心设计目标是快速开发、使用简单、功能强大、轻量级、易扩展。LigerUI 简单而又强大，致力于快速打造 Web 前端界面的解决方案，可应用于 .NET、JSP、PHP 等 Web 服务器环境。LigerUI 是中国人开发的、免费的 UI 框架，其文档较全，但整体感觉不如 EasyUI，其显示界面如图 9.3 所示。

图 9.3　LigerUI

LigerUI 有如下优点：

① 使用简单，轻量级；

② 控件实用性强，功能覆盖面大，可以解决大部分企业信息应用的设计场景；

③ 可实现快速开发，使用 LigerUI 能比传统开发极大地减小代码量；

④ 易扩展，包括默认参数、表单/表格编辑器、多语言支持等；

⑤ 支持 Java、.NET、PHP 等 Web 服务端；

⑥ 支持 Internet Explorer 6＋、Chrome、FireFox 等浏览器；

⑦ 开源，源码框架层次简单易懂。

9.2 以 Bootstrap 为核心的前端框架

Bootstrap 是基于 HTML、CSS、JavaScript 开发的前端开发框架，提供了优雅的 HTML 和 CSS 规范。下面简单介绍一些以 Bootstrap 为核心的前端框架，这些框架都是基于 Bootstrap 并进行了性能优化而来的。

9.2.1 HUI

HUI 前端框架是在 Bootstrap 思想的基础上，基于 HTML、CSS、JavaScript 开发的轻量级 Web 前端框架，其开源免费、简单灵活、兼容性好，可满足大多数中国网站的要求。HUI 分为前端 UI 和后端 UI。目前 HUI 的名气小，用的人少，资料不全，配套组件也不多，其显示界面如图 9.4 所示。

图 9.4 HUI

HUI 有如下优点：
① 为完全响应式布局（支持电脑、平板、手机等所有主流设备）；
② 基于最新版本的 Bootstrap 3＋；
③ 提供 3 套不同风格的皮肤；
④ 支持多种布局方式；
⑤ 使用最流行的扁平化设计。

9.2.2 H＋UI

H＋UI 是一个完全响应式，基于 Bootstrap 3＋最新版本开发的扁平化主题，采用了主流的左右两栏式布局，使用了 HTML 5＋CSS 3 等现代技术，提供了诸多强大的可以重新组合的 UI 组件，集成了最新的 JQuery 版本和很多功能强大、用途广泛的 JQuery 插件,可用于所有 Web 应用程序的开发,如网站管理后台,网站会员中心、以及 CMS、CRM、OA 等,其显示界面如图 9.5 所示。当然,也可以对 H+ UI 进行深度定制,以做出更强的系统。H+ UI 与 Metronic 和 INSPINIA 非常像,插件非常多,但收费。

图 9.5 H+ UI

H+ UI 有如下优点:
① 完全响应式布局(支持电脑、平板、手机等所有主流设备);
② 基于最新版本的 Bootstrap 3+;
③ 提供 3 套不同风格的皮肤;
④ 支持多种布局方式;
⑤ 使用最流行的扁平化设计;
⑥ 提供诸多 UI 组件;
⑦ 集成多款国内优秀插件,诚意奉献;
⑧ 提供盒型、全宽、响应式视图模式;
⑨ 采用 HTML 5 和 CSS 3。

9.2.3 Ace Admin

响应式 Bootstrap 网站后台管理系统模板 Ace Admin 是一个非常不错的轻量级易用的后台管理系统,它基于 Bootstrap 3+,拥有强大的功能组件和 UI 组件,基本能满足后台管理系统的需求,并能根据不同的设备来适配显示;它还有四个主题可供切换,功能强大,组件多,美观,使用了很多不同的插件,其显示界面如图 9.6 所示。

Ace Admin 有如下优点:
① 完全响应式布局(支持电脑、平板、手机等所有主流设备);

图 9.6　Ace Admin

② 基于最新版本的 Bootstrap 3+；
③ 提供 4 套不同风格的皮肤；
④ 支持多种布局方式；
⑤ 使用最流行的扁平化设计；
⑥ 提供诸多 UI 组件。

Ace Admin 兼容的浏览器有：Internet Explorer 8+，以及最新版的 Chrome、Firefox、Opera 和 Safari。

9.2.4　Metronic

Metronic 是一个商业框架，是国外的基于 HTML、JavaScript 等技术的 Bootstrap 开发框架的整合。它整合了很多 Bootstrap 的前端技术和插件的使用，是一个非常不错的前端框架，构成一套精美的响应式后台管理模板，其显示界面如图 9.7 所示。

Metronic 拥有简洁优雅的 Metro UI 风格界面，有 6 种颜色可选，76 个模板页面，包括图表、表格、地图、消息中心、监控面板等后台管理项目所需的各种组件。Metronic 的体系庞大，而且对 IE 的兼容性不太好。

Metronic 有如下优点：
① 支持 HTML 5 和 CSS 3。
② 可自适应，基于响应式 Bootstrap 框架，同时面向桌面电脑、平板、手机等终端。
③ 整合了 AngularJS 框架。

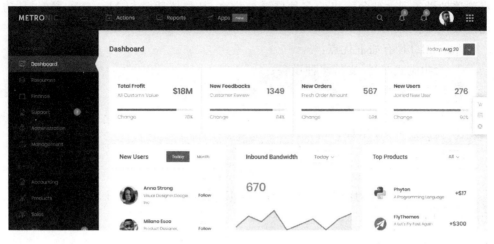

图 9.7 Metronic

④ 可自定义管理面板,包括灵活的布局、主题、导航菜单、侧边栏等。
⑤ 提供了部分电子商务模块,如 CMS、CRM、SAAS。
⑥ 具有多风格的简洁扁平风格设计。
⑦ 拥有 1 000 多个网页模板,1 500 多个 UI 小组件,100 多个集成插件。
⑧ 有详细的说明文档。

9.2.5　AdminLTE

AdminLTE 是一个基于 Bootstrap 3.x 的免费开源框架,也是一个完全响应式管理模板,其显示界面如图 9.8 所示。AdminLTE 的高度可定制,易于使用,适合从

图 9.8　AdminLTE

小型移动设备到大型台式机的多种屏幕分辨率,其整体感觉与 Metronic 类似,且功能强大,UI 精致。

AdminLTE 有如下优点:

① 为响应式布局,支持多种设备;

② 可打印增强;

③ 具有丰富可排序的面板组件;

④ 提供诸多 UI 组件;

⑤ 轻量、快速;

⑥ 支持 Glyphicons、Fontawesome 和 Ion 图标。

AdminLTE 兼容的浏览器有:Internet Explorer 9+,以及最新版的 Chrome、Firefox、Opera 和 Safari。

9.2.6　INSPINIA

INSPINIA 是一个商业用的框架,是具有平面设计理念的管理模板,采用扁平化设计,使用 Bootstrap 3+ 框架,以及 HTML 5 和 CSS 3 等技术开发而成的。它有很多可重用的 UI 组件,并集成了最新的 JQuery 插件,可用于所有类型的 Web 应用程序的开发,如自定义管理面板、项目管理系统、管理仪表板、应用程序的后端、CMS 或 CRM,其显示界面如图 9.9 所示。

图 9.9　INSPINIA

INSPINIA 有如下优点:

① 为响应式布局(支持台式机、平板电脑和移动设备);

② 搭配了新的 Bootstrap 4+;

③ 拥有平面 UI 和干净的审美风格；

④ 使用了 HTML 5 和 CSS 3 技术；

⑤ 拥有 7 个不同的图表库；

⑥ 可定制模态视图；

⑦ 具有表单验证功能；

⑧ 拥有登录、注册和错误页面。

INSPINIA 兼容的浏览器有：Internet Explorer 9＋，以及最新版的 Chrome、Firefox、Opera 和 Safari。

9.3　以 ExtJS 为核心的前端框架

ExtJS 是一个 JavaScript 框架，用于创建前端用户界面，是一个与后台技术无关的前端框架。下面简单介绍一个基于 ExtJS 的、应用于 ASP.NET 的前端框架——FineUI。

FineUI 是基于 JQuery/ExtJS 的 ASP.NET 控件库，用于创建 No JavaScript、No CSS、No UpdatePanel、No ViewState、No WebServices 的网站应用程序。FineUI 完全遵循 ASP.NET 的命名习惯和开发框架，控件的命名、属性、方法和事件与原生的 ASP.NET 控件一模一样，Ajax 交互也尽量与 ASP.NET 的开发习惯一致，因此无须了解 ExtJS 的知识。每个控件的开发都以服务实际项目需要为目的，不会追求大而全的封装，但会在 80％的常见功能上进行细致入微的设计和思考。

FineUI 包含很多简单而实用的创新，支持原生的 Ajax；具有轻量级的数据传输；基于 IFrame 的页面框架，使开发人员专注于业务逻辑的实现而非技术细节。

FineUI 基于 C♯ 和 ASP.NET 2.0 开发模式，使用 C♯ 代替传统的 JavaScript，从而不需要 JavaScript 代码就能完成无刷新的 Web 应用。超过 50 个专业的 ASP.NET 控件不仅可以极大地减少开发时间和降低开发成本，而且让页面更加生动美观，并可在所有主流浏览器下流畅运行，方便维护升级。FineUI 内置多种主题，同时还允许自定义主题，满足项目的个性化需求。

所有的页面回发都不会导致整个页面的刷新，而只是部分页面得到更新。这个 Ajax 过程对开发人员完全透明，开发人员可以像往常一样在服务器端改变控件的属性，该修改会立即更新到前台页面，而无需任何额外的代码。

FineUI 是中国人的作品，开源、免费，界面美观、控件多，帮助说明详细。但因 FineUI 是为 ASP.NET 量身定做的，故有一定的局限性。此外，FineUI 本身较大，并且 ExtJS＋ASP.NET 页面中的 ViewState 体积庞大，其显示界面如图 9.10 所示。

在前面所述的 WebUI 框架中，EasyUI 整体来说比较小，功能强大，而且可免费使用，兼容性好，帮助内容详细，容易入门，使用的人多，生态好，资源较多。

图 9.10　FineUI

参考文献

[1] Lander Rich,Chen Jason. .NET Core 指南.(2018-08-01)[2018-08-05]. https://docs.microsoft.com/zh-cn/dotnet/core.

[2] Roth Daniel,Anderson Rick,Luttin Shaun. ASP.NET Core 简介.(2019-04-07) [2019-04-11]. https://docs.microsoft.com/zh-cn/aspnet/core.

[3] Anderson Rick. ASP.NET Core 基础知识.(2019-03-31)[2019-04-04]. https://docs.microsoft.com/zh-cn/aspnet/core/fundamentals.

[4] Anderson Rick,Luo John,Smith Steve. Cache in-memory in ASP.NET Core. (2019-04-11).[2019-04-12]. https://docs.microsoft.com/en-us/aspnet/core/performance/caching/memory.

[5] Latham Luke,Smith Steve. Distributed caching in ASP.NET Core.(2019-03-30)[2019-04-05]. https://docs.microsoft.com/en-us/aspnet/core/performance/caching/distributed.

[6] Smith Steve. ASP.NET Core Mvc 概述.(2018-01-08)[2018-10-04]. https://docs.microsoft.com/zh-cn/aspnet/core/mvc/overview.

[7] Anderson Rick. ASP.NET Core 中的标记帮助程序.(2019-03-18)[2019-03-23]. https://docs.microsoft.com/zh-cn/aspnet/core/mvc/views/tag-helpers/intro.

[8] Anderson Rick,Latham Luke,Mullen Taylor. ASP.NET Core 的 Razor 语法参考.(2018-10-26)[2018-11-18]. https://docs.microsoft.com/zh-cn/aspnet/core/mvc/views/razor.

[9] Lee Terry G. 安装 Visual Studio.(2019-02-11)[2019-02-16]. https://docs.microsoft.com/zh-cn/visualstudio/install/install-visual-studio.

[10] Warren Genevieve. 欢迎使用 Visual Studio IDE.(2019-03-19)[2019-03-22]. https://docs.microsoft.com/zh-cn/visualstudio/get-started/visual-studio-ide?view=vs-2017.

[11] Nandwani Karan. Nuget 简介.(2018-01-10)[2018-10-06]. https://docs.microsoft.com/zh-cn/nuget/what-is-nuget.

[12] Autofac 入门.[2018-10-12]. https://autofaccn.readthedocs.io/zh/latest/

getting-started/index.html.

[13] Autofac 注册组件.[2018-10-12].https://autofaccn.readthedocs.io/zh/latest/register/index.html.

[14] ASP.NET Core 中集成 Autofac.[2018-10-12].https://autofaccn.readthedocs.io/zh/latest/integration/aspnetcore.html.

[15] Miller Rowan.Entity Framework Core.(2016-10-27)[2018-10-03].https://docs.microsoft.com/zh-cn/ef/core/index.

[16] EntityFrameworkCore Tutorial.[2018-10-03].https://www.entityframeworktutorial.net/efcore/entity-framework-core.aspx.

[17] EntityFramework 6 Core-First Tutorial.[2018-10-03].https://www.entityframeworktutorial.net/code-first/what-is-code-first.aspx.

[18] EasyUI 教程.[2018-10-07].http://www.jeasyui.net/tutorial.

[19] 张果.后台管理 UI 的选择.(2016-08-18)[2018-10-05].https://www.cnblogs.com/best/p/5782891.html.

[20] 罗志超.ASP.NET Core 2.0：六.举个例子来聊聊它的依赖注入.(2018-03-08)[2018-10-02].https://www.cnblogs.com/FlyLolo/p/ASPNETCore2_6.html.

[21] 罗志超.ASP.NET Core 2.0：七.一张图看透启动背后的秘密.(2018-03-22)[2018-10-02].http://www.cnblogs.com/FlyLolo/p/ASPNETCore2_7.html.

[22] Lock Andrew.How to register a service with multiple interfaces in ASP.NET CoreDI.(2018-09-18)[2018-10-12].https://andrewlock.net/how-to-register-a-service-with-multiple-interfaces-for-in-asp-net-core-di/.

[23] 蒋金楠..NET Core 的日志[1]：采用统一的模式记录日志.(2016-08-17)[2018-09-11].https://www.cnblogs.com/artech/p/logging-for-net-core-01.html.